Special Issue Dedicated to
the 60 Years of the Laboratory
of Analytical Chemistry of
the School of Chemistry of the
Aristotle University of Thessaloniki

Special Issue Dedicated to the 60 Years of the Laboratory of Analytical Chemistry of the School of Chemistry of the Aristotle University of Thessaloniki

Editors

Paraskevas D. Tzanavaras
Victoria Samanidou

Basel • Beijing • Wuhan • Barcelona • Belgrade • Novi Sad • Cluj • Manchester

Editors

Paraskevas D. Tzanavaras
Chemistry
Aristotle University
Thessaloniki
Greece

Victoria Samanidou
Chemistry
Aristotle University
Thessaloniki
Greece

Editorial Office
MDPI
St. Alban-Anlage 66
4052 Basel, Switzerland

This is a reprint of articles from the Special Issue published online in the open access journal *Molecules* (ISSN 1420-3049) (available at: www.mdpi.com/journal/molecules/special_issues/242153H0PK).

For citation purposes, cite each article independently as indicated on the article page online and as indicated below:

Lastname, A.A.; Lastname, B.B. Article Title. *Journal Name* **Year**, *Volume Number*, Page Range.

ISBN 978-3-7258-1382-7 (Hbk)
ISBN 978-3-7258-1381-0 (PDF)
doi.org/10.3390/books978-3-7258-1381-0

© 2024 by the authors. Articles in this book are Open Access and distributed under the Creative Commons Attribution (CC BY) license. The book as a whole is distributed by MDPI under the terms and conditions of the Creative Commons Attribution-NonCommercial-NoDerivs (CC BY-NC-ND) license.

Contents

About the Editors vii

Preface ix

Maria Antoniadou, Valentin Schierer, Daniela Fontana, Jürgen Kahr and Erwin Rosenberg
Development of a Multiplexing Injector for Gas Chromatography for the Time-Resolved Analysis of Volatile Emissions from Lithium-Ion Batteries
Reprinted from: *Molecules* **2024**, *29*, 2181, doi:10.3390/molecules29102181 1

Pedro Araujo, Sarah Iqbal, Aleksander Arnø, Marit Espe and Elisabeth Holen
Validation of a Liquid–Liquid Extraction Method to Study the Temporal Production of D-Series Resolvins by Head Kidney Cells from Atlantic Salmon (*Salmon salar*) Exposed to Docosahexaenoic Acid
Reprinted from: *Molecules* **2023**, *28*, 4728, doi:10.3390/molecules28124728 18

Myrto Kakarelidou, Panagiotis Christopoulos, Alexis Conides, Despina P. Kalogianni and Theodore K. Christopoulos
Fish DNA Sensors for Authenticity Assessment—Application to Sardine Species Identification
Reprinted from: *Molecules* **2024**, *29*, 677, doi:10.3390/molecules29030677 30

Dimitrios Tsikas
Application of the Bland–Altman and Receiver Operating Characteristic (ROC) Approaches to Study Isotope Effects in Gas Chromatography–Mass Spectrometry Analysis of Human Plasma, Serum and Urine Samples
Reprinted from: *Molecules* **2024**, *29*, 365, doi:10.3390/molecules29020365 42

Dimitrios Tsikas
Mass Spectrometry-Based Evaluation of the Bland–Altman Approach: Review, Discussion, and Proposal
Reprinted from: *Molecules* **2023**, *28*, 4905, doi:10.3390/molecules28134905 57

Dimitrios Tsikas
GC-MS Studies on Nitric Oxide Autoxidation and S-Nitrosothiol Hydrolysis to Nitrite in pH-Neutral Aqueous Buffers: Definite Results Using ^{15}N and ^{18}O Isotopes
Reprinted from: *Molecules* **2023**, *28*, 4281, doi:10.3390/molecules28114281 87

Electra Mermiga, Varvara Pagkali, Christos Kokkinos and Anastasios Economou
An Aptamer-Based Lateral Flow Biosensor for Low-Cost, Rapid and Instrument-Free Detection of Ochratoxin A in Food Samples
Reprinted from: *Molecules* **2023**, *28*, 8135, doi:10.3390/molecules28248135 99

Anthi Panara, Evagelos Gikas, Anastasia Koupa and Nikolaos S. Thomaidis
Longitudinal Plant Health Monitoring via High-Resolution Mass Spectrometry Screening Workflows: Application to a Fertilizer Mediated Tomato Growth Experiment
Reprinted from: *Molecules* **2023**, *28*, 6771, doi:10.3390/molecules28196771 114

Maria Filopoulou, Giorgios Michail, Vasiliki Katseli, Anastasios Economou and Christos Kokkinos
Electrochemical Determination of the Drug Colchicine in Pharmaceutical and Biological Samples Using a 3D-Printed Device
Reprinted from: *Molecules* **2023**, *28*, 5539, doi:10.3390/molecules28145539 129

Tatiana G. Choleva, Charikleia Tziasiou, Vasiliki Gouma, Athanasios G. Vlessidis and Dimosthenis L. Giokas
In Vitro Assessment of the Physiologically Relevant Oral Bioaccessibility of Metallic Elements in Edible Herbs Using the Unified Bioaccessibility Protocol
Reprinted from: *Molecules* **2023**, *28*, 5396, doi:10.3390/molecules28145396 140

Ioanna Dagla, Evagelos Gikas and Anthony Tsarbopoulos
Two Fast GC-MS Methods for the Measurement of Nicotine, Propylene Glycol, Vegetable Glycol, Ethylmaltol, Diacetyl, and Acetylpropionyl in Refill Liquids for E-Cigarettes
Reprinted from: *Molecules* **2023**, *28*, 1902, doi:10.3390/molecules28041902 150

Ourania Oikonomidou, Margaritis Kostoglou and Thodoris Karapantsios
Structure Identification of Adsorbed Anionic–Nonionic Binary Surfactant Layers Based on Interfacial Shear Rheology Studies and Surface Tension Isotherms
Reprinted from: *Molecules* **2023**, *28*, 2276, doi:10.3390/molecules28052276 160

Thomai Mouskeftara, Olga Deda, Grigorios Papadopoulos, Antonios Chatzigeorgiou and Helen Gika
Lipidomic Analysis of Liver and Adipose Tissue in a High-Fat Diet-Induced Non-Alcoholic Fatty Liver Disease Mice Model Reveals Alterations in Lipid Metabolism by Weight Loss and Aerobic Exercise
Reprinted from: *Molecules* **2024**, *29*, 1494, doi:10.3390/molecules29071494 175

Thomas Meikopoulos, Helen Gika, Georgios Theodoridis and Olga Begou
Detection of 26 Drugs of Abuse and Metabolites in Quantitative Dried Blood Spots by Liquid Chromatography–Mass Spectrometry
Reprinted from: *Molecules* **2024**, *29*, 975, doi:10.3390/molecules29050975 193

Natalia Manousi, Abuzar Kabir, Kenneth G. Furton and Aristidis Anthemidis
Sol-Gel Graphene Oxide-Coated Fabric Disks as Sorbents for the Automatic Sequential-Injection Column Preconcentration for Toxic Metal Determination in Distilled Spirit Drinks
Reprinted from: *Molecules* **2023**, *28*, 2103, doi:10.3390/molecules28052103 206

Argyro G. Gkouliamtzi, Vasiliki C. Tsaftari, Maria Tarara and George Z. Tsogas
A Low-Cost Colorimetric Assay for the Analytical Determination of Copper Ions with Consumer Electronic Imaging Devices in Natural Water Samples
Reprinted from: *Molecules* **2023**, *28*, 4831, doi:10.3390/molecules28124831 222

Christina Papatheocharidou and Victoria Samanidou
Two-Dimensional High-Performance Liquid Chromatography as a Powerful Tool for Bioanalysis: The Paradigm of Antibiotics
Reprinted from: *Molecules* **2023**, *28*, 5056, doi:10.3390/molecules28135056 233

About the Editors

Paraskevas D. Tzanavaras

Dr. Tzanavaras is an associate professor of Analytical Chemistry at the Laboratory of Analytical Chemistry of the School of Chemistry (Aristotle University of Thessaloniki, Greece). He has published more than 150 research and review articles in the field of automation, flow chemistry, chromatography, and optical sensors. He has also edited numerous Special Issues for international journals and co-authored seven book chapters in international collective volumes. His current research interests include liquid chromatography coupled to on-line post column derivatization and instrument-free optical sensors for the quality control of pharmaceuticals and bioanalysis.

Victoria Samanidou

Dr. Victoria Samanidou is full professor and director of the Laboratory of Analytical Chemistry in the School of Chemistry, Aristotle University of Thessaloniki, Greece. Since 2022, she has also been the vice president of the School of Chemistry. Her research interests focus on the development of sample preparation methods using sorptive extraction prior to chromatographic analysis, in accordance with green chemistry demands. She has co-authored 220 original research articles in peer-reviewed journals, 69 reviews, 90 editorials/opinions/commentaries, and 61 chapters in scientific books (h-index 48). She is an editorial board member of more than 31 scientific journals and guest editor of more than 36 special issues. In 2016, she was included in the top 50 power list of women in Analytical Science, as proposed by Texere Publishers. In 2021, she was included in the "The Analytical Scientist" power list of the top 100 influential people in analytical science. In 2023, she was included in the power list created in honour of The Analytical Scientist Magazine's 10-year anniversary, acknowledging excellence and impact over the past decade. In recent years, she has been included among the World Top 2% Scientists published in PLOS Biology, for single years and career, based on citations from SCOPUS. In 2023, she was awarded the Social Service Award by the Faculty of Sciences of Aristotle University of Thessaloniki. She is also vice president of the Committee of Ethics in the Research Committee of The Aristotle University of Thessaloniki. She was also a leader of the Working Group 1 Science and Fundamentals of EuChemS-DAC Sample Preparation Study Group and Network (2021). From 2016 onwards, she has been elected as president of the Steering Committee of the Association of Greek Chemists-Regional Division of Central & Western Macedonia.

Preface

Dear Colleagues,

We are pleased to present the printed version of the Special Issue dedicated to the 60-year anniversary of the foundation of the Laboratory of Analytical Chemistry (School of Chemistry, Aristotle University of Thessaloniki, Greece). Our intention was to launch a Special Issue in which contributions were not restricted to our laboratory alone. We, therefore, invited colleagues and friends from the other labs of our School of Chemistry and from other departments of Chemistry in Greece and abroad. Our goal was to gather high-quality research/review articles to celebrate 60 years since the foundation of our lab.

Due to contributions from almost all invited colleagues and the assistance of MDPI, we launched a successful Special Issue of high scientific quality. It contains seventeen research and review articles covering modern aspects of chemistry and/or analytical science. The "geography" of the SI is rather impressive, taking into account its anniversary character. Besides Thessaloniki, we received articles from colleagues and friends from Athens, Ioannina, Patras, Austria, Germany, and even Norway.

We are confident that interested readers will enjoy the contents of this SI. For example, Gikas and Tsarbopoulos reported GC-MS Methods for the analysis of Refill Liquids for E-Cigarettes, Kokkinos et. al. developed a 3D-printed device for the electrochemical determination of Colchicine, Christopoulos and co-workers reported DNA sensors for food authenticity studies, and Rosenberg et. al. developed a Multiplexing Injector for GC towards the analysis of Volatile Emissions from Lithium-Ion Batteries.

Once again we would like to express our gratitude to all contributing colleagues for the quality of the submitted research and to MDPI who provided the opportunity to launch this SI in a top-rated journal such as *Molecules*.

Paraskevas D. Tzanavaras and Victoria Samanidou
Editors

Article

Development of a Multiplexing Injector for Gas Chromatography for the Time-Resolved Analysis of Volatile Emissions from Lithium-Ion Batteries

Maria Antoniadou [1], Valentin Schierer [2], Daniela Fontana [3], Jürgen Kahr [2] and Erwin Rosenberg [1,*]

1. Institute of Chemical Technologies and Analytics, Vienna University of Technology, Getreidemarkt 9/164, A-1060 Vienna, Austria
2. Electric Drive Technologies, Electromobility Department, Austrian Institute of Technology GmbH, Giefinggasse 2, A-1210 Vienna, Austria; valentin.schierer@tuwien.ac.at (V.S.); juergen.kahr@ait.ac.at (J.K.)
3. FAAM Research Centre, Strada del Portone 13, I-10137 Torino, Italy
* Correspondence: erwin.rosenberg@tuwien.ac.at

Abstract: Multiplex sampling, so far mainly used as a tool for S/N ratio improvement in spectroscopic applications and separation techniques, has been investigated here for its potential suitability for time-resolved monitoring where chromatograms of transient signals are recorded at intervals much shorter than the chromatographic runtime. Different designs of multiplex sample introduction were developed and utilized to analyze lithium-ion battery degradation products under normal or abuse conditions to achieve fast and efficient sample introduction. After comprehensive optimization, measurements were performed on two different GC systems, with either barrier discharge ionization detection (BID) or mass spectrometric detection (MS). Three different injector designs were examined, and modifications in the pertinent hardware components and operational conditions used. The shortest achievable sample introduction time was 50 ms with an interval of 6 s. Relative standard deviations were lower than 4% and 10% for the intra- and inter-day repeatability, respectively. The sample introduction system and column head pressure had to be carefully controlled, as this parameter most critically affects the amount of sample introduced and, thus, detector response. The newly developed sample introduction system was successfully used to monitor volatile degradation products of lithium-ion batteries and demonstrated concentration changes over the course of time of the degradation products (e.g., fluoroethane, acetaldehyde and ethane), as well as for solvents from the battery electrolyte like ethyl carbonate.

Keywords: gas chromatography; multiplexing; multiplex sampling; time-resolved measurements; lithium-ion batteries

1. Introduction

Gas chromatography is the primary technique for the analysis of volatile and semi-volatile organic compounds in a large variety of environmental, industrial, biological or food/flavor-related samples. Its versatility is a consequence of the numerous options this technique offers for sample introduction, separation and detection that allow the customization of a GC system to the particular requirements [1]. Depending on whether qualitative or quantitative analysis is the aim, specific detectors such as mass spectrometry (MS) or universal detectors such as flame ionization (FID), thermal conductivity (TCD) or barrier discharge-ionization detection (BID) can be used. This latter detector, being a relatively recent addition to the range of gas chromatographic detectors [2], is an attractive alternative to both the FID and the TCD in that it offers sensitive, non-destructive and universal detection even of analytes that do not or hardly respond in flame ionization detection. It has thus been used in a variety of applications [3–9], including the analysis of volatile degradation products of the organic electrolyte of lithium-ion batteries [10,11].

Separation techniques are typically used where several analytes have to be determined simultaneously in complex matrices or where the interference-free determination of one or more analytes requires their separation from the matrix. Since separation is achieved in time (rather than in space), chromatography is a technique that—in its classical format—cannot be used for continuous monitoring of process streams. Various strategies have therefore been developed to overcome this shortcoming. Among the more common approaches are the use of short columns or miniaturized separation systems [12], separations under vacuum outlet (low pressure) conditions [13], low thermal mass/directly resistively heated systems and very recently, also negative thermal gradient GC systems [14].

A totally different approach that provides high(er) temporal resolution for chromatographic analyses is the use of multiplexed or multiplexing chromatography [15]. The core of this technique is that a sample is introduced into the separation system at predefined intervals. The samples are separated, and as the duration of a single separation is longer than the interval between sample injections, the individual chromatograms overlap, leading to a complex signal as a response. This signal must then be deconvoluted to arrive again at the individual chromatograms or an average chromatogram (Supplementary Materials, Figure S1).

The conditions that are to be fulfilled for successful multiplexing chromatography are the following:

(a) The sample is introduced at irregular intervals into the separation system according to a pre-defined so-called 'pseudo-random binary sequence' (PRBS) consisting of only "0" and "1" values where the former codes no sample introduction, while the latter stands for the introduction of sample;

(b) The interval of sample introduction (I) is much shorter than the interval between injections, the period T, and also the width of each individual signal (Figure 1);

(c) Chromatographic conditions must be stationary so that each injected sample is exposed to exactly the same separation conditions. This means in practice that separations have to be performed in the isocratic (for HPLC) or in the isothermal mode (for GC), which limits its practical applicability to samples with a relatively narrow polarity range (HPLC) or boiling point distribution (GC).

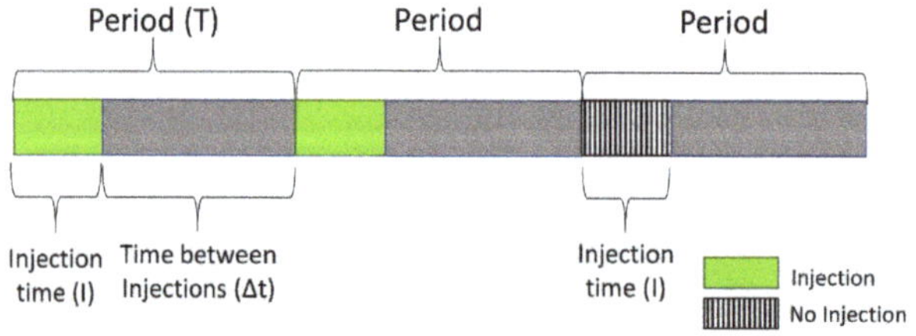

Figure 1. Schematic illustration of the terminology related to modulated injection sequences.

When these conditions are fulfilled, and additionally, the number of data points that have been acquired can be expressed as 2^n (ranging from $0\ldots m = 2^n - 1$), then it is possible to deconvolute the data using the Hadamard transform, or rather, the corresponding back-transformation. The Hadamard transformation is, similar to the Fourier transformation, an ideally loss-free linear transformation from one into another data space. It is typically used for data compression, signal processing and error correction [16]. It explains the formation of the convoluted chromatogram (represented by a vector of dimensionality m) by the product of the so-called convolution matrix $[S]$ (a $m \times m$ matrix derived from the PRBS (Equation (1)) [17]

$$[S] \times \begin{bmatrix} deconvoluted \\ chromatogram \end{bmatrix} = \begin{bmatrix} convoluted \\ chromatogram \end{bmatrix} \quad (1)$$

The deconvoluted chromatogram was obtained by multiplying the inverse convolution matrix with the convoluted chromatogram, as shown in Equation (2) (inverse Hadamard transformation):

$$[S]^{-1} \times \begin{bmatrix} convoluted \\ chromatogram \end{bmatrix} = \begin{bmatrix} deconvoluted \\ chromatogram \end{bmatrix} \quad (2)$$

The most important motivation for using multiplexing chromatography is to improve the signal-to-noise ratio of weak signals. While the signal increases n times with an n-fold performance of a measurement, the noise increases only by a factor of $\sqrt{n}$, therefore also improving the signal-to-noise ratio by a factor of $\sqrt{n}$ (known as 'Fellgett advantage', particularly in spectroscopy).

Generally, the application of multiplexing chromatography to improve time resolution rather than the signal-to-noise ratio is less considered but equally useful. In that case, injections are made from a sample stream whose concentration changes during the measurement sequence, and the objective is to represent these temporal changes with a better time resolution than the cycle time, which, particularly in temperature-programmed operation, can be considerably longer than the chromatographic run time [18]. Depending on the complexity of the resulting chromatogram, and particularly the degree of overlap of the individual signals, different strategies can be applied to deconvolute the data and to derive individual chromatograms in which not only the number of peaks and their retention time is correctly determined, but also their intensity or peak area. In case of a strong overlap, deconvolution of the convoluted chromatogram is required according to the inverse Hadamard transformation outlined above. In that case, the condition of non-periodicity of the sample introduction must be fulfilled, as otherwise, it is not possible to deconvolute the data. Data evaluation is based on the calculation of the average chromatogram as an intermediate result [19]. This chromatogram is then used to calculate the concentration profiles of the individual analytes over time. In the case of only partly overlapping or non-overlapping signals resulting from multiplex injection, there is no need for an inverse Hadamard transformation for the deconvolution of the data, and simpler algorithms can be applied. In the simplest case, the peaks do not overlap and can be evaluated directly if their position is known. To this end, it is no longer a requirement that the sample is introduced in a non-periodic sequence (e.g., the previously discussed PRBS), but on the contrary, it largely simplifies data processing if the sample is introduced in periodic intervals, which makes it easier to determine the relevant peak position.

The algorithm proposed in this work is more universally applicable: It relies on fitting the peak profiles of the individual peaks to the convoluted chromatogram. Provided the chromatographic system is not overloaded, the peak width remains essentially constant with the peak height scales and the concentration. This scaling factor is determined in the process of peak fitting, and the resulting peak is subtracted, while the scaling factor is a measure of its relative concentration. The positions at which the scaled peak is subtracted from the complex chromatogram are given by the retention time of the particular analyte in the first chromatogram under consideration of the time interval between injections (and whether an injection was completed or not in the case of non-periodic sample introduction sequences). This process is repeated for each signal in the initial chromatogram. Evidently, all analytes to be determined must already be present in the first chromatogram (but they may be absent in subsequent chromatograms). The algorithm was shown to work satisfactorily when overlapping peaks had a resolution of at least $R = 1$ and when the peak shape of individual peaks did not change during the multiplexed chromatogram as an effect of overload or column wear (see Figure S2 in the Supplementary Materials) [20].

Multiplexing chromatography requires particular consideration of how the sample introduction is achieved. Ideally, sample introduction is undertaken in a narrow injection pulse with a rectangular peak profile. This can be effected more easily in liquid phase separation (high-performance liquid chromatography, HPLC, and capillary electrophoresis, CE), where the sample is typically introduced through an injection valve with a sample loop of appropriate dimensions (HPLC [21,22]) or through either hydrodynamic or electrokinetic injection in CE [23]. In gas chromatography, the situation is more demanding: In the case of liquid sample introduction, ultrafast introduction and evaporation of the sample must be achieved, which calls for dedicated injector designs [24]. Although with gaseous samples, the use of a six-port valve with a gas sampling loop also appears an attractive option, there are two obstacles: First, multiplexing chromatography with high time resolution ($\Delta t = 2\ldots 3$ s) could mean a chromatographic run of a half hour up to 900 switching cycles, which is a huge number of switching cycles in a very short time, and this could lead to premature leakage and failure of the injection valve. And even then, the typical arrangement of the gas sampling valve upstream of the injector is not suitable for this type of operation if no particular precautions are taken. Depending on the instrument type, the volume of a classical split/spitless injector is in the order of 600–950 µL. Due to this relatively large volume, the injector creates the same effect as if a gaseous sample was introduced into a continuous stirred-tank reactor (CSTR). The characteristics of a CSTR are that with a delta input function (=controlled sample introduction during a short interval), the output function (=transfer of the analytes to the analytical GC column) would have a very steep left edge, while the right edge would follow exponential decay. This is unsuitable for multiplexing injection, which requires narrow, well-defined injection pulses; however, this situation is improving (at the cost of sensitivity) when a high split ratio is used. All these aspects are addressed by the proprietary design of an injector for gaseous samples where the actual location of sample introduction is within the injector in the immediate vicinity of the GC column head. Furthermore, the three designs investigated in this study (design A–C) replace the switching of a six-port valve for sample introduction by the opening and closing of solenoid valves (which are certified to be good for several hundred thousand switching cycles) for the controlled introduction of a defined sample volume. In contrast to this, examples reported in the literature mostly still make use of a six-port valve for sample introduction:

Published examples of multiplex-GC analyses include the detection of volatile organic compounds in indoor air [25], ethanol or toluene in exhaled breath after drinking or smoking [26], acetone in human breath [27] and hexamethyldisiloxane in a wafer cleanroom [24]. Recently, the use of switching valves in combination with column switching techniques has been used for fast sample introduction in the monitoring of catalytic reactions [28,29]. In general, a multiplex sample introduction is interesting for all applications where fast reactions take place and where a better time resolution than one data point per GC cycle time is required. One prominent example is the case of lithium-ion battery degradation products.

Lithium-ion batteries (LIBs) are widely used as an energy storage. To better understand the degradation mechanisms of the organic electrolyte and to improve the safety and performance of LIBs, qualitative and quantitative analysis with high-time resolution is needed. Different chromatographic techniques have already been used for the analysis of electrolyte degradation products [30]. The GC methods use temperature programs that span a wide temperature interval and consequently lead to long cycle times, including the cooling phase. Also, direct MS analysis is not a suitable option in this case, as there are important isobaric interferences that would not be resolved (C_2H_4/CO and CH_3CHO/CO_2, respectively) [31].

The current work addresses the development and investigation of three multiplex injector configurations based on headspace sampling and the introduction of gaseous compounds. Various parameters that can affect the analytical performance are examined and evaluated according to the relative standard deviation (RSD%) and peak area response values. To demonstrate the feasibility of our prototype multiplexing injector, we performed an in situ analysis of lithium-ion battery degradation products and obtained time-resolved chromatograms for a number of relevant compounds.

2. Materials and Methods

2.1. Materials

All reagents had a purity of at least 95%. *n*-heptane was purchased from Fluka (Buchs, Switzerland). Ethanol was obtained from Merck (Darmstadt, Germany), acetonitrile and dimethyl carbonate (DMC) battery grade from Sigma Aldrich (Vienna, Austria) and ethyl methyl carbonate (EMC) Selectilyte from BASF (Ludwigshafen, Germany). Helium, used as the GC carrier and BID discharge gas, was of purity ≥99.999% and was purchased from Messer (Gumpoldskirchen, Austria). All battery cell components were provided by Lithops S.r.l., now FAAM Research Center (Torino, Italy), and the samples were prepared in a commercially available test cell for the in situ analysis of gas species in Li-ion systems, the ECC-DEMS (EL-CELL, Hamburg, Germany).

2.2. Instrumentation

The instrumentation used for the investigation of different multiplex injector configurations was a GC-2010 Plus gas chromatograph with a Tracera BID barrier-discharge ionization detector equipped with an external 6-port switching valve from Shimadzu (Kyoto, Japan). The column was a DB-5MS (5% phenyl-95% methyl-polysiloxane) 30 m × 0.25 mm × 0.25 µm from Agilent J&W. A GC-MS-QP2010 Plus instrument (Shimadzu) equipped with a Rt-Q-BOND (100% divinylbenzene, Restek, Bellefonte, PA, USA) PLOT (porous layer-open tubular) column of 30 m × 0.32 mm × 10 µm dimension equipped with a particle trap and a guard column was used for real sample analysis. For the laboratory-made injector, an Arduino Leonardo ETH board (RS Components, Frankfurt am Main, Germany) with a code written in-house for Arduino Software (IDE) v1.8.9 (https://www.arduino.cc/en/main/software, accessed on 30 December 2023) was used for the control of the solenoid valves. The normally open 2-way, normally closed 2-way and 3-way solenoid valves used for the work were purchased from Bürkert Austria GmbH (Vienna, Austria). Tee connectors were from VICI Valco (Schenkon, Switzerland).

2.3. Analytical Procedure

Sample injection was performed with the laboratory-made injectors as described below. The final method used with the BID detector was an injector temperature of 250 °C, linear velocity of 22.7 cm s^{-1}, split of 10:1, discharge gas flow of 50 mL min^{-1}, column head pressure of 85.6 kPa, total flow of 14.0 mL min^{-1}, column flow of 1.00 mL min^{-1}, purge flow of 3 mL min^{-1}, oven temperature of 35 °C isothermal, and BID temperature of 300 °C. The analysis runtime varied depending on the sequence length from 5 to 100 min as all data were stored in one single data file. The method used on the GC-MS instrument had an injector temperature of 270 °C, split of 8:1, column head pressure of 87 kPa, column flow of 2.22 mL min^{-1}, purge flow of 1 mL min^{-1}, oven temperature program of 100 °C, ion source temperature of 220 °C, interface temperature of 220 °C, and MS scan range of 30 to 300 m/z.

2.4. Multiplex Injector Configurations

Three designs for a multiplex injector were developed and examined. These were the one-valve (A), the two-valve (B) and the three-valve (C) design. Although some of the injector designs could have equally been realized with a six-port switching valve, preference was given to solenoid on/off- or three-way valves due to the longer durability, keeping in mind that a single chromatographic run could require several hundred switching cycles and that during regular operation, tens of thousands of switching cycles could be performed within few days at which regular six-port valves often fail. The general idea of the first two sample introduction device designs was that the gaseous sample was introduced through a fused silica capillary into the GC's injector, which itself was confined within or connected to a stainless-steel capillary. The third introduction device consisted of tubing connections and a stainless-steel sample loop. Depending on the pressure conditions in the injector, at the head of the fused silica capillary and the stainless-steel capillary, and, of course, the valve switching, the sample was or was not introduced into the GC injector and transferred to the GC column.

For design A (Figure 2), a He makeup gas was added to the sample gas stream. Supported by this make-up gas stream, the sample was transported toward the GC through an uncoated capillary column. The capillary was fitted through a Tee connector and this also, into a hypodermic needle, which was inserted through the septum into the GC injection port. The capillary end was close to the needle's end but still confined within the capillary. The other side of the Tee connector was connected to a normally open solenoid valve and through this to the waste. When the sample was not injected, the valve was opened, and the sample that eluted from the fused silica capillary was directed to waste because the GC column head pressure was higher than the pressure at the end of the transfer line, thus preventing the sample from reaching the GC column. In order to inject a portion of the gas stream, the valve was closed, which caused a back-pressure build-up. At one point, the backpressure created by the blocked flow path exceeded the column head pressure, and thus, the sample was injected into the GC column.

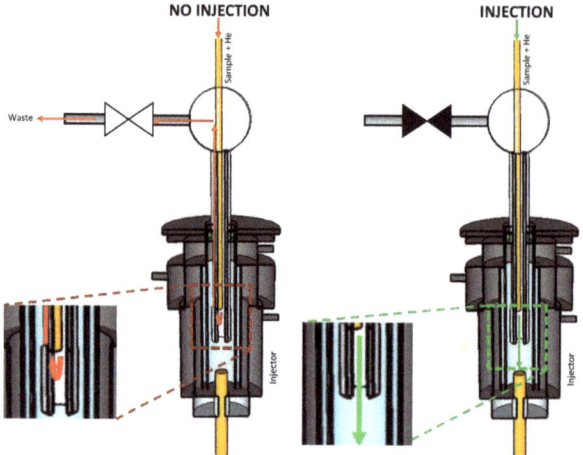

Figure 2. Scheme of multiplex injector design A. The green and red arrows indicate the flow of the carrier gas stream (transporting the sample) into and out of the GC's split/splitless injector when the sample was introduced ('Injection') or not ('No injection'). ▶◀ represents a solenoid valve in the 'closed' state, while ▷◁ represents the valve in the 'open' state.

Figure 3 shows the second design (design B), which is similar to the first, up to the point where the uncoated capillary reaches the Tee connector. The Tee connector was connected to a normally closed valve at the injection port side and a normally open valve at the waste side. When no injection was performed, the normally closed and normally open valves stayed in their default state. The sample that eluted from the capillary was directed to waste because the injection port side was closed. To inject, the normally open and closed valves were now closed and opened, respectively. This means that the flow path to waste was closed, and the sample was forced to the injection port side, which was then open.

Design C (Figure 4) was different from the other two because it was a three-valve set-up that included one 3-way, one 2-way normally open, and one 2-way normally closed solenoid valve. The three-way valve was connected to the He supply from one side and to the sample, which was transported by He make-up gas from the other side. The third port was connected to a length of tubing that defined the sampling volume, similar to the sampling loop of a six-port valve. The tubing was connected via a Tee connector to the normally open valve leading to the waste and to the normally closed valve leading to the injection port. When not injecting, the sample flushed the tubing/loop while connected to the waste line. When injecting the sample, the 3-way valve switched from the sample to the He supply (kept at a higher flow than the sample). The normally open valve closed, and the normally closed valve opened. This led to the sample being injected into the GC column instead of going to waste since the flow path to waste was closed.

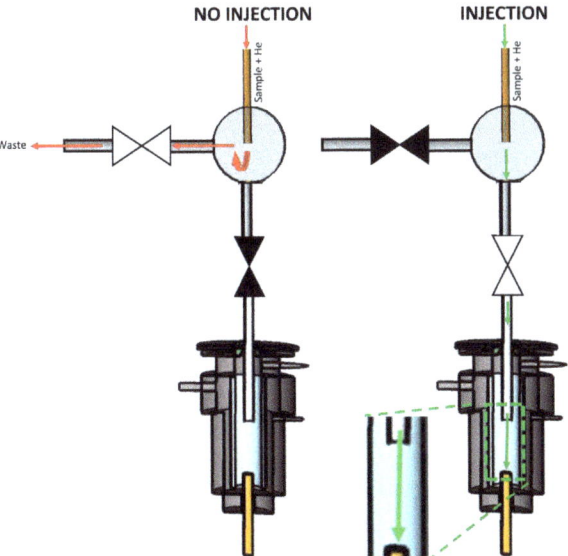

Figure 3. Scheme of multiplex injector design B. The green and red arrows indicate the flow of the carrier gas stream (transporting the sample) into and out of the GC's split/splitless injector when the sample was introduced ('Injection') or not ('No injection'). ▶◀ represents a solenoid valve in the 'closed' state, while ▷◁ represents the valve in the 'open' state.

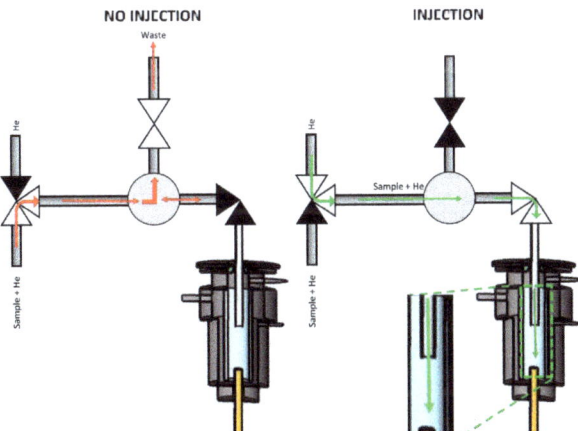

Figure 4. Scheme of multiplex injector design C. The green and red arrows indicate the flow of the carrier gas stream (transporting the sample) into and out of the GC's split/splitless injector when the sample was introduced ('Injection') or not ('No injection'). ▶◀ represents a solenoid valve in the 'closed' state, while ▷◁ represents the valve in the 'open' state. ▷▼◁ represents a three-way solenoid valve with the open symbols representing the open flow path and the closed symbol representing the closed flow path.

3. Results

The three different multiplexing injector designs described in the experimental section were mounted on a GC-BID instrument for further investigations. As a consequence of the multiplexed sample introduction, this system is capable of providing time-resolved chromatograms and improved time resolution. All three designs—denoted 'Design A', 'Design B' and 'Design C' in the following discussion—allow the computer-controlled introduction of a gaseous sample at precisely controlled intervals and sample introduction lengths.

At first, the investigation focused on whether each design was functional or not, and in the second stage, each system was optimized. The compound chosen for the initial testing was heptane, a compound in the middle of the boiling point range with respect to the intended application for LIB emission monitoring, which also shows very good peak shape and detectability with the BID detector. Different parameters and variations of the experimental setup were tested for their influence on the results. Among those were the use of a vacuum pump in the waste line, the inner diameter of the sample introduction capillary, the position of the sample introduction capillary inside the injector, the inner volume of the Tee connector and the length and inner diameter of the tubing connections.

3.1. Investigation of Experimental Setup

Initial experiments with different experimental setups were used with a vacuum pump in the waste flow line. The rationale behind this setup was to create a greater pressure drop between the column head pressure in the GC injector and the pressure at the end of the waste line and thus to efficiently prevent the leakage of sample from the sample introduction capillary into the GC injector during periods of no injection. For testing the usefulness of the pump, the conditions were a 50 °C oven temperature, 1 mL min^{-1} column flow, 20:1 split flow, 50 mL min^{-1} sample flow, 1 s injection time, tested pump flows: 0; 5; 15; 75; and 115 mL min^{-1} (measured at the outlet of the membrane pump). The experimental results showed that the use of the pump leads to irreproducible and somewhat variable amounts of sample being introduced into the GC (Supplementary Materials, Figure S3). The most probable reason for the peak area decreasing with an increasing pump flow rate is that the increasing suction created by the pump affects, i.e., reduces, the amount of sample introduced with each injection. Results were obtained successfully with short tubing connections with smaller inner volumes and without using a vacuum pump in the waste line. Short tubing lengths and small inner volumes were favorable as they resulted in faster sample introduction into the GC and more accurate results. In contrast, greater tubing lengths and larger inner volumes not only resulted in slower sample transfer but also gave rise to carryover between injections. While this problem can be resolved by increasing the sample flow or the injection time, this option is less attractive as it results in high gas and sample consumption. In the subsequent experiments, no pump was used in the waste line for either of the three injector designs. Maximum tubing lengths for designs A and B were 14 cm (inner volume: 90 µL) and 18 cm (inner volume: 100 µL), respectively. For design C, the tubing after the normally closed valve was as short as technically possible and with a small ID (inner volume: 55 µL) to avoid a memory effect from a not adequately flushed sample. The other tubing parts had larger IDs (inner volume: 130 or 530 µL). The three-way valve design mimics the function of a sampling-loop-based injection valve, in contrast to the other two designs, which perform time- and pressure-driven injections. The injected sample volume was calculated from the sample flow and injection time for the time-driven injections. When using lower sample flows, the smaller sample volumes did not disturb the equilibrium between the injector and the instrument, and the injection was achieved more easily. For a loop-based injection (and similarly, for design C, which mimics a loop-based injection), the loop volume defined the injected volume, which means that it had to be chosen adequately.

Normally, in this system, the pressure that comes from the multiplex injector has to be higher or equal to the GC pressure ($P_\text{inj} \geq P_\text{inst}$). This is a prerequisite for a pressure-controlled sample introduction (see Supplementary Materials, Figure S4). With a multiplex injector, the pressure drop (ΔP) between the initial pressure before the sample (P_{init}) and the pressure at the end of the injector (P_{inj}) must be calculated. The pressure drop must not significantly affect the P_{inj}. However, this was also determined by the requirements of the sample introduction. Figure 5 shows the relation between the P_{inj}, the sample flow and the capillary ID. Values for P_{inj} were calculated for different combined flows of sample + He and capillary IDs, depending on the total tubing lengths and IDs according to the Hagen-Poiseuille equation. The parameter range, which is suitable for injection, was identified. The pressures were calculated at sample flows of 1-, 5-, 10- and 20-mL min^{-1} and 0.10-, 0.25-, 0.32- and 0.5-mm ID. As can be seen, there was a specific parameter range where injection could be successfully achieved (highlighted in color in Figure 5). This was always dependent on the column head pressure P_{inst} of the instrument. Thus, the accessible working range can change depending on the P_{inst}.

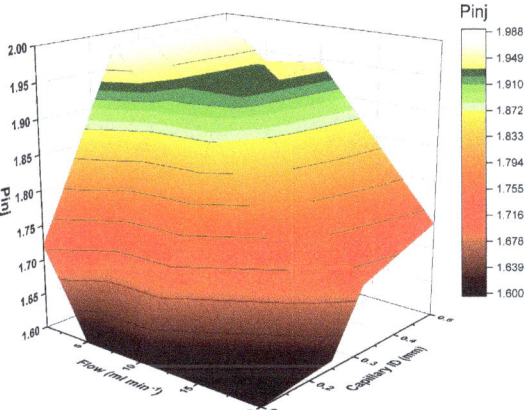

Figure 5. Interrelationship between sample flow rate, capillary ID and required injection pressure P_{inj} (in bar) for successful sample introduction.

To this end, the effect of capillary inner diameter (ID) for the first design was investigated. The investigation reported here was completed under isothermal conditions mentioned in Section 2.3. The tested capillary IDs were 0.10, 0.25 and 0.32 mm. Only the 0.25- and 0.32-mm IDs could be compared because, for the capillary with a 0.1-mm ID and the practically achievable pressure ratio, there was no or only a very small amount of sample injected. A possible reason for this behavior could be the high pressure drop between P_{init} and P_{inj}, resulting in lower P_{inj} from the P_{inst}. The comparison of the analyte signals from the corresponding chromatograms (Figure 6) showed that the 0.25-mm ID capillary produced much higher signal intensities for the same sample. Injections were achieved with injection times as short as 50 ms, but the corresponding sensitivity was low. Of note is that the injected amount increased only after 300 and 500 ms for the 0.25- and 0.32-mm ID capillaries, respectively. The trends looked similar. The only difference was the 0.32-mm ID capillary needed longer injection times to start reaching peak areas similar to those of the 0.25-mm ID column. The difference comes from the pressure drop, which resulted in different sample amounts being injected.

Another important aspect to investigate was the bore size of the Tee connector, which was important for all designs. Tested bore sizes (and corresponding dead volumes, in brackets) were 0.25 mm (0.47 µL), 0.75 mm (4.2 µL), 1 mm (7.5 µL), and 1.5 mm (34.8 µL). The bore sizes affected responses in two ways: With an increasing diameter, the dead volume of the connector increased, making the pressure build up slower and thus also injections slower, and even back-mixing possible. In turn, the pressure drop for the waste line was lower with an increasing bore size, allowing faster removal of the effluent gas stream, which in turn should make injection peaks narrower. Like with the capillary, we could not obtain results with the smallest bore size (0.25 mm), but we managed to compare the remaining ones and selected the 0.75 mm as the most appropriate for achieving higher intensity results for designs B and C (Figure 7) and all three bore sizes (0.75-, 1- and 1.5-mm ID) work successfully with different sample flow rates for the design A. The higher bore sizes worked well, but provided lower sample intensities under the same conditions. Therefore, an increase in the sample flow was needed for those. Unlike the capillary change, which is closely related to the pressure drop, and the structural changes in the bore size affect the flow, which then changes the injected sample amounts.

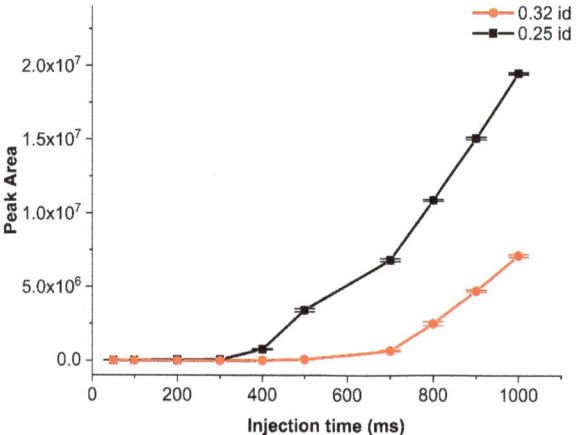

Figure 6. Comparison of the peak area of heptane for two sample introduction capillaries of different inner diameters (0.25 and 0.32 mm) with increasing injection time for design A. The isothermal method is mentioned in Section 3.3 with 60 mL min^{-1} sample flow. Intervals represent one standard deviation around the mean value.

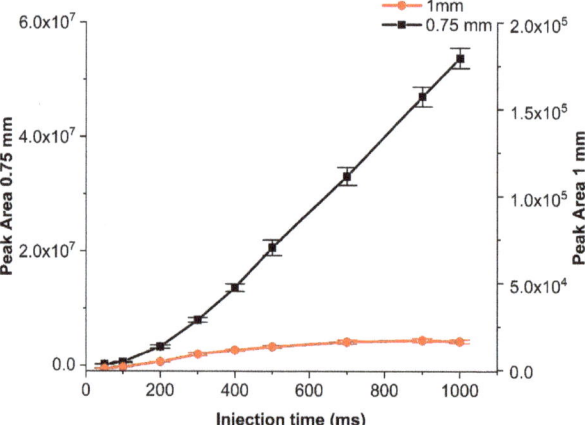

Figure 7. Comparison of the heptane peak area for two Tee connector bore sizes (1 and 0.75 mm) with increasing injection time for design B. The isothermal method is mentioned in Section 3.3 with a 7-mL min^{-1} sample flow. Error bars indicate ±1 s.

3.2. Investigation of Instrumental Parameters

Further instrumental parameters such as split ratio, column flow, sample flow, injection time, and time interval between injections were optimized. The tested split ratios were 10:1, 20:1, 75:1, 100:1, 200:1, 400:1 and 500:1, and column flows of 0.5, 0.75, 1, 1.25, 1.5 mL min^{-1}.

The higher the split ratio, the faster the sample was transferred to the GC column for a 'normal', i.e., liquid phase injection, where a large solvent vapor cloud was formed from the liquid volume injected. In the case of gaseous sample introduction for sub-second periods, this effect was not relevant, and the dilution effect of increasing split ratios was dominant. Hence, and also to reduce carrier gas consumption, a lower split ratio of 10:1 was adopted. Figure 8 demonstrates the effect of the column flow and sample flow on the peak area. Higher column flows moved the sample faster and, as can be seen, resulted in a lower response because of the change in the P_{inst} (Figure 8). The split ratio and the column flow were parameters that could change depending on the P_{inst} we wanted to achieve. In the present work, we needed to work with lower sample flows because our aim was to use the multiplex injector for the investigation of lithium-ion battery emissions, which (due to the small cell volume) allow only very low flow rates. So, the final split ratio was 10:1, and the column flow used was 1 mL min^{-1}. An increase in the column flow or the split flow typically increases the column head pressure. An increased column head pressure will change the pressure situation between the front end of the sample introduction capillary and the GC injector. This may, in the extreme case, lead to a situation where there is no pressure drop toward the GC injector or, if the pressure drop is even inverted, result in no sample being injected. This demonstrates that the optimization of pressure and flow parameters in the multiplexing sample introduction device and the GC injector is delicate and decides whether the sample is successfully introduced or not into the GC. For the optimization of the further parameters, the conditions of GC flow and pressure were the same as reported before, with the only difference that during one run, a sample was injected multiple times (20 times). The investigated sample flows were 50, 80-, 100-, 120- and 140-mL min^{-1} with 50 ms injection time, with time intervals 0.5, 1, 2, 3, 4, 5, 6, 7, 8, 9 and 10 s between samples and the injection times were 10, 20, 40, 50, 100, 200, 300, 400, 500, 600, 700, 800, 900 and 1000 ms. For the first of our injector designs, lower sample flows (below 50 mL min^{-1}) with the use of the 1.5 mm bore Tee union did not reach the necessary P_{inj}. Higher sample flows showed faster sample introduction (due to the fact that the necessary pressure for injection was built up in a much shorter period of time), and thus, lower injection times of 50 ms could be used (Figure 8). With increasing sample flow, the pressure at the end of the sample introduction capillary increased above the column head pressure in the injector ($P_{inj} > P_{inst}$), and more of the sample was introduced. If the P_{inj} is too high, this could result in sample overload. The flow must be high enough to ensure fast sample introduction at the ms time scale. The most appropriate time interval and injection time can be chosen depending on the selected flow. A short time interval (between injections) was desirable as it would determine the temporal resolution of the method. Longer interval times result in fewer data points of the measurement. However, very short interval times can result in significant peak overlap, which will be difficult to deconvolute. In our case, the lowest sampling interval time was 6 s (Table 1). In the case of the injection times, the higher they are, the higher the overload and broadened peak shape of the outcome. This was observed when using a high sample flow of 150 mL min^{-1}. The peak asymmetry was 1.14 ± 0.03 at 50 ms and reached an asymmetry of less than 0.4 at injection times higher than 200 ms. The selected conditions for injector design A were 60–100 mL min^{-1} sample flow (depending on the bore size used), 6 s interval time, and 50 ms injection time. The injection time can vary from 50 ms to 1 s or more as it is a way to increase the injected volume. Similar tests were undertaken for injector designs B and C. With these, we managed to reduce the sample flow to 5–7 and 0.2–1 mL min^{-1}, respectively. This decrease in sample flow rate became possible by the inclusion of an additional normally closed solenoid valve directly before the injector, which isolated the injector from the GC when the sample was not injected.

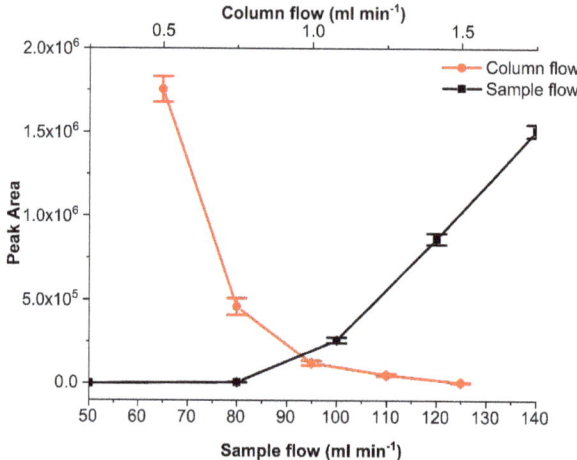

Figure 8. Heptane peak area changes at various columns and sample flows. The isothermal method is mentioned in Section 3.3. Error bars in the graph represent ±1 s.

Table 1. Analytical characteristics of design A for different interval times (Δt).

Interval Time [s]	Peak Area [1]		Peak Height [1]		Asymmetry [2]	
	Mean	RSD%	Mean	RSD%	Mean	RSD%
2	693,516	5.6	338,558	6.7	-	-
3	700,262	2.2	297,074	0.5	-	-
4	698,850	0.9	287,342	0.8	-	-
5	694,628	0.8	286,460	1.2	0.75	1.5
6	672,199	3.0	281,610	1.3	1.13	2.8
7	667,479	1.0	281,128	0.8	1.14	1.5
8	680,908	0.5	285,048	0.4	1.16	0.0
9	681,597	1.0	283,220	1.1	1.13	3.2
10	797,033	2.0	330,935	2.7	1.11	3.0

[1] Mean and RSD% values were calculated from 5 data points. [1] Where asymmetry values are not reported, peaks are not resolved at 10% of the peak height and asymmetry can consequently not be calculated.

Further to the investigation of the used conditions, it was crucial to test whether the retention times were reproducible or not. Reproducible retention times are highly important for the deconvolution of the raw data with the Hadamard transform algorithm and other deconvolution algorithms developed for this work. This was tested after subsequently injecting the same amount of heptane 15 times into the system (three repetitive runs). In all cases, the interval between the injection times of two neighboring peaks was 0.01 min, with relative standard deviations less than 0.01%. For design A, the peak asymmetry was 1.09 ± 0.02, and the relative standard deviation (RSD%) for peak area and height was 3.8 and 2.9%, respectively. These favorable precision data also confirmed the injection time stability (in this case, 50 ms) since the amount of sample introduced was proportional to the injection time. Furthermore, experiments according to a 4-bit PRBS (15 injections) during different days showed RSD% values for the peak area and height of 8.2 and 8.4%. The peak asymmetry average values were 1.07 ± 0.01. For design B, the peaks for repeated injections of the same analyte constantly increased while the analyte's concentration was constant. Even after taking measures against possible memory effects from the preceding injection, the relative standard deviation remained larger than 20%, and this design, therefore, was not used further for the experiments. For design C, the relative standard deviation (RSD%) for peak area and height was 3.3 and 6.2%, respectively. Experiments with random injections during different days showed RSD% values for the peak area and height of 8.4 and 4.2%.

Peak asymmetry was found to be 1.28 ± 0.01. The advantage of design C over design A is that, while the latter, the system is easily affected by changes in the experimental set-up, the former offers the possibility of more accurate determination of the injected volume and, thus, more reproducible and accurate measurements. Additionally, with design C, the instrument was protected from unwanted sample injections during the start-up phase of the system, which was noticed with design A on another GC system.

3.3. Comparison to a 6-Port Switching Valve

Experiments were also performed with the 6-port gas sampling valve provided as a sample introduction option for the GC-BID instrument that was housed in a thermostatted valve box and compared to the designs we had already examined. The first drawback we faced was that all operations of the valve (injection, switching back) were limited by the time base of the instrument's software. The smallest injection/switching time that could be set was 0.01 min (600 ms), which was much longer than the 50 ms achievable with the designs developed in-house. The second drawback consists of the relatively slow switching speed of the electrically actuated switching valve, which did not allow the use of loop filling/injection times of less than 0.02 min to inject the sample into the GC efficiently. It is evident that these problems could be resolved by controlling sample introduction externally, and by using a high-speed switching valve, but these options are not available at the present time. For the experiments that were performed with the switching valve, peak asymmetry was found to be 1.32 ± 0.017, and the RSD values for peak area and height were 0.8% and 1.2%, respectively.

3.4. Application to the Volatile Emissions from Lithium-Ion Batteries

As a realistic test for the versatility of the above-described multiplexing sample introduction system for GC, the developed device was used to study the emissions of a dummy electrochemical cell with GC-BID, while measurements of the degradation products from the electrolyte of a lithium-ion battery by multiplexing-GC/MS were performed for safety reasons at another facility. A dummy cell containing the compounds of interest imitated the transient emission of typical battery electrolyte degradation products. It initially included five compounds: ethanol; ethyl methyl carbonate (EMC); dimethyl carbonate (DMC); acetonitrile; and heptane. Air ($O_2 + N_2$), CO_2 and water leaked into the dummy cell at a later stage or were created from the electrolyte decomposition. The tests were performed with injector design C. Sample injections were performed every 2.02 min in order to obtain a multiplex chromatogram (Figure 9). These were subsequently processed and produced chromatograms, which showed the peaks attributable to seven individual compounds. Figure 10 depicts the peak area change in each compound over time and imitates the battery discharging state where most degradation products show a decrease. The test demonstrated that the multiplex injector and the data processing can successfully handle even complex samples. The investigation was continued by coupling the multiplex injector to a GC-MS instrument for the overcharge experiments (conditions specified in Section 3.3). The tested cell was assembled from am NMC 1:1:1 ($Li_{1-x}(Ni_{0.33}Mn_{0.33}Co_{0.33})O_2$) lithium nickel-manganese-cobalt (1:1:1) oxide cathode prepared vs. a graphite anode in an EL-Cell. During the overcharging experiment, the battery voltage was increased by applying constant current (CC) above its recommended maximum potential. The testing was undertaken after performing a specific constant current–constant voltage (CCCV) formation cycle. This experiment—in which the particular sample introduction device that was previously mounted on the GC-BID system was installed on a GC-MS system—demonstrated that the multiplex injector unit is easily transferable to other instruments and allows comparing the results with the standard sampling method that is usually used. Overcharging experiments were conducted until 250% of the rated capacity. This was achieved by applying the 1C rate for fast overcharging while using multiplex sampling (180 chromatograms were recorded in three hours). In contrast, the monitoring of degradation products by temperature-programmed separation was significantly slower, where the overcharging was produced by a C/3 rate (24 chromatograms in 12 h, Figure 11b). For the 3 h experiments, a drift in the baseline was observed without having a significant effect on the peak identification (Figure 11a). The change is attributed to the high amount of solvents emitted from a Li-ion battery, which continuously accumulate as samples and are constantly introduced into the system. From real battery testing, useful information on the battery degradation products was obtained during overcharging. The compounds detected and investigated for their concentration change over time were CO_2, ethane, water, acetaldehyde, fluoroethane, DMC and methyl formate. The overcharge process resulted in an increase in CO_2, fluoromethane and methyl formate formation.

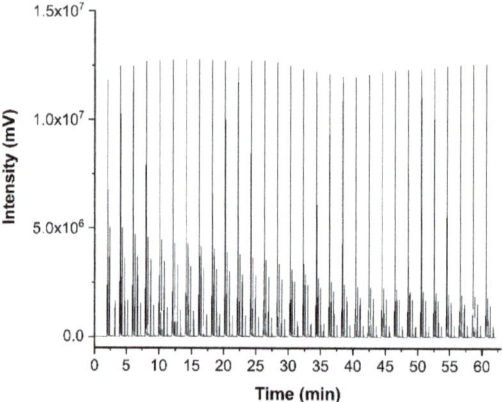

Figure 9. Multiplex chromatogram resulting from a dummy cell measurement with design C. The isothermal method is mentioned in Section 2.3. Seven compounds were analyzed.

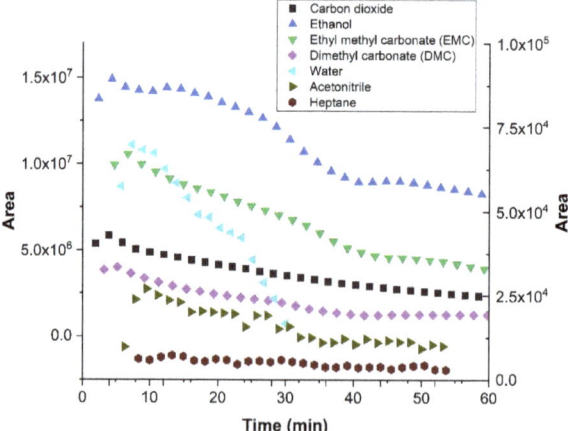

Figure 10. Change in concentration of dummy cell analytes during the experiment (30 injections in 60 min). The isothermal method, as specified in Section 2.3.

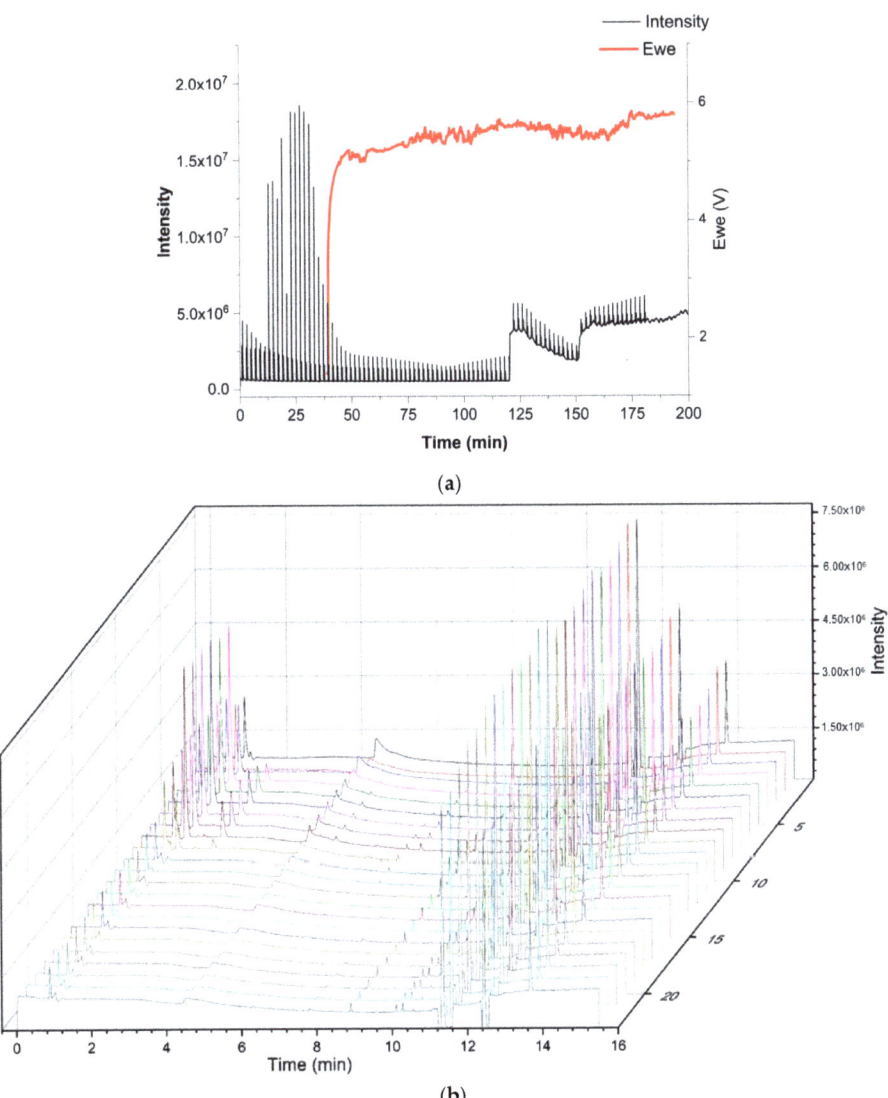

Figure 11. Real lithium-ion battery testing experiment of NMC 111 cell with the design C (**a**) (top) multiplexing and (**b**) (bottom) normal gradient conditions during overcharge. (E_{we}: potential of a working electrode in V).

4. Conclusions

In the current study, we developed three different designs for multiple sample introduction into a GC system. The operational characterization of the three devices showed that two of the designs (A and C) successfully worked after optimization of the respective working conditions. This included the parameters sample flow, column flow and injection time on which the analytical result critically depends, particularly. Another important aspect of the success of the proposed setup was the solenoid valve which had to be fast-switching and had a very small dead volume.

The multiplex system developed in this study provides the advantage of repeated sample introduction in a s short time, which, together with the in-house developed software for data deconvolution, allows the acquisition of a significantly larger number of chromatograms per given time and thus obtains a larger amount of information for a transient process compared to sequential data acquisition. The multiplexing injector is self-confined and can be installed on different instruments, for example, GC with atmospheric pressure detectors such as the BID or even with vacuum detection such as a mass spectrometer (adjustment of the optimum operating parameters is required then).

Finally, GC with multiplex sample introduction was applied in the monitoring of LIB degradation products. Recording the chromatograms in shorter intervals along the transient process allowed for a better understanding of the reactions that take place under normal or abusive conditions and identifying hazardous compounds and their concentration changes.

Supplementary Materials: The following supporting information can be downloaded at: https://www.mdpi.com/article/10.3390/moleculesXXXXX/s1, Figure S1: Graphical representation of the production of a multiplexed chromatogram; Figure S2: Algorithm for the deconvolution of multiplexed chromatograms; Figure S3: Change of peak area response with increase of pump flow in the waste line; Figure S4: Sketch of the different pressures within the GC system in order to achieve a pressure-driven injection into the GC; Figure S5: Construction drawing (left and middle) and photo (right) of the actual set-up of the multiplex injector (Design C), mounted on the rear injection port of a Shimadzu 2010 GC.

Author Contributions: Conceptualization, E.R. and J.K.; methodology, E.R. and M.A.; software, M.A.; validation, M.A. and E.R.; formal analysis, M.A. and V.S.; investigation, M.A. and V.S.; resources, E.R., J.K. and D.F.; data curation, M.A. and V.S.; writing—original draft preparation, M.A.; writing—review and editing, M.A., E.R. and J.K.; visualization, M.A.; supervision, E.R. and J.K.; project administration, E.R. and J.K.; funding acquisition, E.R. and J.K. All authors have read and agreed to the published version of the manuscript.

Funding: This research was funded by the Austrian Research Promotion Agency (FFG) under project number 858298 ("DianaBatt"), which is gratefully acknowledged.

Institutional Review Board Statement: Not applicable.

Informed Consent Statement: Not applicable.

Data Availability Statement: The raw data supporting the conclusions of this article can be made available by the authors on request.

Acknowledgments: The kind support of Wolfgang Tomischko for building the Arduino microprocessor and Andreas Genner for programming the initial algorithm for time-scheduled injection is also acknowledged with thanks.

Conflicts of Interest: The authors declare no conflicts of interest.

References

1. Fowlis, I.A. *Gas Chromatography*, 2nd ed.; Analytical Chemistry by Open Learning; John Wiley & Sons: Chichester, UK, 1995.
2. Shinada, K.; Horiike, S.; Uchiyama, S.; Takechi, R.; Nishimoto, T. *Development of New Ionization Detector for Gas Chromatography by Applying Dielectric Barrier Discharge*; Shimadzu Review: Kyoto, Japan, 2012.
3. Jo, S.H.; Kim, K.H. The applicability of a large-volume injection (LVI) system for quantitative analysis of permanent gases O_2 and N_2 using a gas chromatograph/barrier discharge ionization detector. *Environ. Monit. Assess.* **2017**, *189*, 317. [CrossRef] [PubMed]
4. Frink, L.A.; Armstrong, D.W. The utilisation of two detectors for the determination of water in honey using headspace gas chromatography. *Food Chem.* **2016**, *205*, 23–27. [CrossRef]
5. Weatherly, C.A.; Woods, R.M.; Armstrong, D.W. Rapid Analysis of Ethanol and Water in Commercial Products Using Ionic Liquid Capillary Gas Chromatography with Thermal Conductivity Detection and/or Barrier Discharge Ionization Detection. *J. Agric. Food Chem.* **2014**, *62*, 1832–1838. [CrossRef]
6. Ueta, I.; Nakamura, Y.; Fujimura, K.; Kawakubo, S.; Saito, Y. Determination of gaseous formic and acetic acids by a needle-type extraction device coupled to a gas chromatography-barrier discharge ionization detector. *Chromatographia* **2017**, *80*, 151–156. [CrossRef]

7. Pascale, R.; Caivano, M.; Buchicchio, A.; Mancini, I.M.; Bianco, G.; Caniani, D. Validation of an analytical method for simultaneous high-precision measurements of greenhouse gas emissions from wastewater treatment plants using a gas chromatography-barrier discharge detector system. *J. Chromatogr. A* **2017**, *1480*, 62–69. [CrossRef] [PubMed]
8. Franchina, F.A.; Maimone, M.; Sciarrone, D.; Purcaro, G.; Tranchida, P.Q.; Mondello, L. Evaluation of a novel helium ionization detector within the context of (low-)flow modulation comprehensive two-dimensional gas chromatography. *J. Chromatogr. A* **2015**, *1402*, 102–109. [CrossRef] [PubMed]
9. Antoniadou, M.; Zachariadis, G.A.; Rosenberg, E. Investigating the performance characteristics of the barrier discharge ionization detector and comparison to the flame ionization detector for the gas chromatographic analysis of volatile and semivolatile organic compounds. *Anal. Lett.* **2019**, *52*, 2822–2839. [CrossRef]
10. Schmiegel, J.-P.; Leißing, M.; Weddeling, F.; Horsthemke, F.; Reiter, J.; Fan, Q.; Nowak, S.; Winter, M.; Placke, T. Novel In Situ Gas Formation Analysis Technique Using a Multilayer Pouch Bag Lithium Ion Cell Equipped with Gas Sampling Port. *J. Electrochem. Soc.* **2020**, *167*, 060516. [CrossRef]
11. Leißing, M.; Winter, M.; Wiemers-Meyer, S.; Nowak, S. A method for quantitative analysis of gases evolving during formation applied on $LiNi_{0.6}Mn_{0.2}Co_{0.2}O_2$‖natural graphite lithium ion battery cells using gas chromatography—Barrier discharge ionization detector. *J. Chromatogr. A* **2020**, *1622*, 461122. [CrossRef]
12. Szumski, M.; Buszewski, B. Miniaturization in Separation Techniques. In *Handbook of Bioanalytics*; Buszewski, B., Baranowska, I., Eds.; Springer: Cham, Switzerland, 2022. [CrossRef]
13. Sapozhnikova, Y.; Lehotay, S.J. Review of recent developments and applications in low-pressure (vacuum outlet) gas chromatography. *Anal. Chim. Acta* **2015**, *899*, 13–22. [CrossRef]
14. Boeker, P.; Leppert, J. Flow Field Thermal Gradient Gas Chromatography. *Anal. Chem.* **2015**, *87*, 9033–9041. [CrossRef]
15. Trapp, O. Boosting the Throughput of Separation Techniques by "Multiplexing". *Angew. Chem. Intern. Ed.* **2007**, *46*, 5609–5613. [CrossRef]
16. Graff, D.K. Fourier and Hadamard: Transforms in Spectroscopy. *J. Chem. Educ.* **1995**, *72*, 287–378. [CrossRef]
17. Kaljurand, M.; Küllik, E. Application of the Hadamard transform to gas chromatograms of continuously sampled mixtures. *Chromatographia* **1978**, *11*, 328–330. [CrossRef]
18. Krkošová, Ž.; Kubinec, R.; Jurdáková, H.; Blaško, J.; Ostrovský, I.; Soják, L. Gas chromatography with ballistic heating and ultrafast cooling of column. *Chem. Pap.* **2008**, *62*, 135–140. [CrossRef]
19. Wunsch, M.R.; Lehnig, R.; Trapp, O. Online Continuous Trace Process Analytics Using Multiplexing Gas Chromatography. *Anal. Chem.* **2017**, *89*, 4038–4045. [CrossRef]
20. Antoniadou, M. Development and Evaluation of a Fast Multiplexing Injector for GC Analysis: Application to the Analysis of Lithium Ion Battery Volatiles. Ph.D. Thesis, TU Wien, Vienna, Austria, 2021.
21. Siegle, A.F.; Trapp, O. Development of a Straightforward and Robust Technique to Implement Hadamard Encoded Multiplexing to High-Performance Liquid Chromatography. *Anal. Chem.* **2014**, *86*, 10828–10833. [CrossRef]
22. Siegle, A.F.; Trapp, O. Hyphenation of Hadamard Encoded Multiplexing Liquid Chromatography and Circular Dichroism Detection to Improve the Signal-to-Noise Ratio in Chiral Analysis. *Anal. Chem.* **2015**, *87*, 11932–11934. [CrossRef]
23. Kaneta, T.; Yamaguchi, Y.; Imasaka, T. Hadamard Transform Capillary Electrophoresis. *Anal. Chem.* **1999**, *71*, 5444–5446. [CrossRef]
24. Fan, Z.; Lin, C.-H.; Chang, H.-W.; Kaneta, T.; Lin, C.-H. Design and application of Hadamard-injectors coupled with gas and supercritical fluid sample collection systems in Hadamard transform-gas chromatography/mass spectrometry. *J. Chromatogr. A* **2010**, *1217*, 755–760. [CrossRef]
25. Cheng, Y.-K.; Lin, C.-H.; Kuo, S.; Yang, J.; Hsiung, S.-Y.; Wang, J.-L. Applications of Hadamard transform-gas chromatography/mass spectrometry for the detection of hexamethyldisiloxane in a wafer cleanroom. *J. Chromatogr. A* **2012**, *1220*, 143–146. [CrossRef]
26. Cheng, Y.-K.; Lin, C.-H.; Kaneta, T.; Imasaka, T. Applications of Hadamard transform-gas chromatography/mass spectrometry to online detection of exhaled breath after drinking or smoking. *J. Chromatogr. A* **2010**, *1217*, 5274–5278. [CrossRef]
27. Fan, G.-T.; Yang, C.-L.; Lin, C.-H.; Chen, C.-C.; Shih, C.-H. Applications of Hadamard transform-gas chromatography/mass Spectrometry to the detection of acetone in healthy human and diabetes mellitus patient breath. *Talanta* **2014**, *120*, 386–390. [CrossRef]
28. Wunsch, M.R.; Reiter, A.M.C.; Schuster, F.S.; Lehnig, R.; Trapp, O. Continuous online process analytics with multiplexing gas chromatography by using calibrated convolution matrices. *J. Chromatogr. A* **2019**, *1595*, 180–189. [CrossRef]
29. Wunsch, M.R.; Lehnig, R.; Janke, C.; Trapp, O. Online High Throughput Measurements for Fast Catalytic Reactions Using Time-Division Multiplexing Gas Chromatography. *Anal. Chem.* **2018**, *90*, 9256–9263. [CrossRef]
30. Stenzel, Y.P.; Horsthemke, F.; Winter, M.; Nowak, S. Chromatographic Techniques in the Research Area of Lithium Ion Batteries: Current State-of-the-Art. *Separations* **2019**, *6*, 26. [CrossRef]
31. Geng, L.; Wood, D.L.; Lewis, S.A.; Connatser, R.M.; Li, M.; Jafta, C.J.; Belharouak, I. High accuracy in-situ direct gas analysis of Li-ion batteries. *J. Power Sources* **2020**, *466*, 28211. [CrossRef]

Disclaimer/Publisher's Note: The statements, opinions and data contained in all publications are solely those of the individual author(s) and contributor(s) and not of MDPI and/or the editor(s). MDPI and/or the editor(s) disclaim responsibility for any injury to people or property resulting from any ideas, methods, instructions or products referred to in the content.

Article

Validation of a Liquid–Liquid Extraction Method to Study the Temporal Production of D-Series Resolvins by Head Kidney Cells from Atlantic Salmon (*Salmon salar*) Exposed to Docosahexaenoic Acid

Pedro Araujo [1,*], Sarah Iqbal [1,2], Aleksander Arnø [1,3], Marit Espe [1] and Elisabeth Holen [1]

1 Institute of Marine Research (HI), P.O. Box 1870 Nordnes, N-5817 Bergen, Norway; sarah.iqbal@hi.no (S.I.); aleksander.arno@student.uib.no (A.A.); marit.espe@hi.no (M.E.); elisabeth.holen@hi.no (E.H.)
2 Department of Chemistry, University of Bergen, Allégaten 41, N-5007 Bergen, Norway
3 Department of Biological Sciences, University of Bergen, Thormøhlens Gate 53A, N-5006 Bergen, Norway
* Correspondence: pedro.araujo@hi.no; Tel.: +47-47645029

Abstract: A simple and rapid method for the extraction of D-series resolvins (RvD1, RvD2, RvD3, RvD4, RvD5) released into Leibovitz's L-15 complete medium by head kidney cells from Atlantic salmon and the further determination of liquid chromatography triple quadrupole mass spectrometry is proposed. A three-level factorial design was proposed to select the optimal concentrations of internal standards that were used in the evaluation of the performance parameters, such as linear range (0.1–50 ng mL^{-1}), limits of detection and quantification (0.05 and 0.1 ng mL^{-1}, respectively), and recovery values ranging from 96.9 to 99.8%. The optimized method was used to determine the stimulated production of resolvins by head kidney cells exposed to docosahexaenoic acid, and the results indicated that it is possible that the production was controlled by circadian responses.

Keywords: resolvins; liquid chromatography quadrupole mass spectrometry; Atlantic salmon; head kidney cells; liquid–liquid extraction

Citation: Araujo, P.; Iqbal, S.; Arnø, A.; Espe, M.; Holen, E. Validation of a Liquid–Liquid Extraction Method to Study the Temporal Production of D-Series Resolvins by Head Kidney Cells from Atlantic Salmon (*Salmon salar*) Exposed to Docosahexaenoic Acid. *Molecules* **2023**, *28*, 4728. https://doi.org/10.3390/molecules28124728

Academic Editor: Paraskevas D. Tzanavaras

Received: 26 May 2023
Revised: 9 June 2023
Accepted: 11 June 2023
Published: 13 June 2023

Copyright: © 2023 by the authors. Licensee MDPI, Basel, Switzerland. This article is an open access article distributed under the terms and conditions of the Creative Commons Attribution (CC BY) license (https://creativecommons.org/licenses/by/4.0/).

1. Introduction

Resolvins from the D-series are bioactive oxygenated metabolites of docosahexaenoic acid (22:6n-3; DHA) that were discovered in mice exudate cells treated with aspirin and DHA and were termed resolvins due to their role in dampening and promoting the resolution of inflammation processes [1]. Resolvins from the D-series (RvD) are biosynthesized by the action of 15-LOX on DHA to produce 17S-hydroperoxydocosahexaenoic acid that is converted to various types of resolvin D (RvD): RvD1, RvD2, RvD3, RvD4, RvD5 and RvD6 by the action of 5-LOX [1,2]. RvD1 has therapeutic effects such as slowing the progression of osteoarthritis in joints, preventing neuronal dysfunction in Parkinson's disease, and increasing efferocytosis in the elderly [3]. RvD2 promotes subcellular localized healing, and regenerative and protective effects in burn wounds, such as keratinocyte restoration, muscle regeneration, tissue necrosis restriction, tumor growth inhibition and clearing cellular debris in mice [3,4]. RvD3 has powerful anti-inflammatory effects on leukocytes, decreases the levels of pro-inflammatory cytokines (MCP-1, IL-6, and keratinocyte chemoattractant protein), and eicosanoids (LTB4, PGD2 and TxB2), inhibits neutrophil transmigration and enhances macrophage absorption of microbial particles [5]. RvD4 protects organs in cases of ischemia kidney injury [3,6]. RvD5 improves phagocytosis by increasing neutrophil and macrophage movement, modulates TNF-α and NF-κB in blood, synovial fluid and exudates discharged after a hemorrhage, and, in the late stages of coagulation, prolongs the agglutination process and reduces the likelihood of bleeding [7]. RvD6 increases nerve regeneration and stimulates hepatocyte growth factor genes specifically as upstream regulators and a gene network involved in axon growth and the suppression of neuropathic pain, indicating a novel function of this lipid mediator to maintain cornea integrity and homeostasis after injury [2,8].

In general, resolvins produced from DHA have demonstrated promising therapeutic advantages in terms of cell damage reduction, oxidative stress inhibition, and tumor growth suppression [9]. In mammals, there is growing evidence that resolvins may assist in the resolution of acute inflammation and potently suppress inflammatory and neuropathic pain. Although it is unknown whether this is the case in fish, a study conducted by Ruyter and colleagues showed that greater dietary DHA levels lead to higher concentrations of resolvins in plasma, which may have health benefits in fish [10]. Neuronal deficiencies and developmental issues in larvae have been reported in fish fed with a diet scarce in DHA [11], hence the possible link between resolvins in the neuronal function of fish.

Different approaches have been reported for the determination of resolvins in different kinds of samples. Enzyme immunoassay (EIA) kits have been used for determining the production of RvD_1 in human and fish plasma [10,12] with high sensitivity. However, it is also well known that EIA is prone to cross-reactivity, which in turn causes an overestimation of the levels of specific resolvins. In addition, EIA are limited to just one type of resolvin per commercial kit (e.g., either RvD_1 or RvD_2, which makes the technique remarkably expensive when different resolvins are considered.

High performance liquid chromatography with UV diode array detection (HPLC-DAD) and gas chromatography coupled to mass spectrometry (GC-MS) have been used as alternative techniques to validate the results of liquid chromatography mass spectrometry in tandem mode (LC-MS/MS). For instance, enzymatically generated RvD_1 was determined by LC-MS/MS analysis, and further evidence of its positive identification was obtained using HPLC-DAD to confirm the presence of a conjugated tetraene structure within RvD_1 that is responsible for its characteristic triplet chromophore at a λ_{max} = 301 nm. A subsequent GC-MS analysis was performed after derivatizing RvD_1 with diazomethane to its corresponding trimethylsilyl derivative [9,13]. In a similar way, actively phagocytosing polymorphonuclear neutrophils were converted to RvD_2 and determined by LC-MS/MS analysis, followed by a subsequent GC-MS analysis of derivatized RvD_2 to validate the LC-MS/MS determination [9]. The main drawback associated with HPLC-DAD is the potential coelution of isomeric resolvins with a similar spectrum, which may hinder their discrimination, while the most evident disadvantage of GC-MS is that is restricted to thermally stable volatile compounds, generally prepared by a time-consuming derivatization process [14–16].

Resolvins are commonly found in biological fluids and organs at extremely low concentrations, including peripheral blood, cerebral fluid, placenta, synovial fluids, urine, sputum, spleen, lymph nodes, cell cultures, and others [17]. As a result, successful extraction is essential to wash and clean up the sample, followed by drying up and reconstitution in a small volume to improve the concentration of the analyte. An overview of the literature indicated that solid phase extraction (SPE) is arguably the most popular extraction method for the analysis of resolvins in different kind of samples and LC-MS/MS the preferred quantitative technique for its outstanding sensitivity. A recent article has proposed a cumbersome methodology that combines liquid–liquid extraction (LLE) with chloroform and acetate-water buffered at pH 4 followed by μ-SPE with methanol-water buffered at pH 4 and a final LC-MS/MS analysis for the quantification of resolvins in human keratinocyte cell lysates to obtain recoveries at around 42 and 64% [18]. Unfortunately, SPE alone (not to mention when combined with LLE) is a time-consuming and complicated method that requires multiple steps and different solvents prior to LC-MS/MS analysis. Our current SPE protocol for the analysis of resolvins (RvD_1, RvD_2) in cell culture requires a total of seven different solvents (six solvents for SPE and one solvent for final reconstitution) [14]. Poor recovery and reproducibility, and insufficient cleaning of sample extracts are some of the drawbacks commonly associated with SPE [19]. Different SPE adsorbent materials and LLE have been compared to propose a suitable method for the extraction of RvD_1 from human endothelial cells and further determination by LC-MS, and the best results were obtained using LLE with a solution of methanol containing the internal standard [20]. Although it seems a promising approach, this LLE protocol was validated incorrectly by using spiked plasma samples instead of human endothelial cells. Furthermore, methanol is generally used for protein removal, and therefore LLE with methanol for biological samples requires multiple and time-consuming centrifugation steps to ensure complete precipitation: for instance, centrifugation times of 45 min for metabolome [21] and 70 min for RvD_1 analysis [20], followed by drying methanol with a stream of nitrogen, which is a lengthy operation prior to final reconstitution and LC-MS/MS analysis.

Overall, sample preparation for the analysis of multiple compounds (e.g., resolvins) is the major bottleneck in analytical laboratories. The aim of this study is to propose a simple and rapid LLE procedure to quantify the temporal production of released resolvins (RvD_1, RvD_2, RvD_3, RvD_4, RvD_5) into Leibovitz's L-15 complete medium by head kidney cells from Atlantic salmon exposed to DHA and further liquid chromatography triple quadrupole mass spectrometry (LC-MS/MS). To our knowledge, this is the first validated LLE procedure for quantifying biosynthesized D-series resolvins using cell cultures.

2. Results and Discussion

2.1. Optimal Concentration of Internal Standards

A 3^2-factorial design, where the base 3 represents the number of concentration levels (low, medium, high) and the exponent 2 the number of factors (analyte and internal standard), was used to investigate whether the response factor (RF) remains constant at different internal standard (IS) concentrations over a fixed analytical range. The nine combinations of analytes and internal standards (3^2) suggested by the factorial design (Table 1) were prepared by dissolving the analytical resolvins and internal standards in the L-15 medium. For instance, Experiment 2 represents an L-15 media solution containing all the analytes (RvD_1, RvD_2, RvD_3, RvD_4, RvD_5) at a concentration level of 30 ng/mL and submitted to the extraction protocol by using a mixture of deuterated internal standards (RvD_1-d_5, RvD_2-d_5 and RvD_3-d_5) at 15 ng/mL each. Four blanks, consisting of L-15 medium containing 0, 15, 30 and 45 ng/mL of internal standards, were also prepared. The experiments in Table 1 were were prepared in triplicate and run in random order.

Table 1. Proposed 3^2-factorial design to select the optimal concentrations of internal standards.

Experiment	Resolvin [A] (ng/mL)	Internal Standard [IS] (ng/mL)
1	15	15
2	30	15
3	45	15
4	15	30
5	30	30
6	45	30
7	15	45
8	30	45
9	45	45

The RF for every resolvin at the different concentrations of IS were calculated at every experimental point in Table 1 by the expression RF = [IS]/[A] × (y_A/y_{IS}), where [A] and [IS] represent the analyte and IS concentrations and y_A and y_{IS} their corresponding signals, respectively. This RF approach has been published elsewhere in the selection of an optimal concentration of RvD_2-d_5 for extracting RvD_1 and RvD_2 from fish cells by SPE and further quantification using an LCMS ion-trap instrument [14]. The results are shown in Table 2.

Table 2. Calculated response factors (RF) after implementing the 3^2-factorial design to select the optimal amount of deuterated internal standard for the liquid–liquid extraction of resolvins from L-15 cell culture media. The RF values are expressed as average ± standard error.

[IS]	RvD1-d5	RvD2-d5	RvD3-d5 15 ng/mL	RvD1-d5	RvD1-d5	RvD1-d5	RvD2-d5	RvD3-d5 30 ng/mL	RvD1-d5	RvD1-d5	RvD1-d5	RvD2-d5	RvD3-d5 45 ng/mL	RvD1-d5	RvD1-d5
[A] (ng/mL)	RvD1	RvD2	RvD3	RvD4	RvD5	RvD1	RvD2	RvD3	RvD4	RvD5	RvD1	RvD2	RvD3	RvD4	RvD5
15	2.70	12.41	33.74	0.79	1.00	2.85	12.81	34.63	0.83	0.95	3.32	16.68	39.22	1.00	1.16
15	2.57	13.78	31.24	0.79	0.92	2.77	14.39	32.76	0.87	0.93	3.35	16.73	38.37	0.99	1.15
15	3.04	15.80	36.96	0.93	1.10	2.85	14.05	34.03	0.87	0.96	3.27	16.00	39.58	1.01	1.22
30	3.06	15.07	36.22	0.85	1.07	3.03	14.97	32.95	0.89	1.03	3.12	15.41	35.83	0.95	1.25
30	2.94	14.01	32.97	0.80	1.08	3.24	16.35	37.58	0.98	1.14	3.23	15.43	35.06	0.91	1.26
30	2.98	15.17	33.63	0.86	1.26	3.09	15.24	32.57	0.95	1.00	3.18	16.22	38.42	0.94	1.32
45	2.92	14.63	34.31	0.84	1.11	2.82	14.53	32.44	0.82	1.01	2.92	14.15	34.87	0.83	1.16
45	2.97	14.58	33.71	0.84	1.11	2.87	14.33	34.57	0.87	0.98	2.92	14.86	32.89	0.90	1.09
45	2.96	14.80	36.16	0.82	1.24	2.60	13.35	31.01	0.80	1.05	2.97	14.57	32.29	0.82	1.14
AVG	2.90 ± 0.05	14.43 ± 0.33	34.10 ± 0.61	0.84 ± 0.01	1.08 ± 0.04	2.94 ± 0.06	14.58 ± 0.35	33.94 ± 0.63	0.89 ± 0.02	1.00 ± 0.02	3.17 ± 0.06	15.69 ± 0.31	36.78 ± 0.91	0.94 ± 0.02	1.20 ± 0.02
%CV	1.86	2.27	1.78	1.75	3.25	2.14	2.37	1.85	2.21	2.14	1.79	1.95	2.47	2.48	2.01

%CV = coefficient of variation.

A multiple range test, applied to determine significant differences ($p < 0.05$) between the calculated RF values, revealed that apart from RvD$_5$, the RF remains constant for the concentrations 15 and 30 ng/mL of IS with coefficients of variations around 2%. Hence, 15 ng/mL of RvD$_1$-d$_5$ (for RvD$_1$, RvD$_4$ and RvD$_5$), RvD$_2$-d$_5$ (for RvD$_2$) and RvD$_3$-d$_5$ (for RvD$_3$) that yields average RFs of 2.90 ± 0.05, 14.47 ± 0.33, 34.33 ± 0.61, 0.84 ± 0.01 and 1.01 ± 0.04 for RvD$_1$, RvD$_2$, RvD$_3$, RvD$_4$, and RvD$_5$, respectively, was selected as the optimal concentration level to be used in connection with the proposed extraction protocol and for further quantitative analysis using LC-MS/MS. Although the RF values for RvD$_5$ (1.10 ± 0.04, 1.01 ± 0.02, 1.19 ± 0.02 at 15, 30 and 45 ng/mL of IS, respectively), exhibited statistically significant differences, the RF for 15 ng/mL lay between 30 and 45 ng/mL, with a coefficient of variation (3.2%) similar to those reported elsewhere (4–13% at 1 ng/mL IS) for cell experiments using LCMS [18].

2.2. Analytical Validation

The deuterated internal standards RvD$_1$-d$_5$, RvD$_2$-d$_5$ and RvD$_3$-d$_5$ were dissolved in acetonitrile (15 ng/mL each) and used to extract the resolvins from seven L-15 media preparations containing a mixture of five analytical resolvins (RvD$_1$ to RvD$_5$) at seven different concentrations (0, 1, 5, 15, 30, 45, and 50 ng/mL). The characteristic mass fragments of the five resolvins and three internal standards were extracted from the precursor ions in both unspiked (0 ng/mL) and spiked (1–50 ng/mL) L-15 media. The five resolvins were separated chromatographically, according to their retention times, including RvD$_1$ and RvD$_2$, which have the same precursor (m/z 375) and product (m/z 141) ions. The chromatographic elution order was RvD$_3$ (2.55 min), RvD$_2$ (2.59 min) RvD$_1$ (2.77 min), RvD$_4$ (3.12 min), and RvD$_5$ (3.55 min). The extracted ion chromatograms (EIC) revealed that the proposed extraction protocol allows for detecting unequivocally the resolvins with negligible background interferences (Figure 1), and therefore the analyses were regarded as highly selective towards the five resolvins.

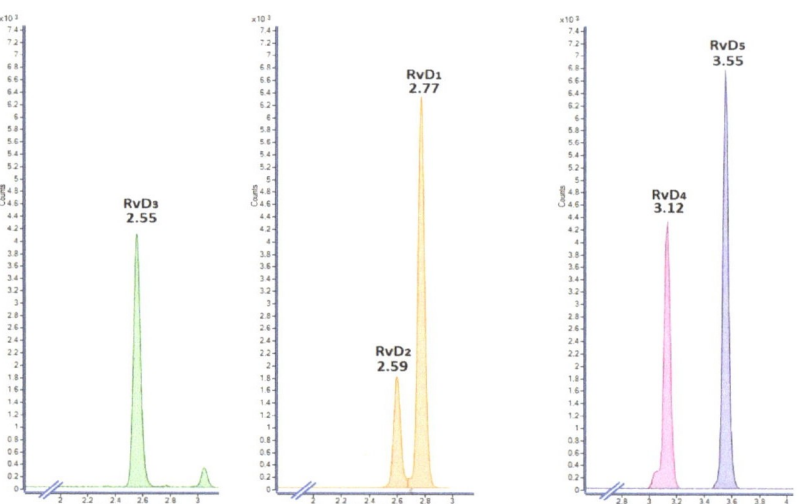

Figure 1. Extracted ion chromatograms, in increasing order of retention times, to indicate the selectivity of the analysis towards the five analyzed resolvins, after implementing the proposed LLE protocol.

The chromatographic peak area ratios RvD$_1$/RvD$_1$-d$_5$, RvD$_2$/RvD$_2$-d$_5$, RvD$_3$/RvD$_3$-d$_5$, RvD$_4$/RvD$_1$-d$_5$ and RvD$_5$/RvD$_1$-d$_5$ were calculated and plotted against the analytical concentrations to compute the function $y_A/y_{IS} = \varphi[A] + \beta$ and obtain the various calibration parameters displayed in Table 3.

Table 3. Analytical performance parameters. The linearity is judged by considering simultaneously the closeness of R^2 to the unity and the comparison of $F_{experimental}$ against the tabulated $F_{critical}$ = 2.958 for 5 and 14 degrees of freedom at the 95% confidence level.

Resolvin (0–50 ng/mL)	Slope (φ)	Intercept (β)	R^2	$F_{experimental}$	LOD	LOQ	Recovery (%)
RvD$_1$	0.197	−0.063	0.997	0.526	0.042	0.127	98.2 ± 1.3
RvD$_2$	0.982	−0.235	0.999	0.342	0.028	0.086	99.5 ± 1.1
RvD$_3$	2.304	−0.318	0.998	0.142	0.032	0.097	99.5 ± 0.9
RvD$_4$	0.055	0.002	0.999	0.123	0.024	0.074	99.8 ± 1.2
RvD$_5$	0.077	−0.048	0.994	0.624	0.059	0.180	96.9 ± 2.3

The degree of linearity of the calibrations was provided by both the regression coefficients (R^2) and the Fisher test, defined as the quotient between the lack-of-fit and the pure error variances (F_{exp} in Table 3). In general, the five resolvins were linear over the studied range of concentrations, as reflected in Table 3, where the R^2 values (between 0.994 and 0.999) indicate that a high proportion of the variance of the calculated y_A/y_{IS} signals are explained by the analytical concentrations [A] in the proposed regression models. This conclusion is also supported by the $F_{experimental}$ values for the five calibration models that were lower than the critical value of 2.958 for 5 and 14 degrees of freedom at the 95% confidence level (Table 3). The LOD (0.028–0.059 ng/mL) and LOQ (0.074–0.180 ng/mL) were within the range of previously reported values for cell cultures [22], and were considered appropriate for all resolvin species. In the present work, the LOQ for resolvins in L-15 complete media by using an Agilent 6495 triple quadrupole are similar to (and in some cases better than) those reported in pure standards by using Sciex QTRAP 6500 [23,24]. For instance, the referred Quadrupole/QTRAP (present/[24]) values for RvD$_1$, RvD$_2$, RvD$_3$, RvD$_4$ and RvD$_5$ are 0.127/0.05, 0.5/0.086, 0.097/0.05, 0.074/0.1 and 0.180/0.1, respectively. Based on the matrix complexity, namely the present L-15 medium versus the pure standard [24], the LOQ values of the present research can be regarded as remarkable. In addition, the proposed extraction protocol in conjunction with the triple quadrupole spectrometer is an outstanding strategy, considering the widespread consensus that QTRAP delivers better data than quadrupole systems [25]. The recovery of the method, expressed as the ratio between found and nominal concentrations ($100 \times [A]_{found}/[A]_{nominal}$) was higher than 95% in all cases, as described in Table 3.

2.3. Analysis of Released Resolvins in L-15 Media by Head Kidney Cells

The proposed LLE protocol and further LC-MS/MS quantification was implemented to study the induced production of resolvins by salmon head kidney cells with and without exposure to exogenous DHA. The production of resolvins was expressed in ng/mL and measured at 6, 12 and 24 h (Table 4). The levels of resolvins in decreasing order of concentration were RvD$_4$ > RvD$_2$ > RvD$_3$ > RvD$_1$ > RvD$_5$ and RvD$_4$ > RvD$_3$ > RvD$_2$ > RvD$_1$ > RvD$_5$ in control and DHA, respectively. These levels agree with previously reported results that were obtained by using SPE and LC-MS/MS to estimate the production of RvD$_1$ and RvD$_2$ by salmon liver cells, and where it was suggested that the production of RvD$_4$ was preferred over RvD$_1$ and RvD$_2$ after exposing the cells to different polyunsaturated fatty acids, including DHA [14].

Table 4. Temporal production of resolvins by head kidney cells from Atlantic salmon (*Salmon salar*) exposed to docosahexaenoic acid.

Group	Time (hours)	RvD$_1$	RvD$_2$	RvD$_3$	RvD$_4$	RvD$_5$
				Concentrations (ng/mL)		
Control	6	0.127 ± 0.022	0.190 ± 0.109	0.186 ± 0.055	0.601 ± 0.090	0.055 ± 0.018
	12	0.074 ± 0.009	0.238 ± 0.141	0.119 ± 0.049	0.382 ± 0.082	0.069 ± 0.009
	24	0.094 ± 0.014	0.154 ± 0.079	0.092 ± 0.018	0.419 ± 0.031	0.017 ± 0.003
DHA	6	3.918 ± 1.798	4.412 ± 1.386	6.849 ± 3.156	10.231 ± 6.342	3.696 ± 2.022
	12	1.388 ± 0.404	2.574 ± 1.143	5.212 ± 2.328	4.737 ± 1.710	1.395 ± 0.361
	24	4.391 ± 2.103	8.095 ± 4.129	9.583 ± 4.682	10.439 ± 5.121	4.256 ± 1.815

A principal component analysis (PCA) indicated that 76.00% of the total data variability was explained by the concentrations of resolvins in control and DHA groups, and 16.24% was explained by the different times (Figure 2).

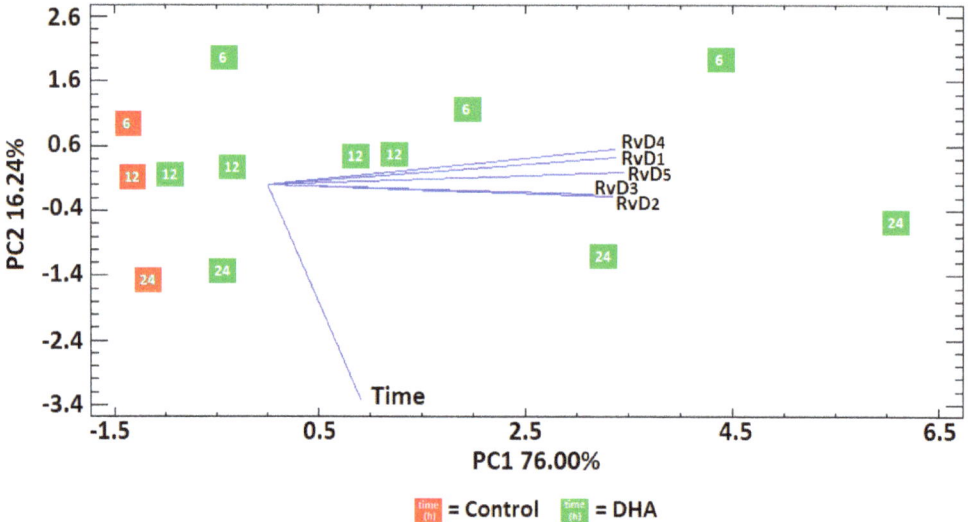

Figure 2. Principal component analysis of the released resolvins in L-15 media by head kidney cells with and without exposure to DHA, after implementing the proposed LLE protocol.

The control and DHA groups were clearly discriminated along the PC1 axis and characterized by negative and positive scores along this axis, respectively. The degree of overlapping within the control and within the DHA group was higher in the former than in the latter group. For instance, the scores for the control at 6 or 12 or 24 h appeared as three separated clusters along PC2 (red squares), while for the DHA the time clusters were separated along PC2 (green squares), but they were widely spread along PC1, indicating a higher dispersion of the DHA data. The PCA also showed that the highest concentrations of resolvins were associated with the DHA group, suggesting that exogenous DHA promoted the production of resolvins. The vectors, time and concentrations of resolvins, were orthogonal; therefore, the changes in concentrations of RvD$_1$, RvD$_2$, RvD$_3$, RvD$_4$ and RvD$_5$ were independent of the time. This lack of correlation between time and concentration was confirmed by studying the within and between variances at the three selected times for every resolvin. The estimated p-values were not significant ($p > 0.05$) for any type of resolving, either in the control or in the DHA group.

The concentration/time relationship was computed from Table 4, and the results revealed a continuous decrease in production over time for the five analyzed resolvins in the control group (Figure 3). In contrast, the DHA group showed higher production at subjective dawn (6 h) and subjective midnight (24 h) than subjective midday (12 h) for all resolvins (Figure 2), suggesting the presence of a circadian clock that may impose a 24 h rhythmicity on the head kidney cells to process the production of resolvins from the added 50 µM of DHA.

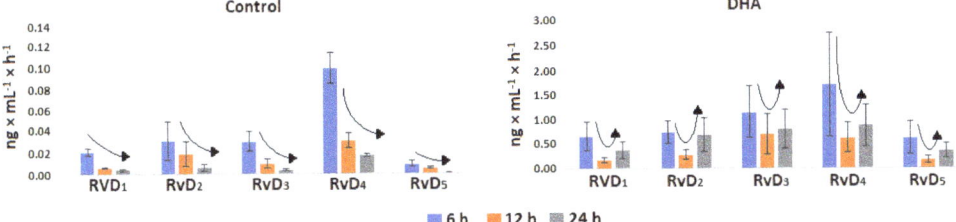

Figure 3. Concentration/time ratio for the different resolvins in the control and DHA group. The former group shows a continuous decrease in production and the latter a plausible regulated production by a circadian clock.

The direct influence of circadian rhythms on resolvin production remains unexplored. However, considering the present findings, it is plausible that there exists an underlying resolvin/15-LOX immune regulating clock that controls the production of resolvins from endogenous DHA. This observation is supported by the results from different studies that have observed higher levels of COX-1 and inflammatory prostaglandins (PGE_2 and $PGF_{2\alpha}$) at midnight than at midday, and concluded that some immunological functions are controlled by circadian responses from a prostaglandin/COX-1 system [26]. Similarly, anti-inflammatory prostaglandin 15d-PGJ_2 has been identified as an entrainment factor aligned with circadian oscillations [27].

3. Materials and Methods

3.1. Reagents

Resolvin D1 (RvD_1, 95%), resolvin D2 (RvD_2, 95%), resolvin D3 (RvD_3, 95%), resolvin D4 (RvD_4, 95%), resolvin D5 (RvD_5, 95%), deuterated resolvin D1 (RvD_1-d_5, 95%), deuterated resolvin D2 (RvD_2-d_5, 95%) and deuterated resolvin D3 (RvD_3-d_5, 95%). Acetonitrile (99.8%) and formic acid (98%) were purchased from Sigma-Aldrich (St. Louis, MO, USA). 2-propanol (HPLC grade, 99.9%) from Merck (Darmstadt, Germany). Chloroform (HPLC grade, 99.8%) was obtained from Merck (Darmstadt, Germany). A Millipore Milli-Q system was used to produce ultra-pure water 18 MΩ (Millipore, Milford, CT, USA). Cis-4,7,10,13,16,19-docosahexaenoic acid (DHA, ≥98%) were purchased from Sigma-Aldrich (Oslo, Norway). Leibovitz's L-15 medium was from Sigma-Aldrich (St. Louis, MO, USA). Fetal bovine serum (FBS, cat# 14-801F) was from BioWhittaker (Petit Rechain, Belgium). The glutaMaxTM 100× (Gibco-BRL, cat# 35056) was from Gibco-BRL (Cergy-Pontoise, France).

3.2. Head Kidney Cells

For each fish, the head kidneys were directly sampled and added to PBS at 5 °C and then cut with scissors and squeezed through a 40 µM Falcon cell strainer. The cells were transferred to tubes and centrifuged in a Hettich Zentrifugen, 320 R, at 400× g for 5 min at 4 °C. The cell pellets were resuspended in PBS and layered carefully on top of equal amounts of diluted Percoll in a density of 1.08 g/mL. The tubes were centrifuged at 800× g for 30 min at 4 °C. The cell layer in the interface containing the head kidney leukocytes was collected and the cells were pelleted by centrifugation, 400× g for 5 min at 4 °C. An additional washing step in PBS was performed. The cells were counted using a Bürker chamber and 0.4% trypan blue solution, and the viability was above 85%.

3.3. Cell Cultures

A L-15 complete (cL-15) medium was supplemented with 10% foetal bovine serum (FBS), 2% pen/strep and 2% glutamax^{TM} 100×, and used to prepare cL-15 solutions containing DHA that was diluted to a concentration of 50 µM and a control solution containing ethanol (the solvent used to dissolve the DHA). Approximately 1×10^7 salmon head kidney cells were cultured into each well (control and DHA). The cell culture plates were incubated in a normal atmosphere incubator (Sanyo Electric Company Ltd. Osaka, Japan) at 9 °C for 6, 12, and 24 h, under dark conditions. The two suspensions of cells (control and DHA) were prepared in pentaplicate. The head kidney cells were centrifuged at $50 \times g$ for 5 min at 4 °C, and the medium collected and stored at −80 °C until extraction, followed by the LC-MS/MS analysis.

3.4. Optimal Concentrations of the Internal Standards

A 3^k factorial design, where 3 represents the number of concentration levels (low, medium, high) and k the number of factors (analyte and internal standard), was used to study variations in the response factor (RF) when the concentrations of both the internal standards (IS) and the analytical resolvins varied between 15 and 45 ng/mL. The optimal IS concentrations should yield a stable RF over the explored analytical range.

3.5. Extraction Protocol

The extraction protocol has been described elsewhere for the determination of arachidonic and eicosapentaenoic acid metabolites from the LOX pathway [28], with some minor modifications. Briefly, two successive aliquots of acetonitrile (500 µL) containing the mixture of internal standards at the concentration levels indicated in Table 1 and chloroform (500 µL) were added successively into an Eppendorf tube containing 200 µL of the mixture of resolvins, at the concentration levels of every experiment in Table 1. The Eppendorf tube was vortex-mixed for 30 s (Bandelin RK 100 ultra mixer, Berlin, Germany) and centrifuged at $1620 \times g$ for 3 min (Eppendorf AG centrifuge, Hamburg, Germany), the top phase was removed, and the extraction procedure repeated in the remaining phase using acetonitrile without internal standards and chloroform. After removing the chloroform phase, the remaining solution was vacuum-dried at room temperature (Labconco vacuum drier system, Kansas, MO, USA), diluted to 50 µL with methanol, transferred to an autosampler vial, and submitted to LC-MS/MS analysis.

3.6. Analytical Performance

The parameters used to assess the analytical performance of the extraction method in conjunction with LC-MS/MS were selectivity, limit of detection (LOD), limit of quantification (LOQ), calibration range and recovery. The selectivity of the method was evaluated by comparing the chromatograms obtained after injection of L-15 medium samples with and without the analytes. The calibration curves for the resolvins in the L-15 medium were prepared between 0 and 50 ng/mL and extracted as described above, by using the optimal concentration of internal standards suggested by the 3^k factorial design. The linearity was judged by computing both the variance ratio of the lack-of-fit to pure error and the coefficient of regression, as suggested by the Analytical Method Committee [29] and the International Council for Harmonisation guidance for the validation of analytical procedures [30]. The ratio of the standard deviation (σ) to the slope (φ) of the regression curves for every resolvin was used to determine the LOD ($3.3 \times \sigma/\varphi$) and LOQ ($10 \times \sigma/\varphi$), as described elsewhere [31]. The percentage of recovery was assessed by comparing the degree of agreement between the experimental and nominal concentrations, as acknowledged by the ICH [30,31].

3.7. Liquid Chromatography Mass Spectrometry

An Agilent ultra-high performance liquid chromatography (UHPLC), coupled to a 6495 QQQ triple quadrupole (Agilent Technologies, Waldbronn, Germany) with an electrospray ionization (ESI) interface and iFunnel ionization, was used to quantify the eicosanoids. The UHPLC system was equipped with a Zorbax RRHD Eclipse Plus C18, 95Å, 2.1 × 50 mm, 1.8 μm chromatographic column. The mobile phase delivered at 0.4 mL/min in gradient mode consisted of ultra-pure water with 0.1% formic acid (solution A) and an equal-volume mixture of acetonitrile and methanol with 0.1% formic acid (solution B). The solvent gradient was as follows: solution A was reduced from 60 to 5% from 0.00 to 4.00 min, kept at 5% between 4.00 and 5.50 min, increased to 60% between 5.50 and 5.51 min and kept at 60% between 5.51 and 10.00 min. Mass spectrometric detection was performed by multiple reactions monitoring (MRM) in negative mode. The monitored transitions in percentage of ion counts (%) were: m/z 375 → 141 for RvD_1 and RvD_2; m/z 375 → 147 for RvD_3; m/z 375 → 101 for RvD_4; m/z 359 → 199 for RvD_5; m/z 380 → 141 for RvD_1-d_5; m/z 380 → 141 for RvD_2-d_5; and m/z 380→147 for RvD_3-d_5. The ESI parameters were gas temperature (120 °C), gas flow rate (19 L/min), nebulizer pressure (20 psi), sheath gas temperature (300 °C), sheath gas flow (10 L/min), capillary voltage (3500 V) and nozzle voltage (2000 V). The integration of the chromatograms was performed using the MassHunter Qualitative Navigator software (version 10.0). The levels of resolvins were estimated by means of the internal standards, and expressed in ng/mL units.

3.8. Statistics

Statgraphics Centurion XV Version 15.2.11 (StatPoint Technologies, Inc., Warrenton, VA, USA) was used for the statistical analyses.

4. Conclusions

This is the first validated liquid–liquid extraction method for resolvins released by head kidney cells from Atlantic salmon in cL-15 media with further quantification by LC-MS/MS using the internal standard calibration method. The small amount of sample (200 μL), the low solvent consumption, the fast extraction times, the high sample throughput (40–50 samples/day) and LOD and LOQ similar to those reported by using SPE and pure standards are important features that make the present approach highly attractive for evaluating the production of resolvins by cell cultures challenged by polyunsaturated fatty acids, such as DHA.

In light of the present findings, it is plausible that production of resolvins by head kidney cells from endogenous DHA is controlled in part by circadian responses.

Author Contributions: Conceptualization, P.A.; Methodology, S.I., A.A., M.E. and E.H.; Validation, P.A., S.I. and A.A.; Formal analysis, P.A., S.I. and A.A.; Investigation, P.A., S.I., M.E. and E.H.; Resources, M.E. and E.H.; Writing—original draft, P.A. and S.I.; Writing—review & editing, P.A., S.I., M.E. and E.H.; Visualization, S.I.; Supervision, P.A., M.E. and E.H.; Project administration, P.A. All authors have read and agreed to the published version of the manuscript.

Funding: The work was funded by the Institute of Marine Research (NuFiMo project. 15473).

Institutional Review Board Statement: Not applicable.

Informed Consent Statement: Not applicable.

Data Availability Statement: All the used data have been provided in the text.

Acknowledgments: We gratefully acknowledge The European Commission, in the context of the Erasmus Mundus Program, for financial support of S.I.

Conflicts of Interest: The authors declare no conflict of interest.

Sample Availability: Not applicable.

References

1. Serhan, C.N.; Dalli, J.; Colas, R.A.; Winkler, J.W.; Chiang, N. Protectins and maresins: New pro-resolving families of mediators in acute inflammation and resolution bioactive metabolome. *Biochim. Biophys. Acta Mol. Cell Biol. Lipids* 2015, *1851*, 397–413. [CrossRef] [PubMed]
2. Pham, T.L.; Kakazu, A.H.; He, J.; Nshimiyimana, R.; Petasis, N.A.; Jun, B.; Bazan, N.G.; Bazan, H.E.P. Elucidating the structure and functions of Resolvin D6 isomers on nerve regeneration with a distinctive trigeminal transcriptome. *FASEB J.* 2021, *35*, e21775. [CrossRef] [PubMed]
3. Chiang, N.; Serha, C.N. Specialized pro-resolving mediator network: An update on production and actions. *Essays Biochem.* 2020, *64*, 443–462. [CrossRef] [PubMed]
4. Dyall, S.C.; Balas, L.; Bazan, N.G.; Brenna, J.T.; Chiang, N.; da Costa Souza, F.; Dalli, J.; Durand, T.; Galano, J.M.; Lein, P.J.; et al. Polyunsaturated fatty acids and fatty acid-derived lipid mediators: Recent advances in the understanding of their biosynthesis, structures, and functions. *Prog. Lipid Res.* 2022, *86*, 101165. [CrossRef]
5. Dalli, J.; Winkler, J.W.; Colas, R.A.; Arnardottir, H.; Cheng, C.Y.C.; Chiang, N.; Petasis, N.A.; Serhan, C.N. Resolvin D3 and Aspirin-Triggered Resolvin D3 Are Potent Immunoresolvents. *Chem. Biol.* 2013, *20*, 188. [CrossRef] [PubMed]
6. Winkler, J.W.; Orr, S.K.; Dalli, J.; Cheng, C.-Y.C.; Sanger, J.M.; Chiang, N.; Petasis, N.A.; Serhan, C.N. Resolvin D4 stereoassignment and its novel actions in host protection and bacterial clearance. *Sci. Rep.* 2015, *6*, srep18972. [CrossRef]
7. Kalyanaraman, C.; Tourdot, B.E.; Conrad, W.S.; Akinkugbe, O.; Freedman, J.C.; Holinstat, M.; Jacobson, M.P.; Holman, T.R.; Perry, S.C.; Kalyanaraman, C.; et al. 15-Lipoxygenase-1 biosynthesis of 7S,14S-diHDHA implicates 15-lipoxygenase-2 in biosynthesis of resolvin D5[S]. *J. Lipid Res.* 2020, *61*, 1087–1103. [CrossRef]
8. Pham, T.L.; Kakazu, A.H.; He, J.; Jun, B.; Bazan, N.G.; Bazan, H.E.P. Novel RvD6 stereoisomer induces corneal nerve regeneration and wound healing post-injury by modulating trigeminal transcriptomic signature. *Sci. Rep.* 2020, *10*, 4582. [CrossRef]
9. Serhan, C.N.; Krishnamoorthy, S.; Recchiuti, A.; Chiang, N. Novel Anti-Inflammatory-Pro-Resolving Mediators and Their Receptors. *Curr. Top. Med. Chem.* 2012, *11*, 629–647. [CrossRef]
10. Ruyter, B.; Bou, M.; Berge, G.M.; Mørkøre, T.; Sissener, N.H.; Sanden, M.; Lutfi, E.; Romarheim, O.H.; Krasnov, A.; Østbye, T.K.K. A dose-response study with omega-3 rich canola oil as a novel source of docosahexaenoic acid (DHA) in feed for Atlantic salmon (Salmo salar) in seawater; effects on performance, tissue fatty acid composition, and fillet quality. *Aquaculture* 2022, *561*, 738733. [CrossRef]
11. Shields, R.J. Larviculture of marine finfish in Europe. *Aquaculture* 2001, *200*, 55–88. [CrossRef]
12. Fedirko, V.; McKeown-Eyssen, G.; Serhan, C.N.; Barry, E.L.; Sandler, R.S.; Figueiredo, J.C.; Ahnen, D.J.; Bresalier, R.S.; Robertson, D.J.; Anderson, C.W.; et al. Plasma lipoxin A4 and resolvin D1 are not associated with reduced adenoma risk in a randomized trial of aspirin to prevent colon adenomas. *Mol. Carcinog.* 2017, *56*, 1977–1983. [CrossRef] [PubMed]
13. Serhan, C.N.; Hong, S.; Gronert, K.; Colgan, S.P.; Devchand, P.R.; Mirick, G.; Moussignac, R.L. Resolvins: A Family of Bioactive Products of Omega-3 Fatty Acid Transformation Circuits Initiated by Aspirin Treatment that Counter Proinflammation Signals. *J. Exp. Med.* 2002, *196*, 1025. [CrossRef]
14. Lucena, E.; Yang, Y.; Mendez, C.; Holen, E.; Araujo, P. Extraction of Pro-and Anti-Inflammatory Biomarkers from fish Cells Exposed to Polyunsaturated Fatty Acids and Quantification by Liquid Chromatography Tandem Mass Spectrometry. *Curr. Anal. Chem.* 2018, *1*, 1–9.
15. Bordet, J.C.; Guichardant, M.; Lagarde, M. Arachidonic acid strongly stimulates prostaglandin I3 (PGI3) production from eicosapentaenoic acid in human endothelial cells. *Biochem. Biophys. Res. Commun.* 1986, *135*, 403–410. [CrossRef] [PubMed]
16. Tsikas, D.; Schwedhelm, E.; Gutzki, F.M.; Frölich, J.C. Gas chromatographic-mass spectrometric discrimination between 8-iso-prostaglandin E2 and prostaglandin E2 through derivatization by O-(2,3,4,5,6-Pentafluorobenzyl)hydroxyl amine. *Anal. Biochem.* 1998, *261*, 230–232. [CrossRef]
17. Li, C.; Wu, X.; Liu, S.; Shen, D.; Zhu, J.; Liu, K. Role of Resolvins in the Inflammatory Resolution of Neurological Diseases. *Front. Pharmacol.* 2020, *11*, 1–12. [CrossRef] [PubMed]
18. Fanti, F.; Oliva, E.; Tortolani, D.; Di Meo, C.; Fava, M.; Leuti, A.; Rapino, C.; Sergi, M.; Maccarrone, M.; Compagnone, D. μSPE followed by HPLC–MS/MS for the determination of series D and E resolvins in biological matrices. *J. Pharm. Biomed. Anal.* 2021, *203*, 114181. [CrossRef] [PubMed]
19. Andrade-Eiroa, A.; Canle, M.; Leroy-Cancellieri, V.; Cerdà, V. Solid-phase extraction of organic compounds: A critical review. part ii. *TrAC Trends Anal. Chem.* 2016, *80*, 655–667. [CrossRef]
20. Dufour, D.; Khalil, A.; Nuyens, V.; Rousseau, A.; Delporte, C.; Noyon, C.; Cortese, M.; Reyé, F.; Pireaux, V.; Nève, J.; et al. Native and myeloperoxidase-oxidized low-density lipoproteins act in synergy to induce release of resolvin-D1 from endothelial cells. *Atherosclerosis* 2018, *272*, 108–117. [CrossRef]
21. Yang, Y.; Cruickshank, C.; Armstrong, M.; Mahaffey, S.; Reisdorph, R.; Reisdorph, N. New sample preparation approach for mass spectrometry-based profiling of plasma results in improved coverage of metabolome. *J. Chromatogr. A* 2013, *1300*, 217–226. [CrossRef]
22. Masoodi, M.; Mir, A.A.; Petasis, N.A.; Serhan, C.N.; Nicolaou, A. Simultaneous lipidomic analysis of three families of bioactive lipid mediators leukotrienes, resolvins, protectins and related hydroxy-fatty acids by liquid chromatography/electrospray ionisation tandem mass spectrometry. *Rapid Commun. Mass Spectrom.* 2008, *22*, 75–83. [CrossRef] [PubMed]

23. Kutzner, L.; Rund, K.M.; Ostermann, A.I.; Hartung, N.M.; Galano, J.M.; Balas, L.; Durand, T.; Balzer, M.S.; David, S.; Schebb, N.H. Development of an Optimized LC-MS Method for the Detection of Specialized Pro-Resolving Mediators in Biological Samples. *Front. Pharmacol.* **2019**, *10*, 169. [CrossRef]
24. Norris, P.; Kapil, S.; Gorti, K. Targeted Profiling of Lipid Mediators. *SCIEX Present* **2017**, 1–6. Available online: https://sciex.com/tech-notes/life-science-research/lipidomics/targeted-profiling-of-lipid-mediators (accessed on 10 June 2023).
25. Agilent 6470B or Sciex 7500 Triple Quad—Chromatography Forum. Available online: https://www.chromforum.org/viewtopic.php?t=110834 (accessed on 19 May 2023).
26. De Zavalía, N.; Fernandez, D.C.; Sande, P.H.; Keller Sarmiento, M.I.; Golombek, D.A.; Rosenstein, R.E.; Silberman, D.M. Circadian variations of prostaglandin E2 and F2α release in the golden hamster retina. *J. Neurochem.* **2010**, *112*, 972–979. [CrossRef]
27. Nakahata, Y.; Akashi, M.; Trcka, D.; Yasuda, A.; Takumi, T. The in vitro real-time oscillation monitoring system identifies potential entrainment factors for circadian clocks. *BMC Mol. Biol.* **2006**, *7*, 5. [CrossRef]
28. Araujo, P.; Janagap, S.; Holen, E. Application of Doehlert Uniform Shell Designs for Selecting Optimal Amounts of Internal Standards in the Analysis of Prostaglandins and Leukotrienes by Liquid Chromatography-Tandem Mass Spectrometry. *J. Chromatogr. A* **2012**, *1260*, 102–110. [CrossRef] [PubMed]
29. Analytical Method Committee. Is My Calibration Linear? *Analyst* **1994**, *119*, 2363–2366. [CrossRef]
30. Stefanini-Oresic, L. Validation of analytical procedures: ICH guidelines Q2(R2). *Farm. Glas.* **2022**, *2*, 1–34.
31. Araujo, P. Key aspects of analytical method validation and linearity evaluation. *J. Chromatogr. B Anal. Technol. Biomed. Life Sci.* **2009**, *877*, 2224–2234. [CrossRef]

Disclaimer/Publisher's Note: The statements, opinions and data contained in all publications are solely those of the individual author(s) and contributor(s) and not of MDPI and/or the editor(s). MDPI and/or the editor(s) disclaim responsibility for any injury to people or property resulting from any ideas, methods, instructions or products referred to in the content.

Article

Fish DNA Sensors for Authenticity Assessment—Application to Sardine Species Identification

Myrto Kakarelidou [1], Panagiotis Christopoulos [1], Alexis Conides [2], Despina P. Kalogianni [1,*] and Theodore K. Christopoulos [1,3,*]

[1] Analytical/Bioanalytical Chemistry & Nanotechnology Group, Department of Chemistry, University of Patras, Rio, 26504 Patras, Greece; myrtokakarelidou@gmail.com (M.K.); pkchristop@gmail.com (P.C.)
[2] Hellenic Centre for Marine Research, Institute for Marine Biological Resources, 46.7 km Athens-Sounion, Anavyssos, 19013 Attika, Greece; conides@hcmr.gr
[3] Institute of Chemical Engineering Sciences, Foundation for Research and Technology Hellas (FORTH/ICE-HT), Platani, 26504 Patras, Greece
* Correspondence: kalogian@upatras.gr (D.P.K.); tchrist@upatras.gr (T.K.C.)

Abstract: Food and fish adulteration is a major public concern worldwide. Apart from economic fraud, health issues are in the forefront mainly due to severe allergies. Sardines are one of the most vulnerable-to-adulteration fish species due to their high nutritional value. Adulteration comprises the substitution of one fish species with similar species of lower nutritional value and lower cost. The detection of adulteration, especially in processed fish products, is very challenging because the morphological characteristics of the tissues change, making identification by the naked eye very difficult. Therefore, new analytical methods and (bio)sensors that provide fast analysis with high specificity, especially between closely related fish species, are in high demand. DNA-based methods are considered as important analytical tools for food adulteration detection. In this context, we report the first DNA sensors for sardine species identification. The sensing principle involves species recognition, via short hybridization of PCR-amplified sequences with specific probes, capture in the test zone of the sensor, and detection by the naked eye using gold nanoparticles as reporters; thus, avoiding the need for expensive instruments. As low as 5% adulteration of *Sardina pilchardus* with *Sardinella aurita* was detected with high reproducibility in the processed mixtures simulating canned fish products.

Keywords: authentication; adulteration; mislabeling; traceability; *Sardina pilchardus*; *Sardinella aurita*; gold nanoparticles; rapid test

Citation: Kakarelidou, M.; Christopoulos, P.; Conides, A.; Kalogianni, D.P.; Christopoulos, T.K. Fish DNA Sensors for Authenticity Assessment—Application to Sardine Species Identification. *Molecules* 2024, 29, 677. https://doi.org/10.3390/molecules29030677

Academic Editors: Paraskevas D. Tzanavaras and Victoria Samanidou

Received: 29 December 2023
Revised: 29 January 2024
Accepted: 30 January 2024
Published: 1 February 2024

Copyright: © 2024 by the authors. Licensee MDPI, Basel, Switzerland. This article is an open access article distributed under the terms and conditions of the Creative Commons Attribution (CC BY) license (https:// creativecommons.org/licenses/by/ 4.0/).

1. Introduction

Food fraud is a major problem worldwide. Because food fraud is linked to economic interests and health concerns, it is a high priority for food safety and quality worldwide. In addition, consumer demand for correct food labeling is constantly increasing. Along with other animal products, fishery products are considered to be one of the most adulterated foods [1,2]. In addition, fish are among the most easily adulterated foods due to morphological changes during processing that are not visible to the naked eye [3]. An increase in fish adulteration has been observed in recent years, being difficult to control due to the evolution of fraudulent practices. Adulteration is the substitution of one species of fish with another species of lower price and lower quality. Species swaps comprise the most prevalent fish fraud [4]. Sardines are vulnerable to adulteration as they contain valuable nutrients and are widely consumed worldwide. The only species that can be listed as sardine in canned food is the *Sardina pilchardus*, while if other similar species are used such as *Sardinella aurita*, the trade name of "X sardines", where X is different from *S. pilchardus* species, must be used [5,6].

To control adulteration, various analytical methods have been developed, including spectroscopy, chromatography, and protein-based methods. However, these methods are typically time-consuming, require highly trained personnel, and often use expensive and complex instrumentation and chemometrics for data interpretation [1,3,7]. The latest advances in spectroscopic techniques for authentication of animal-origin food include terahertz spectroscopy, laser-induced breakdown spectroscopy, hyperspectral imaging, nuclear magnetic resonance spectroscopy (NMR), Raman spectroscopy, near-infrared- and mid-infrared spectroscopy, Fourier transform infrared (FT-IR) spectroscopy. UV–Vis and fluorescence spectroscopy have the advantage of being non-destructive, but provide often low distinction ability between closely-related species [4,8–10]. DNA-based methods are preferred for fish identification, especially for canned fish, because DNA is characteristic of each species and resistant to food processing conditions, such as the heat treatment during canning. Therefore, DNA-based methods are considered valuable analytical tools against food fraud [3]. In addition, among the molecular methods, DNA barcoding has gained increased interest, being a successful tool to correctly identify animal species. Target gene selection is crucial to be able to discriminate between closely related species [7]. Mitochondrial DNA is widely used for species identification as it has a high number of gene copies and is ideal especially for highly processed food products or for a small amount of tissue sample. Finally, next-generation sequencing (NGS) has come to the front to overcome the limitations of DNA-barcoding [11].

The molecular methods for identifying sardine species include polymerase chain reaction followed by restriction fragment length polymorphism analysis (PCR-RFLP) [12–14], DNA sequencing [6,15,16], PCR followed by agarose gel electrophoresis [17–19], exon-primed intron-crossing (EPIC) PCR with acrylamide gel electrophoresis [20], real-time PCR with SYBR Green as fluorescent dye to detect the amplicons [21], real-time PCR using double-labelled detection probes with a fluorescent molecule and a quencher [5,22], and real-time PCR followed by melting curve analysis or high-resolution melting curve analysis [23,24]. The above methods either use specialized and expensive instrumentation or are time-consuming and have a low throughput.

In the present work, we have developed fish DNA sensors in a simple rapid-test format, that enable visual identification of the two main sardine species, *Sardina pilchardus* and *Sardinella aurita*. The sensing principle includes species recognition by hybridization of PCR-amplified sequences with specific probes, capture in the test zone of the sensor, and detection by the naked eye using gold nanoparticles as reporters, eliminating the need for expensive instruments. In contrast to other methods, the proposed method provides shorter analysis time, and allows the detection of as low as 1% adulteration in mixtures of PCR products and 5% adulteration in tissue mixtures of processed (canned) samples of both species. To our knowledge, this is the first report of fish DNA sensors for the identification of sardine species.

2. Results

In this study, a molecular rapid test was developed for the detection of fish adulteration for the species *S. pilchardus* with *S. aurita*. For this purpose, DNA was first isolated from the fresh tissue of both fish species, as well as from processed mixtures of the two species. The DNA was subjected to amplification by PCR using a common primer pair for the two species. The PCR products were 181 bp and 204 bp for *S. pilchardus* and *S. aurita*, respectively. The products were analyzed by electrophoresis in a 2% agarose gel using 2 μL of each PCR product. Quantification of the products was performed using the free-online ImageJ-gel Analyzer software (National Institutes of Health and the Laboratory for Optical and Computational Instrumentation, LOCI, University of Wisconsin, WI, USA). The DNA fragments were compared with the commercial ΦX174 DNA-HaeIII DNA marker (500 ng) (Figure S1).

The identification of the two sardine species was performed by using two different species-specific DNA probes that were hybridized to the amplified sequences followed by detection with the fish DNA sensor. The principle of the method is illustrated in Figure 1. The PCR products were biotinylated and the specific detection probes carried a poly-dA tail at one end. The hybrids were captured from the immobilized poly-dT sequences at the test zone of the sensor and detected by antibiotin-AuNPs through interaction of the anti-biotin Ab with the biotin moiety of the amplification product, forming a red line. A second red line was visualized at the control zone of the strip, as the excess of antibiotin-AuNPs was accumulated to the immobilized biotinylated BSA.

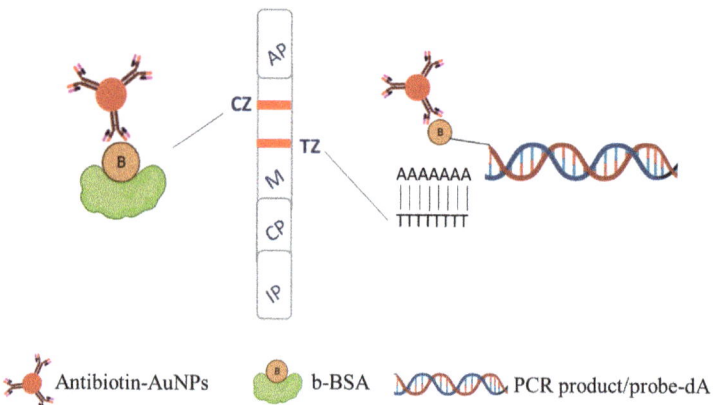

Figure 1. Principle of the fish DNA sensor. B: biotin, AuNPs: gold nanoparticles, IP: immersion pad, CP: conjugate pad, M: membrane, AP: absorbent pad, A: adenine, T: thymine, CZ: control zone, and TZ: test zone.

2.1. Effect of the Treatment of the Sample to the Optical Signal of the Test Zone of the Strip

We initially investigated whether the type of sample (processed and unprocessed) influenced the intensity of the color of the test zone of the sensor. For this purpose, PCR products from both fresh and cooked/processed samples were hybridized at a concentration of 20 nM with the species-specific probes. The fresh samples of both species were found to have only a slightly stronger signal than the processed samples, making the proposed method suitable for detecting adulterations in processed fish samples (Figure 2).

Effect of the treatment of samples on signal intensity

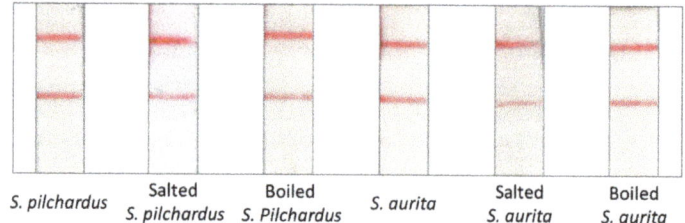

Figure 2. Effect of the treatment of fish samples to the signal of the test zone of the DNA sensor. Samples were treated with salt and boiling in the presence of various ingredients to simulate canned conditions.

2.2. Effect of the Amount of the Amplification Product on the Signal

The effect of the amount of PCR product on the signal at the test zone of the strip was studied as follows: The amplification products from *S. pilchardus* and *S. aurita* were separated by electrophoresis and quantified by imaging of the ethidium bromide-stained agarose gels. Serial two-fold dilutions of the products were then prepared and 5-µL aliquots containing 100, 50, 25, 12.5, 6.25, 3.125, and 1.56 fmol DNA were used for hybridization with the species-specific probes and application to the DNA sensor. Negative samples that contained no target DNA were also included in the study. All samples were analyzed in triplicates. It was observed that the intensity of the color band at the test zone of the sensor increased with increasing amount of the PCR product (fmol) applied to the sensor. As low as 6.25 fmol of PCR product from *S. pilchardus* and 3.13 fmol of the PCR product from *S. aurita* were detectable by the proposed DNA sensor (Figure 3).

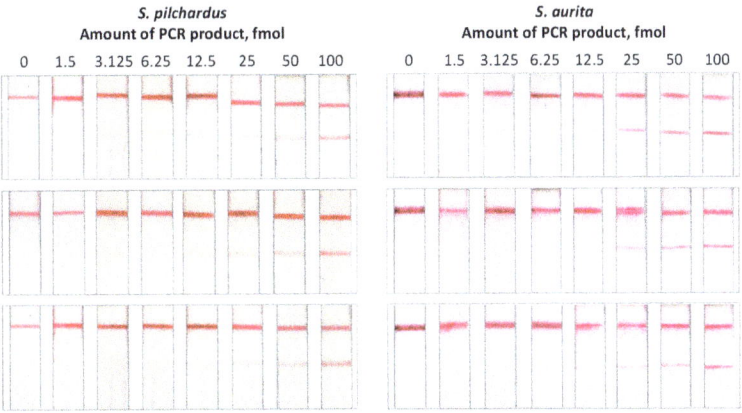

Figure 3. Effect of the amount of the amplification product on the signal of the DNA sensor. Various amounts of PCR product, from *S. pilchardus and S. aurita*, ranging from 0 to 100 fmol were analyzed in triplicate with the DNA sensor.

2.3. Cross-Hybridization Study

In this study, the specificity of the probes was examined, i.e., whether *S. pilchardus* probe hybridized with *S. aurita* DNA and whether the *S. aurita* probe hybridized with *S. pilchardus* DNA. PCR products from *S. pilchardus* and *S. aurita* were hybridized, separately, with both available probes. A-1.5 µL volume (150 fmol) of each PCR product was hybridized to 0.5 pmol of the *S. pilchardus* probe and *S. aurita* probe and the hybrids were detected by the DNA sensor. The results are shown in Figure 4. It was observed that a red line was obtained in the test zone of the sensor only when the PCR product hybridized with its complementary species-specific probe; otherwise, no signal was visible.

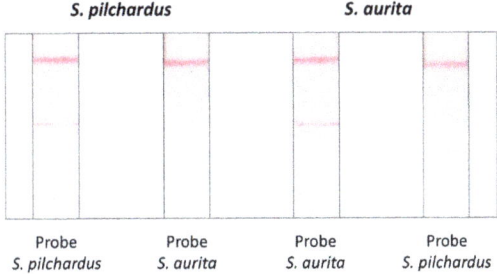

Figure 4. Cross-hybridization study for the experimental confirmation of probe specificity. Amplification products from *S. pilchardus* and *S. aurita* were hybridized, separately, to both probes and analyzed with the DNA sensor.

2.4. Method Performance Evaluation Using Mixtures of Amplification Products

Mixtures of the PCR products were prepared, containing different percentages of the *S. aurita* PCR product (0, 1, 5, 10, 20, 50, and 100%) in the *S. pilchardus* PCR product. The study was conducted in triplicate. The mixtures were hybridized with the specific probe for *S. aurita* and an amount of 100 fmol of the hybrids was applied to the DNA sensor. The results are shown in Figure 5. It was observed that adulteration of *S. pilchardus* with *S. aurita* was detected up to 1%. The mixtures were also tested by hybridization with the *S. pilchardus*-specific probe as a positive control of the entire procedure (see Figure S2). As the percentage of adulteration increased, a slight change in signal intensity was observed, while at a level of adulteration of 100%, no signal was obtained in the test zone of the DNA sensor due to the absence of *S. pilchardus* DNA.

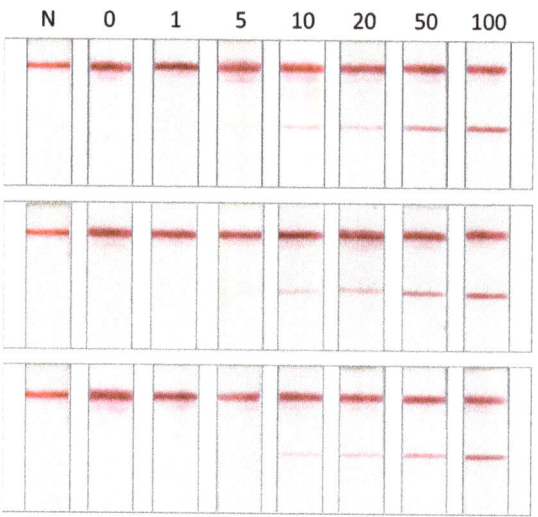

Figure 5. Method performance evaluation using mixtures of amplification products from *S. pilchardus* and *S. aurita*. The samples were analyzed in triplicate with the DNA sensor.

2.5. Method Performance Evaluation Using Mixtures of Processed Samples

Mixtures of the two fish species were prepared by mixing tissues of both species with oil and paprika at different adulteration ratios (0–100%). The mixtures were cooked to simulate canned conditions. DNA was then isolated from the processed samples. The isolated DNA was subjected to PCR and the products were analyzed in triplicate with the proposed DNA sensor. Salted *S. pilchardus* (0% adulteration) and *S. aurita* (100% adulteration) were also analyzed as above. The PCR products of the boiled samples were diluted three times with 1× TE buffer, pH 8.0, while the salted *S. pilchardus* was diluted five times. Hybridization with the species-specific probes was performed, using a 1-µL volume of the diluted PCR product. The DNA sensor results are shown in Figure 6 for the *S. aurita*-specific probe ($n = 3$) and in Figure S3 for the *S. pilchardus*-specific probe. For the processed mixtures (cooked), the adulteration level detected was 5% with high repeatability.

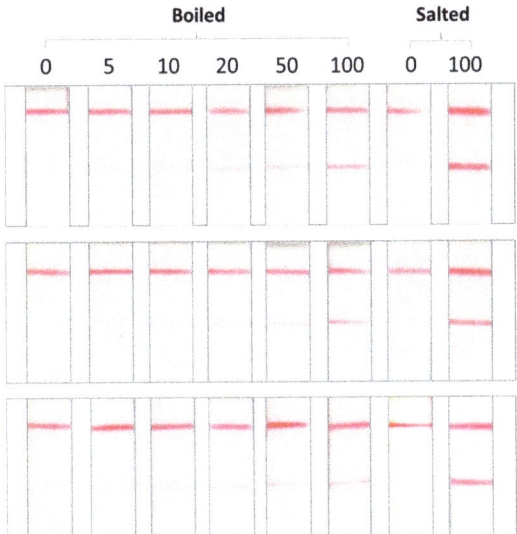

Figure 6. Method performance evaluation using mixtures of processed (canned) samples. The samples were analyzed in triplicate with the DNA sensor.

2.6. Repeatability of the DNA Sensor

Finally, the repeatability of the DNA sensor was assessed. Different amounts of PCR products of both fish species were analyzed in triplicate with the DNA sensor. The results are presented in Figure 7. The images of the strips were obtained using a regular benchtop scanner. After densitometric analysis of the test zones of the strips, the coefficient of variations, $\%CV = \frac{SD}{\bar{x}} \times 100$, were calculated and found to be 3.6, 4.4, and 7.8% for the 6.3, 25, and 100 fmol of *S. pilchardus* and 17.4, 9.6, and 8.1% for the 3.1, 12.5, and 100 fmol of the PCR product of *S. aurita*, proving a very good repeatability of the proposed DNA sensor.

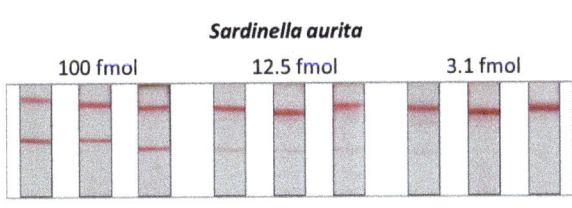

Figure 7. Repeatability of the DNA sensor. Different amounts of PCR products of *S. pilchardus* and *S. aurita* were analyzed in triplicate with the DNA sensor.

3. Discussion

Fish authentication is a major concern worldwide. For these reasons, new methods for fish species identification are of great importance. DNA-based methods are the preferred ones due to DNA stability and enhanced sensitivity. Herein, we have developed a DNA sensor for visual discrimination of *S. pilchardus* from each most common adulterant *S. aurita*.

So far, few methods have been reported for authenticity testing of sardines. More specifically, in 2016, PCR-RFLP and DNA sequencing for confirmation were used and compared by Leonardo et al. for the discrimination of several sardine species in canned products. The cytochrome b mitochondrial gene was used in this study, as well as phylogenetic analysis which was also conducted in the samples. As a conclusion, RFLP was not adequate to allow discrimination of the species and had a low throughput. On the other hand, DNA sequencing allowed unambiguous identification of different sardine species [13]. Also, in 2012 and 2003, a PCR-RLFP method combined with phylogenetic analysis was developed. A fragment of the cytochrome b gene of the mitochondrial DNA was exploited to discriminate *S. pilchardus* from other species in both fresh and canned products [12,14]. Lago et al., in 2011, used PCR, DNA sequencing, and phylogenetic analysis for the identification of different single nucleotide polymorphisms (SNPs) in the cytochrome b mitochondrial gene of *S. pilchardus* and *S. aurita*. This method could be applied to all kinds of canned food, regardless of the treatment that fish samples have undergone. The method was also applied to mixtures of tissues of both species, and down to 5% (w/w) of the tissue of *S. aurita* was detected [6]. Direct sequencing methods were also applied for species identification in fresh and canned products based again on mitochondrial DNA cytochrome b or the cytochrome oxidase I gene. The results were assessed by phylogenetic analysis [15,16]. PCR based on the 16S RNA mitochondrial gene, followed by agarose gel electrophoresis was used for identification of *S. pilchardus* [17,19] and *S. pilchardus*, *S. aurita*, and *S. maderensis* [18]. Exon-primed intron-crossing (EPIC) PCR with acrylamide gel electrophoresis was also exploited for *S. pilchardus* identification [20]. DNA barcoding and real-time PCR based on SYBR Green for detection of amplicons was used for *S. pilchardus* screening in commercial fish products [21]. Taqman probes with real-time PCR was introduced to increase the specificity of the method and applied for the discrimination of *S. pilchardus* from *S. aurita* in commercial products. An amount of 25 pg was adequate for positive results. Processed products had higher Cq values than fresh samples, while Cq remained <30 for all samples [5]. The same method was developed for the discrimination of *S. pilchardus* from anchovy with a detection limit of 0.05 ng of DNA [22]. Finally, real-time PCR with subsequent (high-resolution) melting curve analysis for confirmation of the results was developed for *S. pilchardus* identification [23,24]. All the above methods usually require expensive instrumentation or are highly time-consuming. PCR-RLFP, however, has a lower distinction capability and cannot provide quantitative results. The present work reports the first DNA sensor developed for sardine authenticity testing. It enables rapid and visual discrimination by the naked eye of down to 5% of adulteration of *S. pilchardus* with *S. aurita* with high discrimination capability. However, it can only provide semi-quantitative results. The sensor is easy to develop and use, without the need for special and expensive instrumentation. Therefore, it can be easily adopted by any laboratory and public authorities for authenticity testing. Also, one of the great advantages of the proposed DNA sensor is its universality and versatility. The sensor can be used for the identification of any other fish species as the molecular recognition by species-specific primers and probes is performed prior to the application to the sensor. Finally, the proposed DNA sensor can be modified accordingly to be able to detect more than two different species-specific DNA sequences on the same sensor. A comparative Table S1 is available in the Supplementary Material.

4. Materials and Methods

4.1. Materials

Oligonucleotides, primers, and probes were purchased from Eurofins Genomic (Vienna, Austria). The Nucleospin Tissue kit for DNA extraction was obtained from Macherey-Nagel (Duren, Germany). Kapa 2G Fast polymerase with Buffer A was from Kapa Biosystems (Basel, Switzerland), dNTPs and dUTP from Invitrogen (CarlsBad, CA, USA), FX174 DNA-HaeIII DNA marker and the terminal transferase kit containing the enzyme, the reaction buffer, and the $CoCl_2$ from New England Biolabs (Ipswich, MA, USA), and agarose from Fisher BioReagents (Waltham, MA, USA). Bovine serum albumin (BSA) was purchased from Serva Electrophoresis (Heidelberg, Germany), $NaHCO_3$ from Fisher Scientific (Loughborough, UK), and EZ-Link Sulfo-NHS-LC-LC-Biotin from Thermo Fisher Scientific (Waltham, MA, USA). The 40 nm gold nanoparticles (0.15 nM) were from BBI solutions (Crumlin, UK) and the anti-biotin antibody from Sigma-Aldrich (Saint Louis, MO, USA). EDTA, borax, SDS, and sodium azide were all obtained from Merck (Darmstadt, Germany). For the construction of the strips, Immunopore FP membrane and glass fiber conjugate pad were purchased from GE Healthcare Life Sciences (Buckinghamshire, UK), whereas the absorbent pads were from Schleicher & Schuell (Dassel, Germany). Finally, for the running buffer, glycerol was obtained from Carlo Erba (Barcelona, Spain), the phosphate salts from Lachner (Neratovice, Czech Republic), and Tween-20 from Fluka (Sigma-Aldrich). The $1\times$ phosphate-buffered saline solution (PBS) pH 7.4 consisted of 137 mM NaCl, 2.7 mM KCl, 8 mM Na_2HPO_4, and 2 mM KH_2PO_4 and the $6\times$ saline sodium citrate buffer (SSC) pH 7.0 consisted of 900 mM sodium chloride and 90 mM sodium citrate.

MJ Research PTC-150 Minicycler was used for PCR amplification. The TLC CAMAG Linomat 5 (Muttenz, Switzerland) and UVP Crosslinker CL-3000 by Analytik Jena (Upland, CA, USA) were used for the deposition and the immobilization of the reagents on the membrane, respectively. Finally, a common scanner (EPSON Perfection V600 PHOTO, Seiko Epson Corporation, Suwa, Japan) was used for acquiring the images of the strips.

4.2. Samples

Fresh fish samples of *S. pilchardus* and *S. aurita* were collected from local fish monger markets in Western Greece. The samples were kept at $-20\ °C$ until analysis

4.3. Sample Preparation

Fresh samples were used unprocessed for method development, while processed samples were also prepared to simulate the types of processed fish-based products available in the market. Two fresh samples of each species were processed in two different ways. One sample was boiled with oil and paprika, to simulate canning conditions, and the other sample was left with salt, oil, and vinegar to obtain a salt-preserved product. Mixtures of both species were also prepared using the cooking procedure as described above. The binary mixtures contained different percentages of adulterants (5%, 10%, 20%, 50%, 100%). All samples were stored at $-20\ °C$.

4.4. DNA Extraction

DNA was isolated from fresh samples of *S. pilchardus* and *S. aurita* and from processed mixtures of the two species. Part of the fish dorsal muscle tissue from each sample was cut for DNA isolation. An amount of 25 mg of tissue was used for fresh samples while 24 mg of tissue was cut and weighed directly from processed samples. For the boiled samples, 25 mg of tissue from both species was used for DNA extraction. The Nucleospin Tissue kit was used for DNA isolation according to the manufacturer's instructions. Elution at the final stage was performed with 50 µL of elution buffer and the isolated DNA was stored at $-20\ °C$.

4.5. Primer Design

Ribosomal RNA sequences for *S. pilchardus* (JN590272.1) and *S. aurita* (JN590273.1) were obtained from the NCBI database and sequence alignment was performed using the online free software Clustal Omega (EMBL's European Bioinformatics Institute (EMBL-EBI, Cambridgeshire, UK). The primers and probes sequences were then studied in terms of secondary structures using the Oligo Analyzer Tool (Integrated DNA Technologies, Coralville, IA, USA) and BLAST (NIH, Bethesda, MD, USA) to confirm their specificity to each species. The sequences of the PCR primers and the probes used throughout this work are presented in Table 1.

Table 1. Sequences of primers and probes for *S. pilchardus* and *S. aurita* [4].

	Name	Sequence (5′→3′)
Primers	Sard Forward	Biotin-CGGTGGTMAAACACATG
	Sard Reverse	GTCTGATCTGAGGTCGT
Probes	*S. pilchardus* Probe	CCTTGCTCCAGAGGTCCG
	S. aurita Probe	CCTYGCTCTACGGTCCGG

4.6. Polymerase Chain Reaction (PCR)

PCR was performed using the same primers for both species. The reaction was performed in a final volume of 20 µL and contained 0.8 µM of biotinylated forward primer (Sard Forward), 0.8 µM of reverse primer (Sard Reverse), 1× Kappa 2G buffer A, 0.2 mM each of dATP, dGTP, dCTP, dUTP, 0.25 µL of Kapa 2G Fast polymerase (1.25 U), and 5 µL of isolated DNA from fresh samples or 50 ng of isolated DNA from the mixtures. Amplification was performed as follows: 94 °C for 5 min followed by 35 cycles of 94 °C for 20 s, 52 °C for 20 s, 72 °C for 20 s, and final extension at 72 °C for 7 min.

4.7. Construction of the Sensing Zones of the Device

4.7.1. Biotinylation of Bovine Serum Albumin (b-BSA)

BSA was biotinylated for the construction of the control zone of the sensor. Briefly, 0.2 mg of BSA (200 µL from 1 g/L) was added to 50 µL of 0.5 M NaHCO$_3$, pH 9.1, followed by the addition of 1 mg of sulfo-NHS-LC-LC-biotin and incubation for 1 h, at room temperature.

4.7.2. Insertion of a Poly-dT Tail into the Capture Probe

For the construction of the test zone of the sensor, a poly-dT tail was incorporated into the 3′ end of an oligonucleotide (capture probe) carrying 30 thymine bases (dT$_{30}$). Reaction was carried out in a final volume of 20 µL, containing 1× Reaction TdT Buffer, 0.25 mM CoCl$_2$, 5 mM dTTP, 0.05 mM oligonucleotide dT$_{30}$, and 1.5 U terminal transferase (TdT). The solution was incubated at 37 °C for 1 h. To stop the reaction, 2 µL of a 0.5 M EDTA solution, pH 8.0, was added and the solution was stored at −20 °C.

4.8. Insertion of a Poly-dA Tail into the Detection Probes

For the detection of PCR products with the DNA sensor, a poly-dA tail was inserted into the species-specific detection probes for *S. pilchardus* and *S. aurita*. The reaction conditions were as above except that 2 mM dATP, 0.02 mM of each probe, and 1 U of TdT were used.

4.9. Coupling of Anti-Biotin Antibody to Gold Nanoparticles (Antibiotin-AuNPs)

A-1 mL volume of AuNPs (0.15 nM) was centrifuged at 3300× *g* for 20 min and 500 µL of the supernatant was discarded. The pH was then adjusted to 9.0 with the addition of a proper volume of 200 mM borax. Then, 4.6 µg of anti-biotin antibody, dissolved in 2 mM borax, was added gradually under stirring to a final concentration of 0.01 g/L. The solution was incubated for 45 min in the dark and 100 µL of 100 g/L BSA in 20 mM borax were added for blocking. The solution was incubated for another 10 min at room temperature. This step was followed by centrifugation at $4500\times g$ for 15 min, washing with 500 µL of wash solution (10 g/L BSA in 2 mM borax), centrifugation at $4500\times g$ for 5 min, and redispersing in 100 µL of redispersion solution (1 g/L BSA and 1 g/L NaN_3 in 2 mM borax). Antibiotin-AuNPs were finally stored at 4 °C. A 4-µL aliquot of antibiotin-AuNPs was deposited on the conjugate pad of the strip-type DNA sensor.

4.10. Assembly of the Fish DNA Sensor

The strip-type DNA sensor (4 mm × 70 mm) consisted of a plastic substrate, on which the individual pieces of absorption pad, nitrocellulose membrane, conjugate pad, and immersion pad were assembled. On the nitrocellulose membrane, b-BSA and poly-dT were immobilized to create the control zone and the test zone, respectively. More specifically, a solution containing 10 µM poly-dT in 5% (v/v) MeOH, 2% (w/v) sucrose and 6× SSC pH 7.0 and a solution containing 1.5 or 2.5 mg/L b-BSA in 5% (v/v) MeOH, 2% (w/v) sucrose and 1× PBS buffer pH 7.4 were deposited at specific areas on the nitrocellulose membrane, using the commercial dispenser Linomat 5. The membrane was finally dried in a UV crosslinker at 125 mJ/cm^2 for 15 min. Subsequently, the strip was properly assembled as follows. Double-sided tape was taped onto a plastic substrate and then the nitrocellulose membrane was attached. The absorbent pad was placed on top of the membrane, the conjugate pad was placed in the section below the membrane, and the immersion pad was placed below the conjugate pad, in such a way that there was an overlap between all the pads. Finally, the assembly was cut to 4 mm width per sensor.

4.11. Detection of S. pilchardus and S. aurita

Each PCR product was hybridized to the specific probe carrying a poly-dA tail prior to the application to the sensor. A volume of 1.5 µL of the biotinylated PCR product was mixed with 1.5 µL of 0.5 µM poly-dA specific probe and 1.5 µL of 0.4 M NaOH. The solution was incubated at room temperature for 5 min and 1.5 µL of 0.4 M Tris-HCl and 1.5 µL of 0.4 M HCl were then added. The mixture was finally incubated at 42 °C for 10 min. Subsequently, a 5-µL volume was applied to the conjugate pad and the strip was immersed into 400 µL of the running buffer (1× PBS, 1% (w/v) BSA, 1% (v/v) Tween-20, 1% (v/v) glycerol, 0.35% (w/v) SDS). The strip was removed from the running solution after 15 min and scanned by a regular scanner.

5. Conclusions

We have developed a fish DNA sensor for fish authentication in a rapid-test format. The method involves detecting adulteration of *Sardina pilchardus* with *Sardinella aurita* using species-specific DNA probes. The developed DNA sensor is the first DNA sensor reported for the identification of sardine species and particularly for the detection of adulteration of *S. pilchardus*. Detection is performed visually using gold nanoparticles as reporters. Compared to existing DNA-based fish adulteration methods, the method is simple and rapid. The sensor is cost-effective (<2 €), easy to use, and very practical as it can be developed in the form of a dry-reagent strip-type sensor; no qualified personnel or costly instrumentation is required, while it can be used by authorities for fish authentication control or any laboratory including fish processing industries. Finally, the method was successfully applied to fresh and processed samples treated by cooking and salting in the presence of other ingredients to simulate canning conditions. For mixtures of the two species, as low as 5% adulteration was detected in the processed samples with very good repeatability.

Supplementary Materials: The following supporting information can be downloaded at: https://www.mdpi.com/article/10.3390/molecules29030677/s1, Figure S1: Electropherogram of PCR products for *S. pilchardus* with *S. aurita*.; Figure S2: Mixtures of PCR products. The mixtures contained 0–100% PCR product from *S. aurita* in PCR product from *S. pilchardus* with a total amount of 100 fmol on the strip; Figure S3: Mixtures of processed samples. The mixtures contained 0–100% of *S. aurita* (tissue) in *S. pilchardus*; Table S1: Comparison of methods for sardines' adulteration detection.

Author Contributions: Conceptualization, P.C., A.C., D.P.K. and T.K.C.; Data curation, M.K.; Formal analysis, M.K., P.C., D.P.K. and T.K.C.; Funding acquisition, P.C., A.C., D.P.K. and T.K.C.; Investigation, M.K., P.C., D.P.K. and T.K.C.; Methodology, M.K., P.C., D.P.K. and T.K.C.; Project administration, A.C. and T.K.C.; Resources, P.C. and A.C.; Supervision, D.P.K. and T.K.C.; Validation, M.K., P.C., D.P.K. and T.K.C.; Visualization, M.K., P.C., D.P.K. and T.K.C.; Writing—original draft, M.K., D.P.K. and T.K.C.; Writing—review & editing, A.C., D.P.K. and T.K.C. All authors have read and agreed to the published version of the manuscript.

Funding: The research presented herein was conducted in the frame of the research project "DNASEEKER-Innovative prototype genetic method for the identification of fish species for the Greek fishery products processing industry" (MIS 5072570), co-funded by the Greek Government Operational Program for Fisheries and Sea 2014–2020 and the EU EMFF.

Institutional Review Board Statement: Not applicable.

Informed Consent Statement: Not applicable.

Data Availability Statement: Data are available upon requirement.

Conflicts of Interest: The authors declare no conflicts of interest.

References

1. Smaoui, S.; Tarapoulouzi, M.; Agriopoulou, S.; D'Amore, T.; Varzakas, T. Current state of milk, dairy products, meat and meat products, eggs, fish and fishery products authentication and chemometrics. *Foods* **2023**, *12*, 4254. [CrossRef] [PubMed]
2. Kotsanopoulos, K.V.; Exadactylos, A.; Gkafas, G.A.; Martsikalis, P.V.; Parlapani, F.F.; Boziaris, I.S.; Arvanitoyannis, I.S. The use of molecular markers in the verification of fish and seafood authenticity and the detection of adulteration. *Compr. Rev. Food Sci. Food Saf.* **2021**, *20*, 1584–1654. [CrossRef]
3. Cermakova, E.; Lencova, S.; Mukherjee, S.; Horka, P.; Vobruba, S.; Demnerova, K.; Zdenkova, K. Identification of fish species and targeted genetic modifications based on DNA analysis: State of the Art. *Foods* **2023**, *12*, 228. [CrossRef]
4. Chaudhary, V.; Kajla, P.; Dewan, A.; Pandiselvam, R.; Socol, C.T.; Maerescu, C.M. Spectroscopic techniques for authentication of animal origin foods. *Front. Nutr.* **2022**, *9*, 979205. [CrossRef]
5. Herrero, B.; Lago, F.C.; Vieites, J.M.; Espiñeira, M. Development of a rapid and simple molecular identification methodology for true sardines (*Sardina pilchardus*) and false sardines (*Sardinella aurita*) based on the real-time PCR technique. *Eur. Food Res. Technol.* **2011**, *233*, 851–857. [CrossRef]
6. Lago, F.C.; Herrero, B.; Vieites, J.M.; Espiñeira, M. FINS Methodology to identification of sardines and related species in canned products and detection of mixture by means of SNP analysis systems. *Eur. Food Res. Technol.* **2011**, *232*, 1077–1086. [CrossRef]
7. Dawan, S.J.; Ahn, J. Application of DNA barcoding for ensuring food safety and quality. *Food Sci. Biotechnol.* **2022**, *31*, 1355–1364. [CrossRef]
8. Zaukuu, J.L.Z.; Benes, E.; Bázár, G.; Kovács, Z.; Fodor, M. Agricultural potentials of molecular spectroscopy and advances for food authentication: An overview. *Processes* **2022**, *10*, 214. [CrossRef]
9. Wang, X.Y.; Xie, J.; Chen, X.J. Applications of non-invasive and novel methods of low-field nuclear magnetic resonance and magnetic resonance imaging in aquatic products. *Front. Nutr.* **2021**, *8*, 651804. [CrossRef]
10. Hassoun, A.; Sahar, A.; Lakhal, L.; Aït-Kaddour, A. Fluorescence spectroscopy as a rapid and non-destructive method for monitoring quality and authenticity of fish and meat products: Impact of different preservation conditions. *LWT* **2019**, *103*, 279–292. [CrossRef]
11. Franco, C.M.; Ambrosio, R.L.; Cepeda, A.; Anastasio, A. Fish intended for human consumption: From DNA barcoding to a next-generation sequencing (NGS)-based approach. *Curr. Opin. Food Sci.* **2021**, *42*, 86–92. [CrossRef]
12. Jérôme, M.; Lemaire, C.; Bautista, J.M.; Fleurence, J.; Etienne, M. Molecular phylogeny and species identification of sardines. *J. Agric. Food Chem.* **2003**, *51*, 43–50. [CrossRef]
13. Leonardo, R.; Nunes, R.S.C.; Monteiro, M.L.G.; Conte-Junior, C.A.; Del Aguila, E.M.; Paschoalin, V.M.F. Molecular testing on sardines and rulings on the authenticity and nutritional value of marketed fishes: An experience report in the state of Rio de Janeiro, Brazil. *Food Control* **2016**, *60*, 394–400. [CrossRef]
14. Besbes, N.; Fattouch, S.; Sadok, S. Differential detection of small pelagic fish in tunisian canned products by PCR-RFLP: An efficient tool to control the label information. *Food Control* **2012**, *25*, 260–264. [CrossRef]

15. Jérôme, M.; Lemaire, C.; Verrez-Bagnis, V.; Etienne, M. Direct sequencing method for species identification of canned sardine and sardine-type products. *J. Agric. Food Chem.* **2003**, *51*, 7326–7332. [CrossRef]
16. Labrador, K.; Agmata, A.; Palermo, J.D.; Follante, J.; Pante, M.J. Authentication of processed Philippine sardine products using hotshot DNA extraction and minibarcode amplification. *Food Control* **2019**, *98*, 150–155. [CrossRef]
17. Bottero, M.T.; Dalmasso, A.; Nucera, D.; Turi, R.M.; Rosati, S.; Squadrone, S.; Goria, M.; Civera, T. Development of a PCR assay for the detection of animal tissues in ruminant feeds. *J. Food Prot.* **2003**, *66*, 2307–2312. [CrossRef] [PubMed]
18. Durand, J.D.; Diatta, M.A.; Diop, K.; Trape, S. Multiplex 16S rRNA haplotype-specific PCR, a rapid and convenient method for fish species identification: An application to West African *Clupeiform larvae*. *Mol. Ecol. Resour.* **2010**, *10*, 568–572. [CrossRef] [PubMed]
19. Besbes, N.; Fattouch, S.; Sadok, S. Comparison of methods in the recovery and amplificability of DNA from fresh and processed sardine and anchovy muscle tissues. *Food Chem.* **2011**, *129*, 665–671. [CrossRef] [PubMed]
20. Atarhouch, T.; Rami, M.; Naciri, M.; Dakkak, A. Genetic population structure of sardine (*Sardina pilchardus*) off Morocco detected with intron polymorphism (EPIC-PCR). *Mar. Biol.* **2007**, *150*, 521–528. [CrossRef]
21. Xiong, X.; Yuan, F.; Huang, M.; Xiong, X. Exploring the possible reasons for fish fraud in China based on results from monitoring sardine products sold on Chinese markets using DNA barcoding and real time PCR. *Food Addit. Contam. Part A Chem. Anal. Control Expo. Risk Assess.* **2020**, *37*, 193–204. [CrossRef] [PubMed]
22. Cuende, E.; Mendibil, I.; Bachiller, E.; Álvarez, P.; Cotano, U.; Rodriguez-Ezpeleta, N. A real-time PCR approach to detect predation on anchovy and sardine early life stages. *J. Sea Res.* **2017**, *130*, 204–209. [CrossRef]
23. Armani, A.; Castigliego, L.; Tinacci, L.; Gianfaldoni, D.; Guidi, A. Multiplex conventional and real-time PCR for fish species identification of Bianchetto (juvenile form of *Sardina pilchardus*), Rossetto (*Aphia minuta*), and Icefish in fresh, marinated and cooked products. *Food Chem.* **2012**, *133*, 184–192. [CrossRef]
24. Bréchon, A.L.; Coombs, S.H.; Sims, D.W.; Griffiths, A.M. Development of a rapid genetic technique for the identification of *Clupeid larvae* in the Western English channel and investigation of mislabelling in processed fish products. *ICES J. Mar. Sci.* **2013**, *70*, 399–407. [CrossRef]

Disclaimer/Publisher's Note: The statements, opinions and data contained in all publications are solely those of the individual author(s) and contributor(s) and not of MDPI and/or the editor(s). MDPI and/or the editor(s) disclaim responsibility for any injury to people or property resulting from any ideas, methods, instructions or products referred to in the content.

Article

Application of the Bland–Altman and Receiver Operating Characteristic (ROC) Approaches to Study Isotope Effects in Gas Chromatography–Mass Spectrometry Analysis of Human Plasma, Serum and Urine Samples

Dimitrios Tsikas

Core Unit Proteomics, Institute of Toxicology, Hannover Medical School, 30623 Hannover, Germany; tsikas.dimitros@mh-hannover.de

Abstract: The Bland–Altman approach is one of the most widely used mathematical approaches for method comparison and analytical agreement. This work describes, for the first time, the application of Bland–Altman to study $^{14}N/^{15}N$ and $^{1}H/^{2}H$ (D) chromatographic isotope effects of endogenous analytes of the L-arginine/nitric oxide pathway in human plasma, serum and urine samples in GC-MS. The investigated analytes included arginine, asymmetric dimethylarginine, dimethylamine, nitrite, nitrate and creatinine. There was a close correlation between the percentage difference of the retention times of the isotopologs of the Bland–Altman approach and the area under the curve (AUC) values of the receiver operating characteristic (ROC) approach ($r = 0.8619$, $p = 0.0047$). The results of the study suggest that the chromatographic isotope effects in GC-MS result from differences in the interaction strengths of H/D isotopes in the derivatives with the hydrophobic stationary phase of the GC column. D atoms attenuate the interaction of the skeleton of the molecules with the lipophilic GC stationary phase. Differences in isotope effects in plasma or serum and urine in GC-MS are suggested to be due to a kind of matrix effect, and this remains to be investigated in forthcoming studies using Bland–Altman and ROC approaches.

Keywords: Bland–Altman; chromatography; isotope effects; GC-MS; retention time; ROC

1. Introduction

In gas chromatography (GC) and reversed-phase liquid chromatography (LC), isotopologs differ in their retention times, with deuterated compounds having, as a rule, smaller retention times (t_R) than their protiated analogs [1]. The slightly reduced molecular volume of ^{2}H-labeled compounds compared to their non-labeled analogs is assumed to be responsible for this phenomenon [2]. The t_R of the isotopologs is the main parameter for quantitate H/D isotope effects [3]. One method to calculate the H/D isotope effect (IE) is to divide the t_R of the protiated analyte $t_{R(H)}$ by the t_R of the deuterated analyte $t_{R(D)}$ (Formula (1)). The difference in the retention times can be used to estimate the extent of the isotope effect $\delta_{(H/D)}$ (Formula (2)). The closer the IE value to the unity (1.0000), the lower/weaker the isotope effect between the isotopologs. The smaller the difference $\delta_{(H/D)}$ in the retention times $t_{R(H)}$ and $t_{R(D)}$, the higher/stronger the isotope effect between the isotopologs. These issues are also valid for isotopes of other elements, including ^{14}N and ^{15}N.

$$IE = t_{R(H)} / t_{R(D)} \tag{1}$$

$$\delta_{(H/D)} = t_{R(H)} - t_{R(D)} \tag{2}$$

$$\mu_{(H/D)} = 1/2 \times (t_{R(H)} + t_{R(D)}) \qquad (3)$$

$$\delta\ (\%) = 2 \times ([t_{R(H)} - t_{R(D)}]/[t_{R(H)} + t_{R(D)}]) \times 100 \qquad (4)$$

The Bland–Altman approach is useful for the comparison of methods with comparable analytical performance [4,5]. It is a graphical approach which examines the relationship between the difference (δ) of the values obtained by two methods and the average (μ) of the methods. In a variant of the Bland–Altman method, the percentage difference δ (%) of the two methods is plotted versus the average of these methods (Formulas (3) and (4)). One may expect that the Bland–Altman approach would be useful in calculating the absolute and percentage difference of the retention times of isotopologs as measures of isotope effects. The Bland–Altman method has been sporadically used in this area, including $\delta(^{18}O)$ isotope ratios [6–8]. The receiver operating characteristic (ROC) approach is another graphical plot that is widely used in several disciplines, notably including clinical chemistry [9,10]. The utility of the ROC approach in comparison to methods in analytical chemistry has been demonstrated [5]. The ROC approach is useful for evaluating the agreement/disagreement between the isotopologs.

In the present work, the Bland–Altman and ROC approaches were used for the first time to investigate the H/D and $^{14}N/^{15}N$ isotope effects of endogenous substances in human plasma, serum and urine samples. The analytes considered in the study belong to the L-arginine/nitric oxide pathway [11]. They include nitrite and nitrate, the major metabolites of nitric oxide (NO), L-arginine (Arg) and asymmetric dimethylarginine (ADMA), the endogenous substrate and inhibitor of NO synthase, respectively [11], and dimethylamine (DMA), the major urinary metabolite of ADMA [12]. In addition, creatinine serves as an analyte that is commonly used to correct for the excretion of endogenous substances in urine collected by spontaneous micturition. The above-mentioned analytes were measured by gas chromatography–mass spectrometry (GC-MS) after proper chemical derivatization. The chemical structures of the derivatives of the protiated (unlabeled) and ^{2}H- or ^{15}N-labelled analytes are shown in Scheme 1.

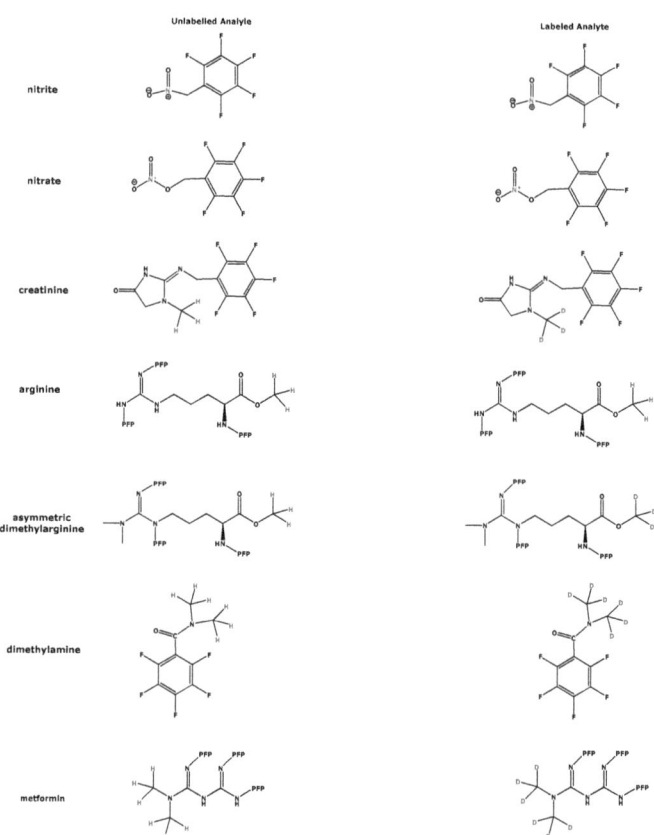

Scheme 1. Chemical structures of the unlabeled analytes (**left**) and their ^{15}N and ^{2}H (D) isotopologs (**right**) investigated in the present work. Nitrite, nitrite and creatinine were derivatized with pentafluorobenzyl bromide in aqueous acetone (e.g., 60 min at 50 °C). Arginine and asymmetric dimethylarginine were first methylated in 2 M HCl in CH$_3$OH or 2 M HCl in CD$_3$OD (e.g., 30 min at 80 °C) and then acylated by pentafluoropropionic anhydride in ethyl acetate (e.g., 60 min at 65 °C). Metformin was derivatized with pentafluoropropionic anhydride only. Dimethylamine was derivatized with pentafluorobenzoyl chloride at room temperature. PFB, pentafluorobenzyl; PFP, pentafluoropropionyl. Blue color indicates ^{14}N and ^{1}H isotopes; red color indicates ^{15}N and ^{2}H (D) isotopes.

2. Methods

2.1. GC-MS Analyses in Human Plasma, Serum and Urine Samples

The human plasma, serum and urine samples analyzed in the present work were collected in previous clinical studies of the author's group and cooperating groups after approval by the local ethics committees [13–19]. The studies were conducted in line with the ethical principles of the Declaration of Helsinki [20]. The COVID-19 study [17] was approved by the local ethics committees (9948_BO_K_2021 Hannover Medical School; 29/3/21 University Medical Center Göttingen). The ASOS study was approved by the Health Research Ethics Committee of North-West University (NWU-00007-15-A1) [18].

Nitrite and nitrate were analyzed simultaneously by GC-MS as described previously, using commercially available ^{15}N-nitrite and ^{15}N-nitrate as internal standards, respectively [13]. Creatinine was analyzed via GC-MS as described elsewhere using commercially available [*methylo*-^{2}H$_3$] creatinine as internal standard [14]. Arg and ADMA were analyzed by GC-MS, using in situ prepared trideuteromethyl ester as described previously [15]. DMA was analyzed by GC-MS using [*dimethylo*-$^2H_3$]DMA (d_6-DMA) as the internal standard [16]. Metformin (METF) was analyzed by GC-MS using [*dimethylo*-$^2H_3$]metformin (d_6-METF) as the internal standard [19]. The isotopic purity in the stable-isotope-labeled analogs was at least 99% at ^2H and ^{15}N. The concentrations of the internal standards were 1 µM for ADMA in plasma and serum; 20 µM in urine; 50 µM for Arg in plasma and serum; 10 mM and 1 mM for creatinine in urine and serum, respectively; 1 mM for DMA in urine; 4 µM for nitrite in plasma and serum and 8 µM in urine; 40 µM for nitrate in plasma and 800 µM in urine.

GC-MS analyses were performed on the apparatus model ISQ from ThermoFisher (Dreieich, Germany), which was equipped with a fused silica capillary OPTIMA-17 column (15 m × 0.25 mm, 0.25 µm film thickness) from Macherey-Nagel (Düren, Germany). Helium and methane were, respectively, used as carrier and reagent gases for negative-ion chemical ionization. Analyses were performed by selected-ion monitoring (SIM) for nitrite, nitrate, creatinine, DMA, ADMA and Arg. Aliquots of 1 µL toluene (for amino acids, DMA and metformin) or ethyl acetate (for nitrite, nitrate and creatinine) extracts of the derivatives were injected in the splitless mode. Ions were detected after conversion to electrons by using an electron multiplier. Different oven temperature programs were used, starting either at 40 °C or 70 °C (for nitrite, nitrate and creatinine).

2.2. Calculations

Isotope effect values were calculated using Formula (1). The difference (in min) between the retention times was calculated by Formula (2). The difference in the retention times was multiplied by 60 to obtain the outcome in seconds. In the Bland–Altman approach, the percentage difference (δ (%)) was plotted versus the average ($\mu_{(H/D)}$). The term bias (%) in the regular Bland–Altman approach corresponds to δ (%). The receiver operating characteristic (ROC) approach was used to determine the area under the curve (AUC) values by using the retention times of the isotopologs. The ratios of the peak areas (PAR) of endogenous analytes and the respective internal standards were calculated and used to test for potential correlations between the difference δ and the PAR values. The Wilcoxon matched-pairs signed-rank test was used to test statistical differences in the retention times of isotopologs.

2.3. Statistical Analyses and Data Presentation

GraphPad Prism Version 7 for Windows (GraphPad Software, San Diego, CA, USA) was used for statistical analyses and preparation of graphs, including the Bland–Altman and ROC plots. The ROC approach was used to calculate AUC values and evaluate agreement/disagreement between the isotopologs. AUC values are reported as mean with standard error. The Wilcoxon matched-pairs signed-rank test was used in two-tailed paired analyses. A *p*-value of <0.05 was considered significant. Chemical structures of the investigated derivatives of the isotopologs were drawn using ChemDraw 15.0 Professional (PerkinElmer, Germany).

3. Results

3.1. Bland–Altman and ROC Approaches to Study Isotope Effects in GC-MS in Biological Samples

The primary results of the present study are listed in Table 1. The secondary results obtained from the application of the Bland–Altman and ROC approaches are summarized in Table 2 and shown in Figure 1.

The retention times of all derivatives were measured with high precision (Table 1). The ^2H and ^{15}N isotopologs had smaller retention times than their ^1H and ^{14}N counterparts. Yet, the differences in the retention times were larger for the ^2H analogs. The IE values for the ^{15}N derivatives of nitrite and nitrate in plasma and urine were practically 1.0000, and the difference in the retention times was not higher than 0.18 s. The highest IE and δ values were observed for the PFBz derivative of d_6-DMA, i.e., 1.007 and 1.5 s, respectively. The concentrations of the stable-isotope labeled analogs, which were added to the plasma and urine samples, were all relevant for the respective biological samples. This is indicated by the measured PAR values for all analytes, which ranged between 0.2 and 2.4. Weak correlations after Spearman between the PAR and δ values were observed for some analytes, indicating a very weak dependence of δ upon the endogenous analyte concentrations in the plasma and urine samples analyzed.

Table 1. Summary of the results of the GC-MS (ISQ) analyses of the investigated analytes in human plasma and urine samples in the present work. The coefficients of variation are reported in parentheses. Me, methyl; PAR, peak area ratio; PFB, pentafluorobenzyl; PFBz, pentafluorobenzoyl; PFP, pentafluoropropionyl; P, plasma; U, urine. The dwell time was 50 ms or 100 ms. The Wilcoxon test p-values were <0.0001. See also: Scheme 1.

Analyte	Derivative	m/z	Retention Time (min)	IE	Δ (s)	PAR	Spearman PAR vs. δ
^{14}N-Nitrite-U	PFB	46	4.517 (0.05)	1.00000 (0.05)	0.07839 (115)	0.227 ± 0.232	none
^{15}N-Nitrite-U	PFB	47	4.516 (0.00)				
^{14}N-Nitrate-U	PFB	62	4.325 (0.02)	1.00000 (0.05)	0.10450 (165)	0.947 ± 0.575	$r = 0.265$ $p = 0.037$
^{15}N-Nitrate-U	PFB	63	4.323 (0.05)				
^{14}N-Nitrite-P	PFB	46	4.521 (0.02)	1.00000 (0.03)	0.0228 (342)	0.234 ± 0.175	$r = 0.217$ $p = 0.030$
^{15}N-Nitrite-P	PFB	47	4.520 (0.03)				
^{14}N-Nitrate-P	PFB	62	4.328 (0.06)	1.00100 (0.06)	0.1794 (82)	0.69 ± 0.42	none
^{15}N-Nitrate-P	PFB	63	4.325 (0.01)				
^{1}H-Creatinine-U	PFB	112	6.913 (0.02)	1.00100 (0.02)	0.6022 (13)	0.602 ± 0.081	$r = 0.316$ $p = 0.019$
^{2}H$_3$-Creatinine-U	PFB	115	6.903 (0.01)				
^{1}H-Arg-P	d$_0$Me-PFP	586	5.729 (0.05)	1.00200 (0.03)	0.699 (13)	2.14 ± 1.27	none
^{2}H$_3$-Arg-P	d$_3$Me-PFP	589	5.718 (0.05)				
^{1}H-ADMA-P	d$_0$Me-PFP	634 → 378	10.85 (0.02)	1.00100 (0.02)	0.9168 (12)	0.493 ± 0.96	none
^{2}H$_3$-ADMA-P	d$_3$Me-PFP	637 → 378	10.84 (0.02)				
^{1}H-ADMA-U	d$_0$Me-PFP	634 → 378	10.97 (0.09)	1.00100 (0.02)	0.8538 (13)	2.35 ± 1.26	none
^{2}H$_3$-ADMA-U	d$_3$Me-PFP	637 → 378	10.95 (0.09)				
^{1}H-DMA-U	PFBz	240	3.520 (0.17)	1.007 (0.14)	1.502 (20)	0.436 ± 0.348	$r = -0.183$ $p = 0.041$
^{2}H$_6$-DMA-U	PFBz	246	3.495 (0.17)				

Table 2. Summary of the results of the GC-MS (ISQ) analyses of the investigated analytes in human plasma and urine samples in the present work as obtained by the (A) Bland–Altman and (B) AUC-ROC approaches. P, plasma; U, urine; LOA, limit of agreement; SD, standard deviation; SE, standard error.

	(A) Bland–Altman				(B) ROC
Analyte	δ (%)	SD (%)	95% Lowest LOA	95% Highest LOA	AUC (Mean ± SE)
DMA-U ($n = 62$)	0.714	0.143	0.4333	0.9946	0.9951 ± 0.0024 $p < 0.0001$
Arginine-P ($n = 100$)	0.2035	0.02725	0.1501	0.2570	0.9932 ± 0.0054 $p < 0.0001$

Table 2. Cont.

	(A) Bland–Altman				(B) ROC
Analyte	δ (%)	SD (%)	95% Lowest LOA	95% Highest LOA	AUC (Mean ± SE)
Creatinine-U (n = 55)	0.1453	0.01949	0.1071	0.1835	1.000 ± 0.000 p < 0.0001
ADMA-P (n = 50)	0.1409	0.01698	0.1076	0.1742	1.000 ± 0.000 p < 0.0001
ADMA-U (n = 52)	0.1298	0.01630	0.0979	0.1618	0.8299 ± 0.0384 p < 0.0001
Nitrate-P (n = 100)	0.06909	0.05674	−0.0421	0.1803	0.7951 ± 0.033 p < 0.0001
Nitrate-U (n = 62)	0.04029	0.04625	−0.0504	0.1309	0.7177 ± 0.0469 p < 0.0001
Nitrite-U (n = 52)	0.02891	0.04768	−0.0646	0.1224	0.6371 ± 0.050 p = 0.008
Nitrite-P (n = 100)	0.00841	0.02879	−0.0480	0.0648	0.5414 ± 0.041 p = 0.311

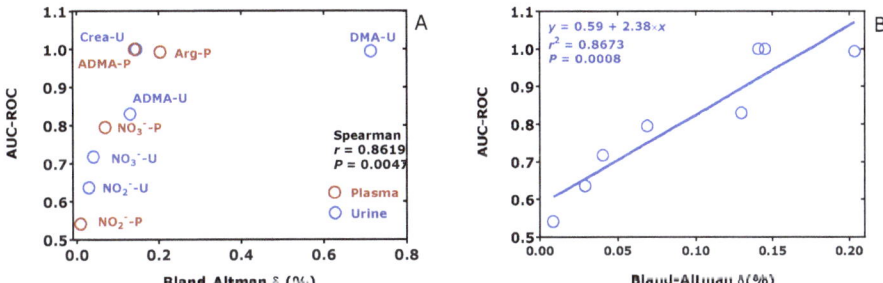

Figure 1. Comparison of the ROC and Bland–Altman approaches. (**A**) The mean AUC-ROC values are plotted versus the percentage differences δ (%) of the retention times of the isotopologs of the indicated analytes in plasma (P) and urine (U). There is a close correlation after Spearman among the two approaches. (**B**) The mean AUC-ROC values are plotted versus the percentage differences δ (%) in the retention times of the isotopologs of the investigated analytes in plasma and urine; the value of DMA was not considered.

The application of the Bland–Altman approach resulted in δ (%) values ranging between 0.714% for DMA in urine and 0.00841% for the PFB derivative of nitrite in plasma (Table 2). The application of the ROC approach resulted in AUC values ranging between 1.0000 for creatinine in urine and ADMA in plasma and 0.5414 for nitrite in plasma. The AUC values for ADMA and nitrate were higher in the plasma compared to urine samples, whereas the AUC value for nitrite was lower in plasma compared to urine. In urine, the δ (%) values (y) increased linearly with the average retention time (x) of the $^{14}N/^{15}N$ isotopologs ($y = -200 + 44 \times x$, $r^2 = 1.000$, $p < 0.0001$), indicating positive proportional error [5].

We tested for potential correlation between the AUC-ROC and Bland–Altman δ (%) values. We found a strong correlation after Spearman between these approaches: $r = 0.862$, $p = 0.005$ (Figure 1A). This Figure also illustrates the AUC-ROC and δ (%) differences for ADMA, nitrate and nitrite in the plasma and urine samples. Omitting the DMA values, a linear regression analysis between the AUC (y) and δ (%) (x) values resulted in a straight line with the regression equation $y = 0.59 + 2.38 \times x$, $r^2 = 0.8673$ (Figure 1B).

In a further, recently performed study [17], we analyzed paired serum and urine samples of 85 volunteers who formerly had COVID-19 or were living with long-COVID-19. Nitrate, nitrite and creatinine were simultaneously analyzed by GC-MS on the apparatus ISQ and a 15 m long GC OPTIMA-17 column, as described previously [18]. The concentrations of the internal standards were 10 mM for creatinine in urine and 100 µM in serum, 4 µM for nitrite in serum and 8 µM in urine as well as 40 µM for nitrate in serum and 800 µM in urine. Instead of toluene [13,14], ethyl acetate [21] was used for the extraction of the PFB derivatives, and 1 µL aliquots of ethyl acetate extracts were injected in the splitless mode. The results of these analyses are summarized in Table 3.

The retention times of the corresponding isotopologs differed statistically significantly from each other. The values of IE, δ and AUC differed for the isotopologs in urine and serum, as well as when compared to urine with serum. The highest IE, δ and AUC values were observed for creatinine and the lowest were observed for nitrite in serum (Table 3). Figure 2 shows the relationship between the AUC-ROC and Bland–Altman values with respect to the retention times of the isotopologs in the serum and urine samples.

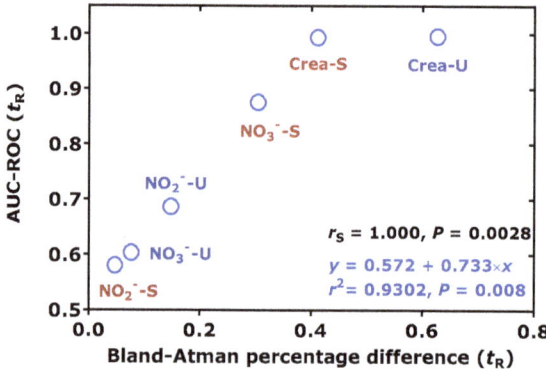

Figure 2. Comparison of the AUC-ROC values and the percentage difference values in the Bland–Altman approach with respect to the retention times in urine (U) and serum (S) samples of the isotopologs. There is a close correlation after Spearman between the two approaches and high linearity when the paired values for creatinine in urine are excluded.

3.2. Isotope Effects as a Measure of Matrix Effects in GC-MS: Proof-of-Concept Studies

Matrix effects are very common in LC-MS/MS, and methods have been proposed, with their measurements implemented in bioanalysis [22–26]. Matrix effects have been sporadically reported in GC-based methods, including GC-MS and GC-MS/MS [27–30]. Matrix-induced ion suppression effects occur both in electron ionization (EI) and NICI, yet the underlying mechanisms have not yet been explained thus far [26]. Stable-isotope-labeled analogs have been used in GC [31] and LC-MS/MS [32] to minimize matrix effects. To the best of our knowledge, isotopologs have not been used to quantify matrix effects in GC-MS or GC-MS/MS [26]. Given the observations of different IE and δ values for some analytes in serum and urine samples in the present study, we tested the utility of isotopologs to quantify matrix effects in GC-MS.

Table 3. Summary of the results of the GC-MS (IS²Q) analyses of the indicated analytes in paired samples in the present work obtained in urine (U) and serum (S) of 85 volunteers with long- or former COVID-19 infection. GC column, OPTIMA-17 (15 m × 0.25 mm I.D., 0.25 µm film thickness).

Analyte	Retention Time (min)	Wilcoxon Test (t_R)	IE	Wilcoxon Test (IE) (U vs. S)	δ(s)	Wilcoxon Test (δ) (U vs. S)	AUC (t_R)	AUC (IE) (U vs. S)	AUC (δ) (U vs. S)	Bland–Altman Percentage (t_R)	Bland–Altman Percentage (IE) (U vs. S)	Bland–Altman Percentage (δ) (U vs. S)
^{14}N-Nitrate-U	3.183 (0.18)	p < 0.0001	1.0010 (0.14)	Nitrate p < 0.0001	0.1482 (176)	Nitrate p < 0.0001	0.6037 ± 0.0432 p = 0.0196	Nitrate 0.8131 ± 0.0352 p < 0.0001	Nitrate 0.7894 ± 0.0352 p < 0.0001	0.078 ± 0.136	Nitrate −0.228 ± 0.257	Nitrate −128 ± 126
^{15}N-Nitrate-U	3.180 (0.18)											
^{14}N-Nitrite-U	3.398 (0.20)	p < 0.0001	1.0010 (0.17)	Nitrite p = 0.0036	0.3025 (117)	Nitrite p < 0.0001	0.6874 ± 0.0271 p < 0.0001	Nitrite 0.6184 ± 0.0437 p = 0.0077	Nitrite 0.6485 ± 0.0422 p = 0.0008	0.149 ± 0.174	Nitrite 0.099 ± 0.200	Nitrite 99 ± 159
^{15}N-Nitrite-U	3.393 (0.19)											
^{14}N-Nitrate-S	3.203 (0.16)	p < 0.0001	1.0030 (0.20)		0.5859 (67)		0.8756 ± 0.0432 p < 0.0001			0.305 ± 0.203		
^{15}N-Nitrate-S	3.193 (0.03)											
^{14}N-Nitrite-S	3.408 (0.13)	p = 0.0005	1.0000 (0.12)		0.0988 (245)		0.5814 ± 0.0434 p = 0.652			0.048 ± 0.118		
^{15}N-Nitrite-S	3.406 (0.15)											
^{1}H-Creatinine-U	7.133 (0.07)	p < 0.0001	1.0060 (0.11)	Creatinine p < 0.0001	2.675 (18)	Creatinine p < 0.0001	0.9959 ± 0.0047 p < 0.0001	Creatinine 0.8727 ± 0.0263 p < 0.0001	Creatinine 0.8694 ± 0.0268 p < 0.0001	0.627 ± 0.114	Creatinine 0.217 ± 0.195	Creatinine 47 ± 51
^{2}H$_{3}$-Creatinine-U	7.089 (0.10)											
^{1}H-Creatinine-S	7.130 (0.02)	p < 0.0001	1.0040 (0.17)		1.751 (41)		0.9943 ± 0.0066 p < 0.0001			0.412 ± 0.167		
^{2}H$_{3}$-Creatinine-S	7.101 (0.17)											

3.2.1. GC-MS Analysis of Dimethyl Amine

We analyzed, via GC-MS, d_0-DMA and d_6-DMA after extractive derivatization with pentafluorobenzoyl chloride/toluene (Scheme 1) [16]. Human urine samples (U, $n = 80$), a 67 mM potassium phosphate buffer of pH 7.0 (B, $n = 33$), a 20 mM Na_2CO_3 solution (C, $n = 33$) and deionized water (W, $n = 33$) were treated as follows:

(1) A total of 10 µL of a 1 mM d_6-DMA solution was introduced into autosampler glass vials;
(2) A total of 10 µL of U, B, C or W was added (B, C and W contained d_0-DMA at 0, 100 and 500 µM);
(3) A total of 90 µL of W was added;
(4) A total of 100 µL of 20 mM Na_2CO_3 was added.

d_6-DMA was used as the internal standard at a fixed final concentration of 1000 µM in all matrices. The d_0-DMA concentrations in B, C and W were 0, 100 and 500 µM. After derivatization, 1 µL aliquots of toluene extracts were injected splitless, and SIM of m/z 240 for d_0-DMA and m/z 246 for d_6-DMA was performed (ISQ apparatus, 15 m long OPTIMA-17 column). The results of this experiment are summarized in Table 4.

Table 4. Summary of the results of the GC-MS (ISQ) analyses of d_0-DMA in urine (U), buffer (B), carbonate (C) and deionized water (W) after extractive derivatization with pentafluorobenzoyl chloride and toluene. GC column: OPTIMA-17 (15 m × 0.25 mm I.D., 0.25 µm film thickness).

Analyte, Matrix	Retention Time (min)	W or M-W Test (t_R)	IE	M-W Test (IE)	δ(s)	M-W Test (δ)	Bland–Altman (δ)	AUC (δ)
d_0-DMA-U	7.166 (0.12)	$p < 0.0001$	1.004 (0.06)		1.765 (16)			
d_6-DMA-U	7.136 (0.16)							
d_0-DMA-B	7.154 (0.07)	$p < 0.0001$	1.004 (0.04)	U vs. B 0.1863	1.690 (10)	U vs. B $p < 0.0001$	U vs. B 5.896 (15) % (−24–36)	U vs. B 0.7287 ± 0.0594 $p < 0.0001$
d_6-DMA-B	7.125 (0.079							
d_0-DMA-C	7.155 (0.07)	$p < 0.0001$	1.004 (0.08)	U vs. C 0.3131	1.891 (18)	U vs. C $p = 0.2984$	U vs. C −4.076 (21) % (−45–36)	U vs. C 0.5577 ± 0.0590 $p = 0.3173$
d_6-DMA-C	7.123 (0.07)							
d_0-DMA-W	7.118 (0.06)	$p < 0.0001$	1.004 (0.04)	U vs. W 0.1413	1.715 (10)	U vs. W $p = 0.0017$	U vs. W 4.493 (14) % (−24–33)	U vs. W 0.6681 ± 0.0621 $p = 0.0036$
d_6-DMA-W	7.090 (0.02)							

The PA of m/z 246 for d_6-DMA-PFBz varied by 7.3%. In the urine samples ($n = 80$), the PAR of m/z 240 for d_0-DMA to m/z 246 for d_6-DMA ranged between 0.1 and 1.6 (mean, 0.608 ± 0.26).

The retention times of the isotopologs differed in all matrices but did not result in different IE values. This parameter was not further investigated. Statistically significant differences with respect to δ were found between urine (U) and buffer (B), as well as between urine (U) and water (W) by the Mann–Whitney test and the ROC approach. The highest δ value was observed for the carbonate solutions of DMA (C). The DMA solutions in B, C and W are more comparable among themselves than with the U samples, which were diluted 10-fold with water and carbonate. The experiment described above is a very simple simulation of potential matrix effects on isotope effects in GC-MS. A modification of this simulation, for instance, by using undiluted urine or urine diluted to varying degrees, would be more meaningful. Whether the Bland–Altman approach or the ROC approach is able to provide more definite results remains to be investigated. The Bland–Altman approach is expected to be more promising because of its higher versatility.

3.2.2. GC-MS Analysis of Metformin

Standard curves were prepared for d_0-metformin (d_0-METF) in human urine (U) and serum (S) samples in relevant metformin concentration ranges, i.e., 0 to 25 mM in urine and 0 to 25 µM in serum, using the internal standard at a fixed concentration of 1000 µM for d_6-metformin (d_6-METF) in urine and 20 µM in serum. SIM of m/z 383 for d_0-METF and m/z 383 for d_6-METF was performed as described previously [19]. The results of this experiment are summarized in Table 5. A typical GC-MS chromatogram from the analysis of metformin in a human serum sample is shown in Figure 3.

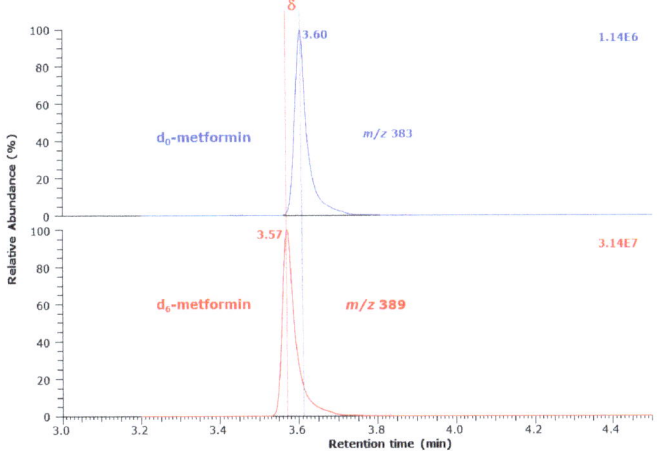

Figure 3. GC-MS chromatogram from the quantitative analysis of metformin in a human serum sample as pentafluoropropionyl (PFP) derivative on an ISQ instrument directly interfaced with a Trace 1310 series gas chromatograph. Selected ion monitoring of m/z 383 for d_0-metformin and m/z 389 for d_6-metformin (20 µM) was performed (dwell time, 100 ms for each ion). Methane (2.4 mL/min) was used as the reagent gas for negative-ion chemical ionization. A fused-silica capillary column Optima 17 (15 m × 0.25 mm I.D., 0.25 µm film thickness) was used. The oven temperature was kept at 40 °C for 0.5 min, then increased to 210 °C at a rate of 15 °C/min and to 320 °C at a rate of 35 °C/min, respectively, and held at 320 °C for 1 min. Helium was the carrier gas at a constant flow rate of 1 mL/min. The GC-MS method for metformin has been reported in detail elsewhere [19]. The symbol δ indicates the difference between the retention times of d_0-metformin (3.60 min) and d_6-metformin (3.57 min).

The PA of m/z 389 for d_6-METF-PFP varied by 31% in U and by 30% in S. The retention times of the isotopologs differed in both matrices. The IE and δ values differed statistically significantly between U and S.

Table 5. Summary of the results of the GC-MS (ISQ) analyses of d_0-metformin (d_0-METF) and d_6-metformin (d_6-METF) in human urine (U) and serum (S) samples after derivatization with pentafluoropropionic anhydride in ethyl acetate. GC column: OPTIMA-17 (15 m × 0.25 mm I.D., 0.25 μm film thickness).

Analyte-Matrix	t_R (min)	W Test (t_R)	IE	M-W Test (IE)	δ(s)	M-W Test (δ)	t_R Bland–Altman	t_R ROC	IE U vs. S	δ U vs. S
d_0-METF-U	3.628 (0.22)	$p = 0.0002$	1.014 (0.42)	U vs. S $p < 0.0001$	3.046 (30)	U vs. S $p = 0.0003$	U 1.409 (0.4157) % (0.59–2.22)	U 1.000 ± 0.000% (1 to 1) $p < 0.0001$	Bland–Altman 0.4889 (0.4291) % (−0.35–1.33)	Bland–Altman 39 (31) % (−23–101)
d_6-METF-U	3.589 (0.23)									
d_0-METF-S	3.630 (0.00)	$p = 0.0005$	1.009 (0.11)		1.90 (12)		S 0.8762 ± 0.1083% (0.66–1.09)	S 1.000 ± 0.000% (1 to 1) $p < 0.0001$	ROC 0.9679 ± 0.0297 $p < 0.0001$	ROC 0.8846 ± 0.0709 $p = 0.0011$
d_6-METF-S	3.598 (0.11)									

4. Discussion

The Bland–Altman approach has been proposed for testing agreement between two measurements [4]. The graphical Bland–Altman approach is frequently used in analytical chemistry to compare two analytical methods for the quantitative determination of analytes in biological samples [5]. The ROC approach is also a graphical plot that is often used to measure differences between two methods of measurement of analytes, although the main aim of this approach is testing disagreement between two approaches, especially in clinical diagnosis [33]. Given the potentially very small differences in the retention times of isotopologs in chromatography [1–3], we investigated, in the present study, the utility of the Bland–Altman and ROC approaches in the GC-MS analyses of selected endogenous analytes in human plasma, serum and urine samples. The focus of the study was on the main members of the L-arginine/nitric oxide pathway [11] and creatinine, which is an important clinical biochemical parameter.

Nitrate and nitrite and the externally added ^{15}N isotopologs were analyzed by GC-MS after derivatization with PFB bromide to their PFB-ONO$_2$ and PFB-NO$_2$ derivatives, respectively (Scheme 1). PFB-ONO$_2$ and PFB-NO$_2$ are separated completely by GC as well as by MS. Being a nitric acid ester, PFB-ONO$_2$ eluted in front of PFB-NO$_2$, which is a nitro derivative [13]. Virtually, both the Bland–Altman and the ROC approach are not able to discriminate ^{14}N/^{15}N isotopologs of PFB-ONO$_2$ and PFB-NO$_2$, respectively. Yet, small differences were detected in plasma, serum and urine samples, independent of the extraction solvent that contained the derivatives, i.e., toluene and ethyl acetate. In contrast, both the Bland–Altman approach and the ROC approach clearly discriminated the respective H/D isotopologs of DMA (PFBz-DMA), ADMA (Me-PFP), Arg (Me-PFP) and creatinine (PFB-creatinine) (Scheme 1), yet with some differences for ADMA between plasma and urine. The strong correlation found between the Bland–Altman and the ROC approaches suggests that both methods are virtually equally suitable to investigate isotope effects in GC-MS.

The two methyl groups of DMA in its PFBz derivative, the methyl group of creatinine in its PFB derivative and the methyl ester groups of Arg and ADMA in the methyl ester PFP derivatives are most likely responsible for the considerably stronger H/D isotope effects compared to the ^{14}N/^{15}N isotope effects observed in PFB-ONO$_2$ and PFB-NO$_2$ derivatives. The greater differences in physical properties between H and D (a 100% increase in mass) compared to the differences between ^{14}N and ^{15}N (a 7% increase in mass) are a likely explanation for the stronger H/D isotope effects.

The charge radius of D is 2.5 times higher compared to the charge radius of H (https://physics.nist.gov/cuu/Constants/index.html, assessed on 10 December 2023). The gravest factor that causes the stronger H/D isotope effects is likely to be a stronger interaction of the methyl groups with the lipophilic stationary phase of the GC column (50% methylpolysiloxane, 50% phenylpolysiloxane) in the present study. H/D effects were observed for non-derivatized methylxanthine isotopologs in GC-MS on a 14% cyanopropy-lphenyl methylpolysiloxane fused silica column [34]. In that study, H/D isotopic effects were found to depend not only on the number of D atoms but also on the position of the CD$_3$ groups in the molecules (IE range, 1.00147 to 1.00668) [34]. The rate of a reaction involving a C–H bond is typically 6–10 times faster than the corresponding C–D bond [35].

In the case of PFB-ONO$_2$ and PFB-NO$_2$, the central N atoms seem to be strongly sterically hindered from interacting with the stationary phase. The differences seen between PFB-ONO$_2$ and PFB-NO$_2$ suggest that the N atom in PFB-ONO$_2$ is somewhat more accessible to interaction with the stationary phase than the N atom in PFB-NO$_2$, which is closer to the PFB group (Scheme 1). This observation demands deeper investigations with nitro and nitric acid derivatives of alkyl/aryl residues.

In the cases of ADMA, nitrate, nitrite and creatinine, which were analyzed both in plasma/serum and in urine, there were some differences in the Bland–Altman δ (%) and ROC-AUC values in plasma or serum compared to urine. ADMA: 0.1409 vs. 0.1298 (1.1-fold); nitrate: 0.06909 vs. 0.04029 (1.7-fold); nitrite: 0.00841 vs. 0.02891 (0.3-fold). These observations may be interpreted as a type of "matrix effect". Matrix effects occur not only in LC-MS/MS but also in GC-MS and GC-MS/MS [22–32]. Several methods have been proposed and used in LC-MS/MS, such as the use of standard line slopes as a measure of relative matrix effects [22]. The results of the present study, including those of the pilot experiment, indicate that the differences in the retention times of isotopologs δ are better suited to quantify matrix effects than the ratio of the retention times IE of d_0-DMA-PFBz and d_6-DMA-PFBz in human urine. IE is a little variable measure but is less sensitive than the more variable measure δ. The Bland–Altman approach seems to be better suited for quantitating isotope effects than the ROC approach.

5. Conclusions

In GC-MS, the Bland–Altman and ROC approaches seem to be suitable for studying H/D and $^{14}N/^{15}N$ isotope effects in the PFB, PFBz and PFP derivatives of endogenous analytes of the L-arginine/nitric oxide pathway and the universal biomarker creatinine. Isotope effects in GC-MS are likely to be caused by differences in the interaction strengths of H/D and $^{14}N/^{15}N$ isotopes in the derivatives with the hydrophobic stationary phase of the GC column. D atoms in the derivatives seem to attenuate the interaction of the skeleton of the molecules with the lipophilic GC stationary phase. Differences in the retention times of isotopologs, i.e., the parameter δ, appear to be a better-suited experimental measure for quantitating matrix effects in GC-MS.

Funding: This research received no external funding.

Institutional Review Board Statement: In this study, material from previously approved human studies was used (see Section 2.1).

Informed Consent Statement: Not applicable (see Section 2.1).

Data Availability Statement: The data presented in this study are available in article.

Acknowledgments: The author is grateful to previous and current members of his group for their contributions.

Conflicts of Interest: The author declares no conflict of interest.

References

1. Biemann, K. *Mass Spectrometry—Organic Chemical Applications*; McGraw Hill: New York, NY, USA, 1962; p. 204.
2. Lehmann, W.D. A timeline of stable isotopes and mass spectrometry in the life sciences. *Mass Spectrom. Rev.* **2017**, *36*, 58–85. [CrossRef]
3. Turowski, M.; Yamakawa, N.; Meller, J.; Kimata, K.; Ikegami, T.; Hosoya, K.; Tanaka, N.; Thornton, E.R. Deuterium isotope effects on hydrophobic interactions: The importance of dispersion interactions in the hydrophobic phase. *J. Am. Chem. Soc.* **2003**, *125*, 13836–13849. [CrossRef]
4. Bland, J.M.; Altman, D.G. Statistical methods for assessing agreement between two methods of clinical measurement. *Lancet* **1986**, *1*, 307–310. [CrossRef]
5. Tsikas, D. Mass Spectrometry-Based Evaluation of the Bland-Altman Approach: Review, Discussion, and Proposal. *Molecules* **2023**, *28*, 4905. [CrossRef]
6. Gucciardi, A.; Cogo, P.E.; Traldi, U.; Eaton, S.; Darch, T.; Simonato, M.; Ori, C.; Carnielli, V.P. Simplified method for microlitre deuterium measurements in water and urine by gas chromatography-high-temperature conversion-isotope ratio mass spectrometry. *Rapid Commun. Mass Spectrom.* **2008**, *22*, 2097–2103. [CrossRef]
7. Wong, W.W.; Roberts, S.B.; Racette, S.B.; Das, S.K.; Redman, L.M.; Rochon, J.; Bhapkar, M.V.; Clarke, L.L.; Kraus, W.E. The doubly labeled water method produces highly reproducible longitudinal results in nutrition studies. *J. Nutr.* **2014**, *144*, 777–783. [CrossRef]
8. Wong, W.W.; Clarke, L.L. Accuracy of δ(18)O isotope ratio measurements on the same sample by continuous-flow isotope-ratio mass spectrometry. *Rapid Commun. Mass Spectrom.* **2015**, *29*, 2252–2256. [CrossRef]
9. Zweig, M.H.; Campbell, G. Receiver-operating characteristic (ROC) plots: A fundamental evaluation tool in clinical medicine. *Clin. Chem.* **1993**, *39*, 561–577. [CrossRef]
10. Tom, F. An Introduction to ROC Analysis. *Pattern Recognit. Lett.* **2006**, *27*, 861–874. [CrossRef]
11. Tsikas, D. A critical review and discussion of analytical methods in the L-arginine/nitric oxide area of basic and clinical research. *Anal. Biochem.* **2008**, *379*, 139–163. [CrossRef]

12. Tsikas, D. Urinary Dimethylamine (DMA) and Its Precursor Asymmetric Dimethylarginine (ADMA) in Clinical Medicine, in the Context of Nitric Oxide (NO) and Beyond. *J. Clin. Med.* **2020**, *9*, 1843. [CrossRef]
13. Tsikas, D. Simultaneous derivatization and quantification of the nitric oxide metabolites nitrite and nitrate in biological fluids by gas chromatography/mass spectrometry. *Anal. Chem.* **2000**, *72*, 4064–4072. [CrossRef]
14. Tsikas, D.; Wolf, A.; Mitschke, A.; Gutzki, F.M.; Will, W.; Bader, M. GC-MS determination of creatinine in human biological fluids as pentafluorobenzyl derivative in clinical studies and biomonitoring: Inter-laboratory comparison in urine with Jaffé, HPLC and enzymatic assays. *J. Chromatogr. B* **2010**, *878*, 2582–2592. [CrossRef]
15. Tsikas, D.; Schubert, B.; Gutzki, F.M.; Sandmann, J.; Frölich, J.C. Quantitative determination of circulating and urinary asymmetric dimethylarginine (ADMA) in humans by gas chromatography-tandem mass spectrometry as methyl ester tri(N-pentafluoropropionyl) derivative. *J. Chromatogr. B* **2003**, *798*, 87–99. [CrossRef]
16. Tsikas, D.; Thum, T.; Becker, T.; Pham, V.V.; Chobanyan, K.; Mitschke, A.; Beckmann, B.; Gutzki, F.M.; Bauersachs, J.; Stichtenoth, D.O. Accurate quantification of dimethylamine (DMA) in human urine by gas chromatography-mass spectrometry as pentafluorobenzamide derivative: Evaluation of the relationship between DMA and its precursor asymmetric dimethylarginine (ADMA) in health and disease. *J. Chromatogr. B* **2007**, *851*, 229–239. [CrossRef]
17. Mikuteit, M.; Baskal, B.; Klawitter, S.; Dopfer-Jablonka, A.; Behrens, G.M.N.; Müller, F.; Schröder, D.; Klawonn, F.; Steffens, S.; Tsikas, D. Amino acids, post-translational modifications, nitric oxide, and oxidative stress in serum and urine of long COVID and ex COVID human subjects. *Amino Acids* **2023**, *55*, 1173–1188. [CrossRef]
18. Bollenbach, A.; Schutte, A.E.; Kruger, R.; Tsikas, D. An Ethnic Comparison of Arginine Dimethylation and Cardiometabolic Factors in Healthy Black and White Youth: The ASOS and African-PREDICT Studies. *J. Clin. Med.* **2020**, *9*, 844. [CrossRef]
19. Baskal, S.; Bollenbach, A.; Henzi, B.; Hafner, P.; Fischer, D.; Tsikas, D. Stable-Isotope Dilution GC-MS Measurement of Metformin in Human Serum and Urine after Derivatization with Pentafluoropropionic Anhydride and Its Application in Becker Muscular Dystrophy Patients Administered with Metformin, l-Citrulline, or Their Combination. *Molecules* **2022**, *27*, 3850. [CrossRef]
20. Carlson, R.V.; Boyd, K.M.; Webb, D.J. The revision of the Declaration of Helsinki: Past, present and future. *Br. J. Clin. Pharmacol.* **2004**, *57*, 695–713. [CrossRef]
21. Hanff, E.; Eisenga, M.F.; Beckmann, B.; Bakker, S.J.L.; Tsikas, D. Simultaneous Pentafluorobenzyl Derivatization and GC-ECNICI-MS Measurement of Nitrite and Malondialdehyde in Human Urine: Close Positive Correlation between These Disparate Oxidative Stress Biomarkers. *J. Chromatogr. B* **2017**, *1043*, 167–175. [CrossRef]
22. Matuszewski, B.K.; Constanzer, M.L.; Chavez-Eng, C.M. Matrix effect in quantitative LC/MS/MS analyses of biological fluids: A method for determination of finasteride in human plasma at picogram per milliliter concentrations. *Anal. Chem.* **1998**, *70*, 882–889. [CrossRef]
23. Matuszewski, B.K.; Constanzer, M.L.; Chavez-Eng, C.M. Strategies for the assessment of matrix effect in quantitative bioanalytical methods based on HPLC-MS/MS. *Anal. Chem.* **2003**, *75*, 3019–3030. [CrossRef]
24. Matuszewski, B.K. Standard line slopes as a measure of a relative matrix effect in quantitative HPLC-MS bioanalysis. *J. Chromatogr. B* **2006**, *830*, 293–300. [CrossRef]
25. Cortese, M.; Gigliobianco, M.R.; Magnoni, F.; Censi, R.; Di Martino, P.D. Compensate for or Minimize Matrix Effects? Strategies for Overcoming Matrix Effects in Liquid Chromatography-Mass Spectrometry Technique: A Tutorial Review. *Molecules* **2020**, *25*, 3047. [CrossRef]
26. Panuwet, P.; Hunter, R.E., Jr.; D'Souza, P.E.; Chen, X.; Radford, S.A.; Cohen, J.R.; Marder, M.E.; Kartavenka, K.; Ryan, P.B.; Barr, D.B. Biological Matrix Effects in Quantitative Tandem Mass Spectrometry-Based Analytical Methods: Advancing Biomonitoring. *Crit. Rev. Anal. Chem.* **2016**, *46*, 93–105. [CrossRef]
27. Erney, D.R.; Gillespie, A.M.; Gilvydis, D.M.; Poole, C.F. Explanation of the Matrix-Induced Chromatographic Response Enhancement of Organophosphorus Pesticides during Open-Tubular Column Gas-Chromatography with Splitless or Hot on-Column Injection and Flame Photometric Detection. *J. Chromatogr. A* **1993**, *638*, 57–63. [CrossRef]
28. Schenck, F.J.; Lehotay, S.J. Does further clean-up reduce the matrix enhancement effect in gas chromatographic analysis of pesticide residues in food? *J. Chromatogr. A* **2000**, *868*, 51–61. [CrossRef]
29. Hajslova, J.; Zrostlikova, J. Matrix effects in (ultra)trace analysis of pesticide residues in food and biotic matrices. *J. Chromatogr. A* **2003**, *1000*, 181–197. [CrossRef]
30. Yu, S.; Xu, X.M. Study of matrix-induced effects in multi-residue determination of pesticides by online gel permeation chromatography-gas chromatography/mass spectrometry. *Rapid Commun. Mass Spectrom.* **2012**, *26*, 963–977. [CrossRef]
31. Colby, B.N.; McCaman, M.W. A comparison of calculation procedures for isotope dilution determinations using gas chromatography mass spectrometry. *Biomed. Mass Spectrom.* **1979**, *6*, 225–230. [CrossRef]
32. Berg, T.; Strand, D.H. C-13 labelled internal standards-A solution to minimize ion suppression effects in liquid chromatography-tandem mass spectrometry analyses of drugs in biological samples? *J. Chromatogr. A* **2011**, *1218*, 9366–9374. [CrossRef] [PubMed]
33. Datta, K.; LaRue, R.; Permpalung, N.; Das, S.; Zhang, S.; Steinke, M.; Bushkin, Y.; Nosanchuk, J.D.; Marr, K.A. Development of an Interferon-Gamma Release Assay (IGRA) to Aid Diagnosis of Histoplasmosis. *J. Clin. Microbiol.* **2022**, *60*, e0112822. [CrossRef] [PubMed]

34. Benchekroun, Y.; Dautraix, S.; Désage, M.; Brazier, J.L. Isotopic effects on retention times of caffeine and its metabolites 1,3,7-trimethyluric acid, theophylline, theobromine and paraxanthine. *J. Chromatogr. B* **1997**, *688*, 245–254. [CrossRef] [PubMed]
35. Laidler, K.J. *Chemical Kinetics*, 3rd ed.; Harper & Row: New York, NY, USA, 1987; ISBN 978-0-06-043862-3.

Disclaimer/Publisher's Note: The statements, opinions and data contained in all publications are solely those of the individual author(s) and contributor(s) and not of MDPI and/or the editor(s). MDPI and/or the editor(s) disclaim responsibility for any injury to people or property resulting from any ideas, methods, instructions or products referred to in the content.

Review

Mass Spectrometry-Based Evaluation of the Bland–Altman Approach: Review, Discussion, and Proposal

Dimitrios Tsikas

Institute of Toxicology, Core Unit Proteomics, Hannover Medical School, 30623 Hannover, Germany; tsikas.dimitros@mh-hannover.de

Citation: Tsikas, D. Mass Spectrometry-Based Evaluation of the Bland–Altman Approach: Review, Discussion, and Proposal. *Molecules* 2023, 28, 4905. https://doi.org/10.3390/molecules28134905

Academic Editors: Paraskevas D. Tzanavaras and Victoria Samanidou

Received: 9 May 2023
Revised: 12 June 2023
Accepted: 16 June 2023
Published: 21 June 2023

Copyright: © 2023 by the author. Licensee MDPI, Basel, Switzerland. This article is an open access article distributed under the terms and conditions of the Creative Commons Attribution (CC BY) license (https:// creativecommons.org/licenses/by/ 4.0/).

Abstract: Reliable quantification in biological systems of endogenous low- and high-molecular substances, drugs and their metabolites, is of particular importance in diagnosis and therapy, and in basic and clinical research. The analytical characteristics of analytical approaches have many differences, including in core features such as accuracy, precision, specificity, and limits of detection (LOD) and quantitation (LOQ). Several different mathematic approaches were developed and used for the comparison of two analytical methods applied to the same chemical compound in the same biological sample. Generally, comparisons of results obtained by two analytical methods yields different quantitative results. Yet, which mathematical approach gives the most reliable results? Which mathematical approach is best suited to demonstrate agreement between the methods, or the superiority of an analytical method A over analytical method B? The simplest and most frequently used method of comparison is the linear regression analysis of data observed by method A (y) and the data observed by method B (x): $y = \alpha + \beta x$. In 1986, Bland and Altman indicated that linear regression analysis, notably the use of the correlation coefficient, is inappropriate for method-comparison. Instead, Bland and Altman have suggested an alternative approach, which is generally known as the Bland–Altman approach. Originally, this method of comparison was applied in medicine, for instance, to measure blood pressure by two devices. The Bland–Altman approach was rapidly adapted in analytical chemistry and in clinical chemistry. To date, the approach suggested by Bland–Altman approach is one of the most widely used mathematical approaches for method-comparison. With about 37,000 citations, the original paper published in the journal *The Lancet* in 1986 is among the most frequently cited scientific papers in this area to date. Nevertheless, the Bland–Altman approach has not been really set on a quantitative basis. No criteria have been proposed thus far, in which the Bland–Altman approach can form the basis on which analytical agreement or the better analytical method can be demonstrated. In this article, the Bland–Altman approach is re-valuated from a quantitative bioanalytical perspective, and an attempt is made to propose acceptance criteria. For this purpose, different analytical methods were compared with *Gold Standard* analytical methods based on mass spectrometry (MS) and tandem mass spectrometry (MS/MS), i.e., GC-MS, GC-MS/MS, LC-MS and LC-MS/MS. Other chromatographic and non-chromatographic methods were also considered. The results for several different endogenous substances, including nitrate, anandamide, homoarginine, creatinine and malondialdehyde in human plasma, serum and urine are discussed. In addition to the Bland–Altman approach, linear regression analysis and the Oldham–Eksborg method-comparison approaches were used and compared. Special emphasis was given to the relation of difference and mean in the Bland–Altman approach. Currently available guidelines for method validation were also considered. Acceptance criteria for method agreement were proposed, including the slope and correlation coefficient in linear regression, and the coefficient of variation for the percentage difference in the Bland–Altman and Oldham–Eksborg approaches.

Keywords: agreement; biomarkers; Bland and Altman approach; comparison; linear regression analysis; mass spectrometry; Oldham; Eksborg; tandem mass spectrometry; validation

1. Introduction

Most likely, nobody knows how many low- and high-molecular-mass chemical compounds are present in biological samples such as blood and urine. Yet, their number is assumed to be very high and to increase with time due to the discovery of natural and the introduction into the environment of new synthetic compounds including drugs. The core mission of *Analytical Chemistry* is both to identify the structure these compounds and to determine their concentration as accurately as possible. Over the years, numerous analytical methods were reported for the quantitative determination of virtually all classes of chemical compounds. Scientific competition, curiosity and striving, often paired with the discovery of novel technologies and improvements in available methodologies, have resulted in, and consistently result in, the development of various analytical methods in part for the same analyte, yet with different analytical performances. The performance of analytical methods can be characterized with a certain degree of objectivity, especially when defined criteria are applied. Generally, an improvement in a current analytical method for a certain analyte is an acceptable justification for the publication of the improved analytical method in a scientific journal, despite its lacking true analytical novelty.

1.1. Method-Comparison Approaches

Method-comparison approaches were proposed, interpreted, discussed, criticized, and improved by several groups [1–36] (in part cited in chronological order). Linear regression analysis (see Formula (1)) of results obtained by two different methods for the same measure, e.g., for an analyte in a biological system, is reportedly the oldest approach of method-comparison. In 1986, Bland and Altman published in *The Lancet* their legendary paper entitled *Statistical methods for assessing agreement between two methods of clinical measurement* [10]. This paper is one of the most frequently cited articles in Life Sciences (Figure S1). Bland and Altman [10] have noted that linear regression analysis, notably the use of the correlation coefficient r, is inappropriate for method-comparison, and they have suggested an alternative approach. Despite the availability of approaches for stronger analytical power, including the Bland–Altman method, linear regression analysis is still, without doubt, the most frequently and routinely used approach in the field of analytical chemistry and in other areas, until this day. Interestingly, the Bland–Altman method seems to be more widespread in the field of clinical chemistry. The consequences of the use of inappropriate or unsatisfactory approaches for method-comparison, such as the sole use of any value of the correlation coefficient r, as long it is associated with a statistically significant p value, i.e., $p \leq 0.05$, may be grave, as is demonstrated in the present study.

$$\tau_{1j} = \alpha + \beta \times \tau_{2j} \tag{1}$$

whereas τ_{1j} and τ_{2j} are the values measured by method 1 and method 2 ($j = 1, 2, \ldots n-1, n$; n = total number of the analyzed samples), respectively; α and β are values of the y-axis intercept and the slope of the straight line, respectively.

The Bland–Altman (BA) method [10] is a rather graphical approach which is still widely used but is less frequently and not routinely applied in analytical chemistry. The Bland–Altman approach examines the relationship between the difference (δ_{BA} or simply δ) of the values obtained by two methods (see Formula (2)) and the mean (μ_{BA} or simply μ) of the methods (see Formula (3)). Usually, in this approach, δ_{BA} is plotted versus the μ_{BA} of the methods.

$$\delta_{BA} = \tau_{1j} - \tau_{2j} \tag{2}$$

$$\mu_{BA} = 1/2 \times (\tau_{1j} + \tau_{2j}) \tag{3}$$

Even if the Bland–Altman approach is steadily used in analytical chemistry, this method-comparison is applied incorrectly, most likely due to the lack of acceptance criteria. Thus, most of the measurements may be within a 95% confidence interval, e.g., the $\pm 1.96 \times$ standard deviation, despite lacking analytically relevant comparability. This is because the 95% confidence interval becomes wider the larger the difference between the methods is (see below). It should be emphasized that Bland and Altman suggested their approach for quite comparable methods [10]. In this respect, both linear regression analysis and the Bland–Altman approach are used arbitrarily and do not provide reliable information about a potentially existing comparability and the extent of agreement.

Oldham [2], and later Eksborg [8], have suggested independently of each other an alternative approach which is based on using the ratio (Λ_{OE} or simply Λ; see Formulas (4a) and (4b)) of the measured values by two methods versus the mean of the values or versus the values τ_2 of method 2 (chosen as the reference method; Formula (4a)) or versus their average (Formula (4b)). In the present work, this method is referred to as OE. Interestingly, and in the opinion of this author, surprisingly, this approach did not find appreciable applications in method-comparison studies until the present day.

$$\Lambda_{OE} = \tau_{1j} : \tau_{2j} \tag{4a}$$

$$\Lambda_{OE} = \tau_{1j} : 0.5 \times (\tau_{1j} + \tau_{2j}) \tag{4b}$$

1.2. Basic Principles of Mass Spectrometry and Tandem Mass Spectrometry

Analytical methods involving the use of mass spectrometers, such as gas chromatography-mass spectrometry (GC-MS) and liquid chromatography-mass spectrometry (LC-MS) apparatus, are based on the separation of inorganic and organic ions produced in the ion-source of the instruments due to their mass-to-charge (m/z) ratio (Scheme S1). This ability provides mass spectrometry (MS)-based approaches with inherent specificity and distinguishes them from other analytical techniques, which are based on the utilization of far less characteristic physicochemical properties such as light absorption, fluorescence, or conductivity. The separation of substances by their m/z values enables the use of stable-isotope-labelled analogues as internal standards (IS) in MS-based methods. This is a unique feature of MS technology and lends MS-based methods high accuracy in quantitative analysis. A quantum jump in specificity and accuracy is represented by tandem mass spectrometry (MS/MS), for instance, as realized in GC-MS/MS and LC-MS/MS instruments. Not without reason, MS/MS-based methods are regarded as the *Reference Methods*, the *Gold Standard*, in the area of analytical chemistry, in basic and clinical research, including clinical chemistry (see for instance Refs. [37–40]).

In biological samples, such as plasma and urine samples, there are myriads of substances that belong to distinctly different classes. In MS-based methods, sample treatment procedures, such as protein precipitation, proper extraction and/or derivatization prior to analysis, generally lead to a considerable reduction in the number of the analytes finally injected into the MS instrument. The number of analytes that may interfere with the analysis of a certain substance may be further reduced by gas chromatographic or liquid chromatographic separation prior to MS separation. Despite a strong reduction in the number of potentially interfering analytes by such steps, the ionization process of co-eluting substances may generate isobaric ions, i.e., structurally different ions which have, however, the same m/z value. This particular situation is illustrated in Scheme S1 for mass spectrometers based on quadrupole (Q) technology.

Commonly, quantification by GC-MS and LC-MS instruments (and by GC-MS/MS and LC-MS/MS instruments operated in the SSQ configuration) is performed in the selected-ion monitoring (SIM) mode, as shown in Scheme S2A. In general, two ions produced in the ion source are selected: one ion for the target analyte A_T and one ion for the corresponding ion of the stable-isotope-labelled analogue A_{IS}, which serves as the IS for the analyte A_T. Quantification by GC-MS/MS and LC-MS/MS instruments is usually carried out in the selected-reaction monitoring (SRM) mode, as illustrated in Scheme S2B. For example, from the ions produced in the ion source of a GC-MS/MS instrument, the first quadrupole Q1 alternately separates the ion with m/z A_T for the target analyte and the corresponding ion m/z A_{IS} for the externally added IS. These precursor ions are fragmented in the collision chamber (second quadrupole) Q2, and the third quadrupole Q3 alternately selects, in general, each one specific product ion (p); for instance, it "filters" m/z P_T for the target analyte and m/z P_{IS} for the IS (Scheme S2B). In the SRM mode, Q3 can also pass product ions of the same m/z value (i.e., m/z P_T = m/z P_{IS}) which are, however, produced from different precursor ions and can therefore be completely discriminated.

A more detailed description of the instrumentation and principles of operation techniques, including ionization techniques, with SSQ and TSQ mass spectrometers and other types of mass spectrometers can be referred to the literature (e.g., Refs. [41–45]). A history of European mass spectrometry is found in Ref. [46]. In the context of quantitative analytical chemistry, it should be emphasized that mass spectrometry is not, per se, a magic bullet, and it does not always guarantee valid data [47]. Yet, it is currently the best available technology in analytical chemistry. It must be used with validated methods, and all findings need to be critically evaluated [48].

1.3. Problem and Aim of the Study

Because of the lack of guide numbers for the correlation coefficient r, the y-axis intercept α and the slope β of the regression equation, the results of linear regression analysis for method-comparison are used rather arbitrarily [10,49]. In particular, the lack of a definition of the acceptance criteria for the correlation coefficient r seduces us into misusing regression analysis, for instance, into suggesting an agreement in doubtful cases or even in missing analytical agreements. Thus, even if the value for the correlation coefficient r is, for instance, only 0.8, a p value below 0.05 is commonly considered satisfactory to claim correlation between the methods tested, irrespective of the y-axis intercept and slope values of regression equations.

In principle, these considerations equally apply to both the Bland–Altman method and the Oldham–Eksborg approach. Actually, these two methods lack a definition of the acceptance criteria for comparability and the validity of the analytical methods being compared. The Bland–Altman approach is useful for comparisons of methods with comparable performances, as originally stated by Bland and Altman in their original work [10]. In cases of considerable disagreement between the methods being compared, the Bland–Altman approach would penalize the method with the better analytical performance, e.g., method 2, in favour of the method with the putatively lower-quality analytical method, e.g., method 1. For example, this could be the case when comparing GC-MS/MS or LC-MS/MS methods with GC-FID or HPLC-UV methods. Application of the Bland–Altman approach to two methods being less comparable would result into a too-large confidence interval. Thus far, no additional established quantitative parameters of this approach have been proposed to value and report the extent of the agreements between methods. Regrettably, and in analogy to regression analysis, the Bland–Altman approach is interpreted incorrectly by many investigators, presumably because of the lack of acceptance criteria. Commonly, the application of this approach to method-comparison is solely restricted to showing the graph and the confidence interval. Eventually, the Oldham–Eksborg approach finds generally very few applications in method-comparisons despite the considerable potential of this method.

As will be shown in this work, the approaches of linear regression analysis, Bland–Altman, and Oldham–Eksborg are linked together. Therefore, one possibility to overcome the flaws of the individual approaches could be the deviation and use of a proper combination of these methods. However, even if this would be profitable, it may not allow us to solve the main, common, and principal problem of method-comparison: the renunciation of the superiority of one method over the other method, or the arbitrariness of defining one of the methods being compared as the *absolute* reference method, the *Gold Standard*. We could extricate ourselves from this dilemma if we accepted that thoroughly validated and proven analytical methods based on the tandem mass spectrometry methodology, such as GC-MS/MS and LC-MS/MS methods, are best-qualified to represent the reference methods [48]. The superiority of tandem mass spectrometry technology over other putatively less reliable analytical techniques is reasonably indisputable in the literature. The examples presented below are supportive of the analytical superiority of the MS/MS methodology over other analytical methodologies.

The aim of the present study was to investigate whether or not defining validated and proven analytical MS/MS-based methods as reference methods may help solve problems associated with method-comparison and may even help to define acceptance criteria for linear regression analysis, the Bland–Altman, and the Oldham–Eksborg approaches. Most currently available guidelines proposed by international associations and analytically oriented journals address exogenous drugs as analytes [50–56] rather than endogenous substances which have special requirements beyond method validation [57]. The present work focuses on the quantitative analysis of endogenous substances in biological samples, which represents a formidable analytical challenge.

2. Methods

2.1. Re-Evaluation of Published Analytical Data

Proceeding

Selected studies published by the author's group and by other investigator groups were examined by three method-comparison approaches: (1) linear regression analysis; (2) the Bland–Altman (BA) method; and (3) the Oldham–Eksborg (OE) approach. The selected studies reported results which allow for a satisfactory re-evaluation. The data reported in the Figures and Tables of this article were reconstituted and re-evaluated by the author to the best of his ability. For simplicity, values in Tables are reported without their respective units. Statistical data from the author's group were generated using GraphPad Prism Version 7 for Windows (GraphPad Software, San Diego, CA, USA). Chemical structures were drawn using ChemDraw 15.0 Professional (PerkinElmer, Germany). The structures of some analytes discussed in the present work are illustrated in Scheme 1. Where applicable, data analysis is reported in the following sections in more detail.

Standard analytical parameters included in the present work are: (1) y-axis intercept α, slope β, and goodness of fit (r^2) from linear regression analysis; (2) the mean of the difference δ_{BA}, the average μ_{BA} and the bias values from the BA approach; and (3) the OE ratio Λ_{OE}. In addition, further statistically relevant parameters, notably the relative standard deviation (RSD) or coefficient of variation (CV) of the absolute difference δ and the percentage difference $\delta(\%)$, and of the ratio Λ_{OE} were included. In the BA approach, linear regression analysis between δ or $\delta(\%)$ versus the average was performed and the goodness of fit (ρ^2) was reported. It is assumed that these measures allow for evaluations of agreement between two methods more effectively and on a quantitative basis as compared to the rather qualitative information provided by the individual approaches. In addition, the receiver operating characteristic (ROC) approach was used and the area under the curve (AUC) values were considered to evaluate agreement/disagreement between two compared methods. The complete set of the results from the meta-analyzed studies is presented in Figures 1–11 and summarized in Table 1.

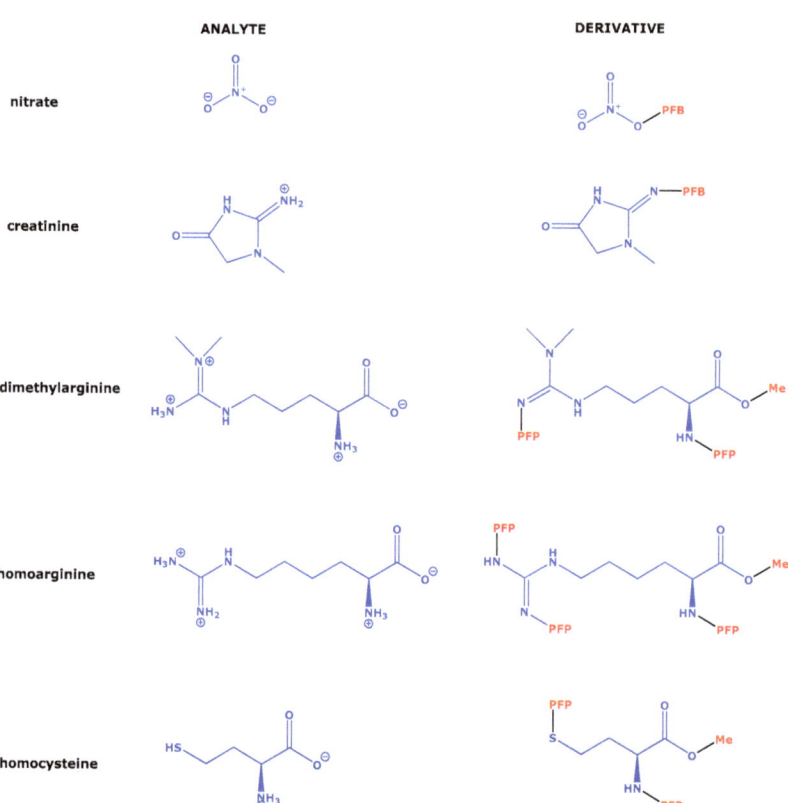

Scheme 1. Chemical structures of the native analytes (**left**) and their chemical derivatives (**right**) discussed in the present work. Nitrate, nitrite, malondialdehyde, and creatinine are derivatized with pentafluorobenzyl bromide in aqueous acetone (e.g., 60 min at 50 °C). Asymmetric dimethylarginine (ADMA), homoarginine and homocysteine are first methylated in 2 M HCl in methanol (e.g., 30 min at 80 °C), and then by pentafluoropropionic anhydride in ethyl acetate (e.g., 60 min at 65 °C). Me, methyl; PFB, Pentafluorobenzyl; PFP, pentafluoropropionyl.

Table 1. Summary of the results from the re-evaluation of published studies by using linear regression analysis, the Bland-Altman and the Oldham-Eksborg method, and the ROC approach between two methods.

No	Mean		n	p Value	Linear Regression			Bland-Altman			Oldham-Eksborg		ROC		Method 1/Method 2	Refs.
	τ_{2u}	τ_{1u}	n	p	α	β	r^2	δ	CV (%)	ρ^2 (δ; δ%)	Λ	CV (%)	AUC ± SE	p		
(1) Nitrate in urine (a, b) and plasma (c) (μM)																
a	1059	1048	20	0.0136	1.24	0.99	0.998	−12 ± 50	1.1 ± 4.7	−1.5 ± 2.7	0.986 ± 0.027	2.8	0.531 ± 0.093	0.735	GC-MS/GC-MS/MS	[58]
b	885	907	240	0.554	−121	1.16	0.973	30 ± 183	3.4 ± 21	n.d.	0.983 ± 0.231	23	n.d.		HPLC/GC-MS	[59,60]
c	41	36	40	0.0002	−21	1.37	0.896	−5.5 ± 8.7	13 ± 21	n.d.	0.822 ± 0.253	31	n.d.		Griess/GC-MS	[61]
(2) Asymmetric dimethylarginine (ADMA) in urine (a) and plasma or serum (aa, b–g) (μM or nM)																
aa	1351	1334	18	0.034	12.7	1.003	0.994	17 ± 1.2	29 ± 180	1.3	1.013 ± 0.0219	2.2	0.549 ± 0.097	0.613	Serum/Plasma	[62]
a	14.9	13.2	10	0.008	0.22	1.100	0.9997	1.16 ± 1.5	12 ± 11	n.d.	1.170 ± 0.097	8.3	n.d.		GC-MS/GC-MS/MS	[63]
b	654	640	10	0.072	2.37	1.020	0.9966	14 ± 21	2.2 ± 3.3	n.d.	1.02 ± 0.03	3.4	n.d.		GC-MS/GC-MS/MS	[63]
c			9		−0.9	0.999	0.991								ELISA/GC-MS	[64]
d			29		0.01	0.85	0.984								ELISA/LC-MS/MS	[64]
e	0.72	0.37	11		0.40	0.34	0.312								ELISA/LC-MS/MS	[65]
f			55												ELISA/HPLC	[66]
g	1.75	0.83	36	1.3×10^{-18}	−0.19	0.86	0.944	0.9 ± 0.3	108 ± 36		2.09 ± 0.24				ELISA/HPLC	[67]
(3) Anandamide (AEA) in plasma (nM)																
	0.844	0.729	305	<0.0001	0.013	0.848	0.821	0.116 ± 0.123	106	0.024; 0.024	0.866 ± 0.136	16	0.626 ± 0.024	<0.0001	GC-MS/MS/LC-MS/MS	[68,69]
(4) Homoarginine (hArg) in plasma (μM or nM)																
a	0.744	0.643	369	<0.0001	0.03	0.821	0.994	0.105 ± 0.070	14.3 ± 27	0.871; 0.174	0.868 ± 0.035	4	0.596 ± 0.021	<0.0001	GC-MS/GC-MS/MS	[70]
b	245	270	79	0.190	186	1.86	0.918	−25 ± 108	13.6 ± 56	0.860; 0.542	0.978 ± 0.439	45	0.508 ± 0.021	0.869	LC-MS/MS/GC-MS/MS	[70,71]
(5) 15(S)-8-iso-PGF2α and other F2-isoprostanes in urine																

Table 1. Cont.

No	Mean $\tau_{2\mu}$	Mean $\tau_{1\mu}$	n	p Value p	Linear Regression α	β	r^2	δ	Bland-Altman CV (%)	ρ^2 (δ; δ%)	Oldham-Eksborg Λ	CV (%)	ROC AUC ± SE	p	Method 1/Method 2	Refs.
a	2.813	2.037	66	0.612	0.814	0.560	0.642	−0.147±0.985	42; 38	0.272; 0.079	1.159 ± 0.691	60	0.510 ± 0.051	0.841	EIA/GC-MS	[72,73]
b	934	417	67	<0.0001	176	1.82	0.752	−517 ± 320		0.725; 0.005	2.412 ± 0.920	38	0.845 ± 0.033	<0.0001	ELISA/LC-MS/MS	[74]
c	3.605	0.825	10	<0.0001	0.996	3.62	0.656	147 ± 985	335	0.906; 0.028	4.513 ± 1.143	25	0.990 ± 0.016	0.0002	GC-MS/LC-MS	[74]
(6) Systolic blood pressure (mmHg)																
	167	178	25	<0.0001	−7	1.11	0.911	10.7 ± 9.0	84	0.203; 0.095	0.943 ± 0.044	5	0.609 ± 0.080	0.187	Apparat. 1/Apparat. 2	[75,76]
(7) Peak expiratory flow rate (mL/min)																
	450	453	17	<0.0001	39	0.917	0.890	2.1 ± 39	1830	0.007; 0.058	0.995 ± 0.114	5	0.509 ± 0.102	0.931	Instrum. 1/Instrum. 2	[10]
Miscellaneous																
(8) Nitrite (a) and nitrate (b) in bronchoalveolar liquid (μM)																
a	0.5	1.71	72	<0.0001	0.07	0.25	0.395	−1.2 ± 0.6	70 ± 35		0.28 ± 0.20	71			CL/GC-MS	[77]
b	6.0	9.5	72	<0.0001	−0.35	0.67	0.946	−3.1 ± 4.0	33 ± 42		0.64 ± 0.21	33			CL/GC-MS	[77]
(9) 3-Nitrotyrosine in plasma (nM)																
	4.8	1.65	18	0.002	3.06	1.04	0.334	3.1 ± 3.7	188 ± 224		3.9 ± 3.7	95			GC-MS/GC-MS/MS	[78]
(10) 2,3-dinor-5,6-dihydro-PGF2α in urine																
	543	502	14	0.434	12	1.06	0.832	40 ± 186	8 ± 37		1.12 ± 0.59	53			GC-MS/GC-MS/MS	[79]
(11) Creatinine in urine (mM)																
a	5.15	5.21	24	0.466	−0.13	1.01	0.998	−0.06 ± 0.40	1.2 ± 7.7		0.97 ± 0.07	7.2			HPLC/GC-MS	[80]
b	4.04	5.20	24		0.29	0.72	0.996	−1.17 ± 1.77	23 ± 34		0.82 ± 0.08	9.8			Jaffé/GC-MS	[80]
(12) Total plasma homocysteine (μM)																
	13.6	12.7	31	<0.001	2.47	0.88	0.990	0.91 ± 1.31	7.2 ± 10		1.13 ± 0.14	12.4			FPIA/GC-MS	[63]

N.A., not available. Abbreviations: CL, chemiluminescence; FPIA, fluorescence polarization immunoassay.

2.2. Measurement of Nitrate in Human Urine-Comparison of GC-MS with GC-MS/MS

Nitrate (Scheme 1) is the major circulating and urinary metabolite of nitric oxide (NO) [81]. Nitrate in urine is a suitable measure of whole-body NO synthesis. Figure 1 shows the results from the re-evaluation of data previously reported by our group (Table 1 of Ref. [58]) regarding validation by GC-MS/MS of a GC-MS method for the quantitative analysis of nitrate in human urine. In the urine samples analyzed, the nitrate concentration ranged between about 100 µM and 4000 µM. The nitrate concentration was measured to be (mean ± SD) 1048 ± 1024 µM (CV, 98%) by GC-MS (method 1) and 1059 ± 1035 µM (CV, 98%) by GC-MS/MS (method 2). The values differed between the methods ($p = 0.014$; two-tailed Wilcoxon test).

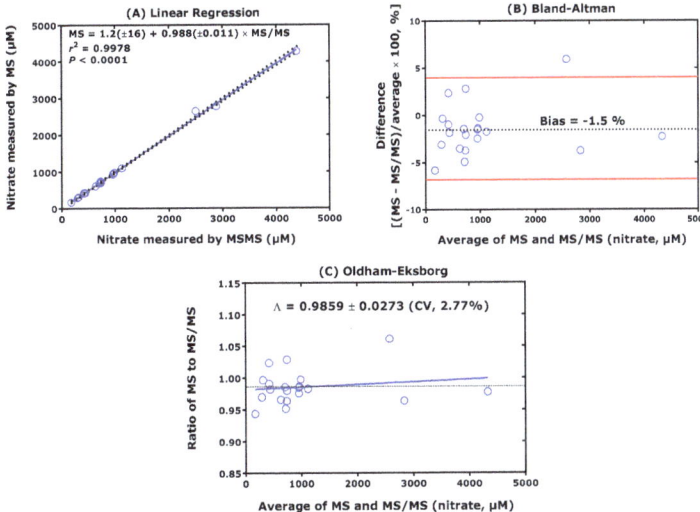

Figure 1. Measurement of nitrate in human urine by GC-MS (i.e., MS, method (1) and GC-MS/MS (i.e., MS/MS, method (2) and their comparison by: (**A**) linear regression; (**B**) Bland–Altman; and (**C**) Oldham–Eksborg. This Figure was constructed by using the data of Table 1 of the article [58]. Samples were analyzed on the instrument TSQ 7000 first by GC-MS in the SIM mode and subsequently by GC-MS/MS in the SRM mode. The closed points in (**A**) indicate the 95% confidence bands. Horizontal solid lines in (**B**) indicate the 95% limits of agreement (±1.96 × SD).

Linear regression analysis between the data obtained by GC-MS and those by GC-MS/MS resulted in a regression equation with a very low y-axis intercept value $\alpha = 1.2$, a slope value $\beta = 0.988$ close to unity, and a very high correlation coefficient ($r^2 = 0.9978$). These observations suggest a very tight agreement between the GC-MS (method 1) and the GC-MS/MS (reference method 2) (Figure 1A).

The Bland–Altman approach revealed a very low difference between the two methods $\delta_{BA} = -12 \pm 50$ µM according to a percentage difference δ (%) of -1.5 ± 2.7% (mean ± SD), which is only a very small portion of the mean concentration of nitrate measured in the whole concentration range (Figure 1B). Neither the difference ($\rho^2 = 0.05$) nor the percentage difference ($\rho^2 = 0.02$) correlated with the average concentration. Thus, the findings argue for a close agreement between the GC-MS and GC-MS/MS methods for nitrate in human urine.

The approach according to Oldham and Eksborg gave a concentration-independent ratio $\Lambda_{OE} = 0.986 \pm 0.027$ which is very close to the unity and has a low CV value of only 2.8% (Figure 1C). In the Bland–Altman approach, the ratio of the two methods can also be plotted against the average of the two methods. It provided a value of 0.9858 ± 0.027 which is identical to the ratio Λ_{OE}. The third approach used to compare the GC-MS method with the GC-MS/MS method for urinary nitrate strongly indicates that the GC-MS method is as suitable as the GC-MS/MS method for the accurate quantitative determination of nitrate in human urine.

The ROC approach on these data resulted in the AUC value of 0.531 ± 0.093 and a p value of $p = 0.735$. This data can be interpreted as having a high agreement between the two methods.

2.3. Measurement of Asymmetric Dimethylarginine (ADMA) in Human Plasma and Serum

Asymmetric dimethylarginine (ADMA) is an endogenous inhibitor of NO synthase (NOS), which catalyzes the conversion of L-arginine to NO and is a cardiovascular risk factor [81]. ADMA circulates in blood and is excreted in the urine. Several methods were developed for the measurement of ADMA mainly in plasma and serum [63–67,82–84]. The concentration of ADMA in serum and heparinized plasma of humans was reported by many groups using HPLC, GC-MS, GC-MS/MS and LC-MS/MS methods and found not to differ significantly, with deviations being in the order of 1% [82–84].

We have utilized this feature of ADMA and quantitated ADMA by GC-MS/MS in heparinized plasma and serum samples generated from blood samples of a patient suffering from end-stage kidney disease before, during, and after extended haemodialysis for 8 h [62]. The ADMA concentration (mean $\pm$ SD) was measured to be 1351 ± 386 nM (CV, 29%) in serum (method 2) and 1334 ± 383 nM (CV, 29%) in plasma (method 1). The values differed between the methods ($p = 0.034$; two-tailed Wilcoxon test). The results of the methods-comparison are shown in Figure 2.

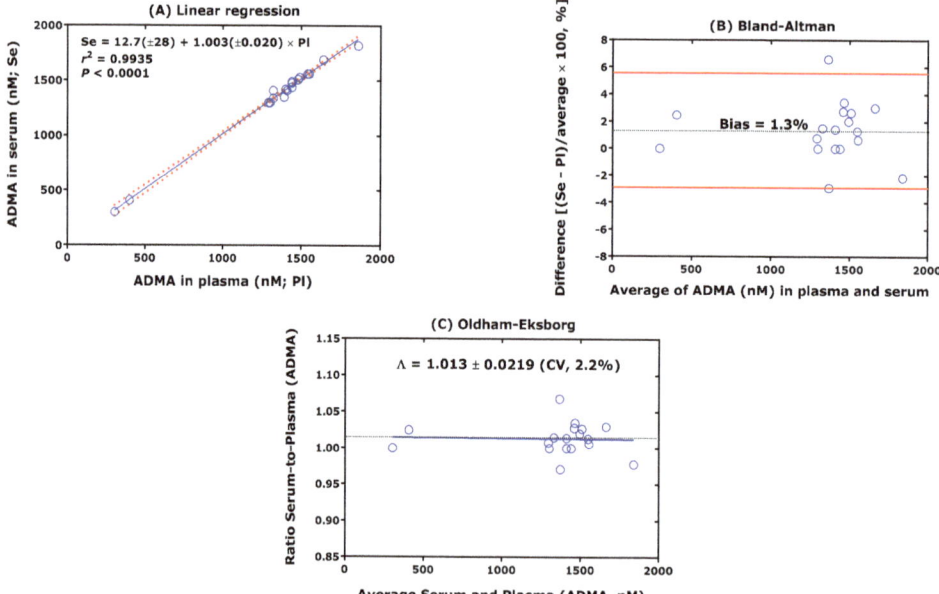

Figure 2. Comparison of measurements of asymmetric dimethylarginine (ADMA) in human plasma (method 1) and serum (method 2) of one patient with acute renal failure before, during and after extended haemodialysis for 8 h: (**A**) linear regression analysis; (**B**) Bland–Altman; and (**C**) Oldham–Eksborg. This Figure was constructed by using the data of Table 1 of a previous article [62]. Samples were analyzed for ADMA on the instrument TSQ 7000 by GC-MS/MS in the SRM mode as reported elsewhere [63]. Horizontal solid lines in (**B**) indicate the 95% limits of agreement.

Linear regression analysis between the data measured in serum (method 2) and those in plasma (method 1) resulted in a regression equation with a very low y-axis intercept value $\alpha = 12.7$, a slope value $\beta = 1.003$ very close to the unity, and a very high correlation coefficient ($r^2 = 0.9935$). These observations suggest a very high agreement between the serum and the plasma ADMA levels (Figure 2A).

The Bland–Altman approach revealed a very low difference between the two methods (17.2 ± 1.2 nM) according to a percentage difference of 1.3 ± 2.1% (bias), which is only a very small portion of the mean concentration of ADMA measured in the whole concentration range (Figure 2B). Neither the difference ($\rho^2 = 0.007$) nor the percentage difference ($\rho^2 = 0.001$) correlated with the average concentration. These findings argue for a close agreement between the plasma and serum "methods" for ADMA.

The approach according to Oldham and Eksborg gave $\Lambda_{OE} = 1.013 \pm 0.0219$ which is very close to the unity and has a CV value of only 2.2% (Figure 2C). Thus, the third approach that was used to compare the serum and plasma levels of ADMA strongly indicates that ADMA can be measured equally accurately in human serum and plasma samples by GC-MS/MS.

The ROC approach on these data resulted in the AUC value of 0.549 ± 0.097 ($p = 0.613$), suggesting no difference, i.e., a high extent of agreement between the two methods.

2.4. Measurement of Anandamide in Human Plasma by GC-MS/MS and LC-MS/MS

Anandamide (AEA) is an endogenous cannabinoid, an endocannabinoid, and is mainly measured in human plasma and serum at concentrations in the upper pM-to-the lower nM range [68,69,85,86]. For the measurement of AEA in human plasma, we developed GC-MS/MS [68] and LC-MS/MS [69] methods, which utilize stable-isotope-labelled AEA as the internal standard. In the GC-MS/MS method, AEA is derivatized, while in the LC-MS/MS method AEA is analyzed without derivatization. We compared these methods by parallel measurements of AEA in the plasma of healthy humans. In this comparison, we considered the GC-MS/MS method as the reference method.

The AEA concentration (mean ± SD) was measured to be 0.844 ± 0.289 nM (CV, 34%) by GC-MS/MS (method 2) and 0.729 ± 0.270 nM (CV, 37%) by LC-MS/MS (method 1). The values differed between the methods ($p < 0.0001$; two-tailed Wilcoxon test). The results of the methods-comparison are shown in Figure 3.

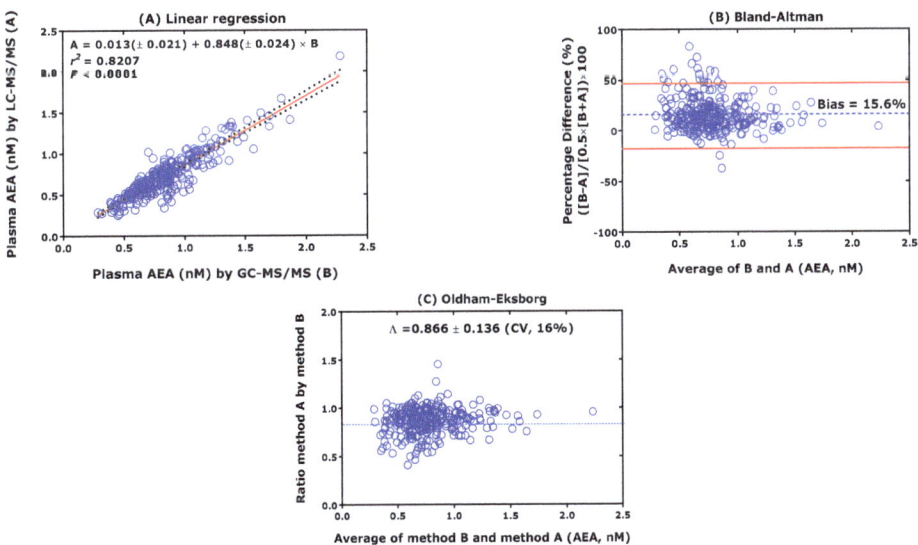

Figure 3. Comparison of measurements of anandamide (AEA) in 277 human plasma samples by LC-MS/MS (method 1) and by GC-MS/MS (method 2): (**A**) linear regression analysis; (**B**) Bland–Altman; and (**C**) Oldham–Eksborg. Samples were analyzed for AEA on the instrument TSQ 7000 by GC-MS/MS [68] and by Xevo LC-MS/MS [69] as reported in these references. Horizontal solid lines in (**B**) indicate the 95% limits of agreement.

The linear regression analysis between the AEA concentrations measured by LC-MS/MS and those measured by GC-MS/MS resulted in a regression equation with a y-axis intercept value $\alpha = 0.013$, a slope value $\beta = 0.848$, and a relatively small correlation coefficient ($r^2 = 0.8207$). These observations suggest a weak agreement between LC-MS/MS and GC-MS/MS (Figure 3A).

The Bland–Altman approach revealed a low but considerably varying difference between the two methods (0.116 ± 0.123 nM; CV, 106%) according to a percentage difference of 15.6% (bias) (Figure 3B). Neither the difference ($\rho^2 = 0.024$) nor the percentage difference ($\rho^2 = 0.024$) correlated with the average concentration. These findings argue for a weak agreement between the LC-MS/MS and GC-MS/MS methods regarding AEA measurement in human plasma samples.

The approach according to Oldham and Eksborg resulted in the ratio $\Lambda_{OE} = 0.866 \pm 0.136$ which is not close to the unity and has a moderate variation (CV, 16%) (Figure 3C). Thus, the third OE-approach confirms the results of the linear regression and Bland–Altman method.

The ROC approach yielded an AUC value of 0.626 ± 0.024 and a p value < 0.0001, suggesting some extent of agreement between the two methods.

In summary, GC-MS/MS and LC-MS/MS are suitable for the measurement of AEA in human plasma but yield considerably different results.

2.5. Measurement of Homoarginine in Human Plasma by GC-MS and GC-MS/MS-Real Data and a Simulation

L-Homoarginine (hArg) is a non-proteinogenic amino acid. Low plasma and urinary hArg concentrations are considered to be risk markers for cardiovascular and renal diseases [87]. For the quantitative determination of hArg in human plasma, serum and urine samples, we have developed GC-MS and GC-MS/MS methods using L-[^{2}H$_3$]homoarginine (d$_3$-hArg) as the internal standard [86]. In healthy humans, plasma, serum, and urine concentrations of hArg are on the order of 2–3 µM [87].

The hArg concentration (mean $\pm$ SD) was measured to be 0.747 ± 0.367 µM (CV, 49%) by GC-MS/MS (method 2) and 0.643 ± 0.302 µM (CV, 47%) by GC-MS (method 1). The values differed between the methods ($p < 0.0001$; two-tailed Wilcoxon test). The results of the methods-comparison are shown in Figure 4.

Linear regression analysis between the hArg concentrations measured by GC-MS (method 1, DSQ) and those measured by GC-MS/MS (method 2, TSQ) resulted in the regression equation with a y-axis intercept value $\alpha = 0.030$, a slope value $\beta = 0.821$, and a correlation coefficient ($r^2 = 0.9943$). These observations suggest a close correlation between GC-MS and GC-MS/MS, with the GC-MS method providing constantly lower hArg concentrations than TSQ (Figure 4A).

The Bland–Altman approach revealed a relatively low difference between the two methods (0.105 ± 0.070 µM) according to a percentage difference of $14.3 \pm 27\%$ (bias) (Figure 4B1). The percentage difference (y) correlated very weakly ($\rho^2 = 0.1739$, $p < 0.0001$) with the average concentration of hArg (x): $y = 10.9 + 4.9 \times x$. The difference correlated more strongly ($\rho^2 = 0.8701$, $p < 0.0001$) with the average concentration of hArg: $y = -0.03 + 0.195 \times x$ (Figure 4B2). These findings argue for a considerable agreement between the GC-MS and GC-MS/MS methods regarding hArg measurement in human plasma samples, with the GC-MS method resulting in constantly lower hArg concentrations.

The approach according to Oldham and Eksborg resulted in the ratio $\Lambda_{OE} = 0.868 \pm 0.035$ (CV, 4%) which is not close to the unity (Figure 4C). Linear regression between the Λ_{OE} ratio (y) and the average (x) resulted in the regression equation $y = 0.897 - 0.043 \times x$ ($r^2 = 0.173$, $p < 0.0001$), indicating weak concentration-dependency.

The ROC approach yielded and AUC value of 0.596 ± 0.021 and a p value < 0.0001, suggesting some extent of agreement between the two methods.

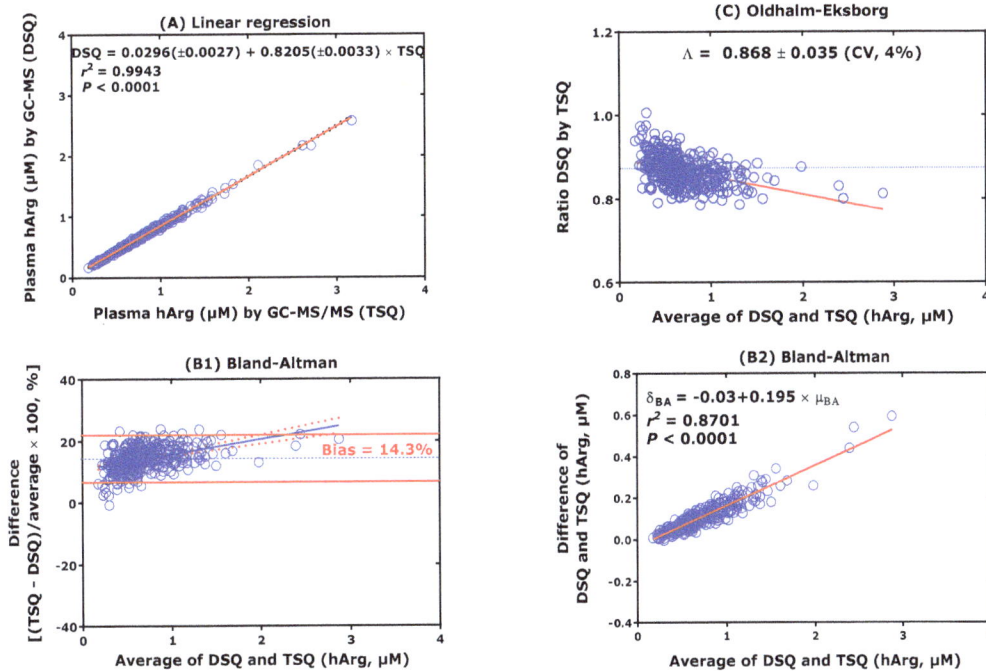

Figure 4. Comparison of measurements of homoarginine (hArg) in plasma by GC-MS (method 1) and by GC-MS/MS (method 2) in 369 plasma samples of pregnant women: (**A**) linear regression analysis; (**B1,B2**) Bland–Altman; and (**C**) Oldham–Eksborg. Within (**B1,B2**): linear regression analysis between percentage difference and average. This Figure was constructed by using the data published in a previous article [70]. Note that two different apparatus were used: Samples were analyzed for hArg on the instrument TSQ 7000 by GC-MS/MS in the SRM mode and on the instrument DSQ in the SIM mode as reported elsewhere [70]. Horizontal solid lines in (**B1**) indicate the 95% limits of agreement. Note that the difference between the methods is used in percentages ((**B1**), %) and in absolute concentrations ((**B2**), μM).

The hArg concentrations measured by DSQ (i.e., by GC-MS) were changed by multiplication to reach higher (DSQ × 1.2) and lower (DSQ × 0.8, DSQ × 0.6) concentrations. All methods of comparison were used to compared unchanged and changed hArg concentrations with those measured by TSQ (i.e., by GC-MS/MS). The results of these analyses are summarized in Table 2.

Table 2. Linear regression, Bland–Altman, Oldham–Eksborg and ROC analysis using simulated homoarginine concentrations. The original concentrations measured by DSQ (GC-MS, i.e., DSQ × 1.0) were multiplied by 1.2, 0.8 and 0.6 and compared with those measured by TSQ (GC-MS/MS).

DSQ-Fold	Linear Regression			Bland–Altman			Oldham–Eksborg	ROC
DSQ × 1.2	$y = 0.04 + 0.9846x$	$r^2 = 0.9943$	$\delta = -0.0234 \pm 0.0281$	CV = 117%	$\rho^2 = 0.0271$	δ (%) = -3.94 ± 3.98 CV = 100%	1.041 ± 0.041 CV = 4%	0.5268 $p = 0.2081$
DSQ × 1.0	$y = 0.03 + 0.8205x$	$r^2 = 0.9943$	$\delta = 0.1046 \pm 0.0698$	CV = 67%	$\rho^2 = 0.8701$	δ (%) = 14.3 ± 3.9 CV = 28%	0.868 ± 0.034 CV = 4%	0.5961 $p < 0.0001$
DSQ × 0.8	$y = 0.02 + 0.6564x$	$r^2 = 0.9943$	$\delta = 0.2331 \pm 0.1276$	CV = 55%	$\rho^2 = 0.9699$	δ (%) = 36.2 ± 3.8 CV = 11%	0.694 ± 0.028 CV = 4%	0.7285 $p < 0.0001$
DSQ × 0.6	$y = 0.02 + 0.4923x$	$r^2 = 0.9943$	$\delta = 0.3617 \pm 0.1871$	CV = 51%	$\rho^2 = 0.9903$	δ (%) = 63.2 ± 3.6 CV = 6%	0.520 ± 0.021 CV = 4%	0.8595 $p < 0.0001$

The best agreement between TSQ and DSQ were observed between plasma hArg concentrations measured by TSQ and by DSQ × 1.2:β = 0.985, a very small difference δ and bias (δ%) in the Bland–Altman method with no linearity between the difference and the average (ρ^2 = 0.027), a weakly (CV, 4%) varying Oldham–Eksborg ratio of 1.04, and an AUC value very close to 0.5, indicating complete agreement between the methods. These results suggest that r^2 alone is not a useful measure of agreement between two methods.

2.6. Measurement of Homoarginine in Mouse Plasma by GC-MS/MS and LC-MS/MS

hArg was measured in the plasma of mice by GC-MS/MS [86] and LC-MS/MS [71]. The hArg concentrations were measured as 245 ± 105 nM (CV, 43%) by the GC-MS/MS and 270 ± 204 nM (CV, 76%) by the LC-MS/MS. These values did not differ from each other (p = 0.190; two-tailed Wilcoxon test). The results of the methods-comparison are shown in Figure 5.

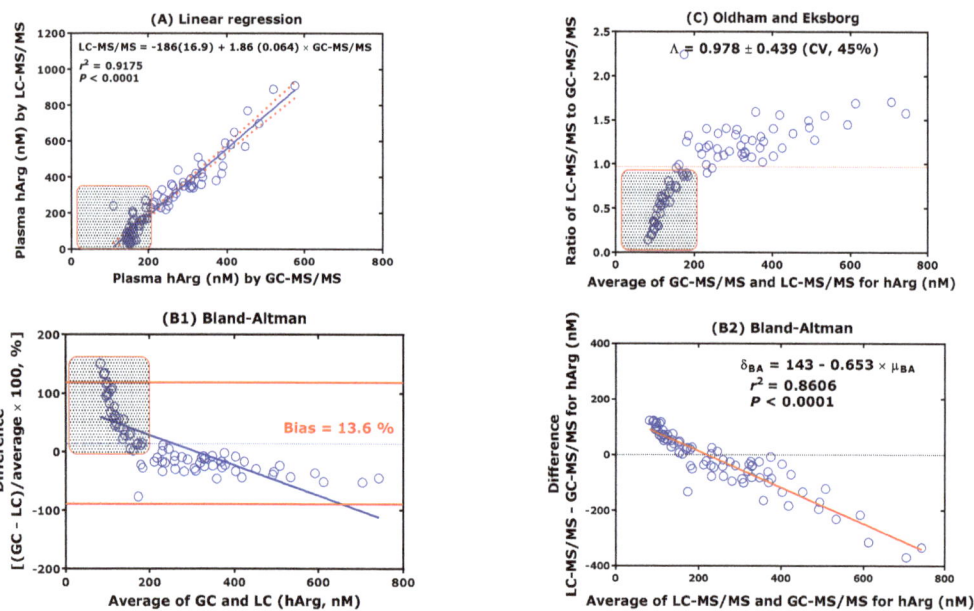

Figure 5. Comparison of measurements of homoarginine (hArg) in mouse plasma by GC-MS/MS (method A) and by LC-MS/MS (method B) in 79 plasma samples: (**A**) linear regression analysis; (**B1,B2**) Bland–Altman; and (**C**) Oldham–Eksborg. Within (**B1,B2**): linear regression analysis between percentage difference and average. This Figure was constructed by using the data published in a previous article [71]. Note that two different apparatus were used: Samples were analyzed for hArg on the instrument TSQ 7000 by GC-MS/MS in the SRM mode and on the instrument Varian 1200 L Triple Quadrupole MS in the SRM mode as reported elsewhere [71]. Horizontal solid lines in (**B1,B2**) indicate the 95% limits of agreement. Shaded insets indicate ranges of maximum disagreement.

Linear regression analysis between the hArg concentrations measured by the LC-MS/MS (method 1) and those measured by the GC-MS/MS (method 2) resulted in a regression equation with a y-axis intercept value α = −186, a slope value β = 1.86, and a correlation coefficient (r^2 = 0.9175). These observations suggest a good correlation between LC-MS/MS and GC-MS/MS, yet with the LC-MS/MS providing higher values for hArg in mouse plasma than the GC-MS/MS (Figure 5A).

The Bland–Altman approach revealed a moderate difference between the two methods (−25 ± 108 nM) according to a percentage difference of 13.6 ± 25% (bias) (Figure 5B1). The percentage difference (y) correlated weakly (ρ^2 = 0.5421, p < 0.0001) with the average concentration (x): $y = 80 - 0.26 \times x$. The difference correlated more strongly ($\rho^2 = 0.8601$, $p < 0.0001$) with the average concentration of hArg: $y = 80 - 0.258 \times x$ (Figure 5B2). These findings argue for a moderate agreement between the GC-MS/MS and LC-MS/MS methods regarding hArg measurement in mouse plasma samples.

The approach according to Oldham and Eksborg resulted in the ratio $\Lambda_{OE} = 0.978 \pm 0.439$ (CV, 45%) which is close to unity, but is considerably variable (Figure 5C).

The ROC approach resulted in an AUC value of 0.508 ± 0.021 and a p value of 0.8688, suggesting a good agreement between the two methods.

In summary, the Bland–Altman and the Oldham–Eksborg approaches indicate considerable disagreement between the GC-MS/MS and LC-MS/MS methods for hArg measurement in mouse plasma. Disagreement is especially visible at hArg concentrations lower than 200 nM (Figure 5), presumably because of the lower sensitivity of the LC-MS/MS method in terms of a higher limit of quantitation (LOQ).

The LC-MS/MS method for hArg was compared with an ELISA method for this amino acid in human plasma [71]. The linear regression analysis of the hArg concentrations measured by ELISA (y) correlated with those measured by LC-MS/MS: $y = 0.04 + 0.76 \times x$, $r^2 = 0.78$ [71]. Thus, LC-MS/MS yielded consistently higher hArg values than ELISA ($p < 0.001$). Analysis by the Bland–Altman approach resulted in a considerable difference of 0.50 ± 0.39 μM hArg between ELISA and LC-MS/MS. The data indicate a considerable disagreement between LC-MS/MS and ELISA for hArg measurement in human plasma.

2.7. Measurement of F_2-Isoprostanes in Biological Samples

Free-radical-induced peroxidation of arachidonic acid and other polyunsaturated fatty acids esterified to lipids and subsequent hydrolysis generates prostaglandin-like compounds including isoprostanes and neuroprostanes. These compounds are accessible for quantitative determination in tissue, plasma and urine. F_2-Isoprostanes have emerged as markers of lipid peroxidation in vivo in humans. Among them, 8-*iso*-prostaglandin (PG) $F_{2\alpha}$, also known as 8-*iso*-$PGF_{2\alpha}$, 8-*epi*-$PGF_{2\alpha}$, 15-F_{2t}-IsoP, i$PF_{2\alpha}$-III, and its major urinary metabolites, i.e., 2,3-dinor-4,5-dihydro-8-*iso*-$PGF_{2\alpha}$ and 2,3-dinor-8-*iso*-$PGF_{2\alpha}$, were the subjects of extensive investigation [72–74,79,88–91]. Analytical approaches for F_2-isoprotanes include chromatographic techniques such as thin-layer chromatography (TLC), high-performance liquid chromatography (HPLC), and GC, particularly in combination with MS. F_2-Isoprostanes are extracted from biological samples by solid-phase extraction (SPE) or immunoaffinity column chromatography (IAC). HPLC and TLC are used for the isolation of particular F_2-isoprostanes.

2.7.1. Comparison between EIA and GC-MS

Figure 1 of the article by Devaraj et al. [73] shows the relationship for "F2Isoprostanes" in urine as measured by the authors' group by using a commercially available EIA (i.e., method 1) and as measured by the group of Roberts by using a GC-MS method (i.e., method 2) after Morrow and Roberts [72].

The F_2-isoprostanes concentration (mean ± SD) was measured as being 2.183 ± 1.623 ng/mg (CV, 74%) by the GC-MS (method 2) and 2.037 ± 1.135 ng/mg (CV, 56%) by EIA (method 1). The values did not differ between the methods ($p = 0.612$; two-tailed Wilcoxon test). The results of the methods-comparison are shown in Figure 6.

The linear regression analysis between the F_2-isoprostanes concentrations measured in urine by EIA (method 1) and those measured by GC-MS (method 2) resulted in a regression equation with a y-axis intercept value $\alpha = 0.814$, a slope value $\beta = 0.560$, and a correlation coefficient ($r^2 = 0.6422$). These observations suggest a weak correlation between EIA and GC-MS (Figure 6A). On the basis of the data of Figure 6A, and on the assumption that the GC-MS method is the reference method, one may, at first glance, conclude that the EIA method is valid for the intended analyte. However, the calculated correlation coefficient r^2 value of 0.64 across the whole concentration range is rather low. Nevertheless, such a value is frequently considered to be high enough throughout the literature.

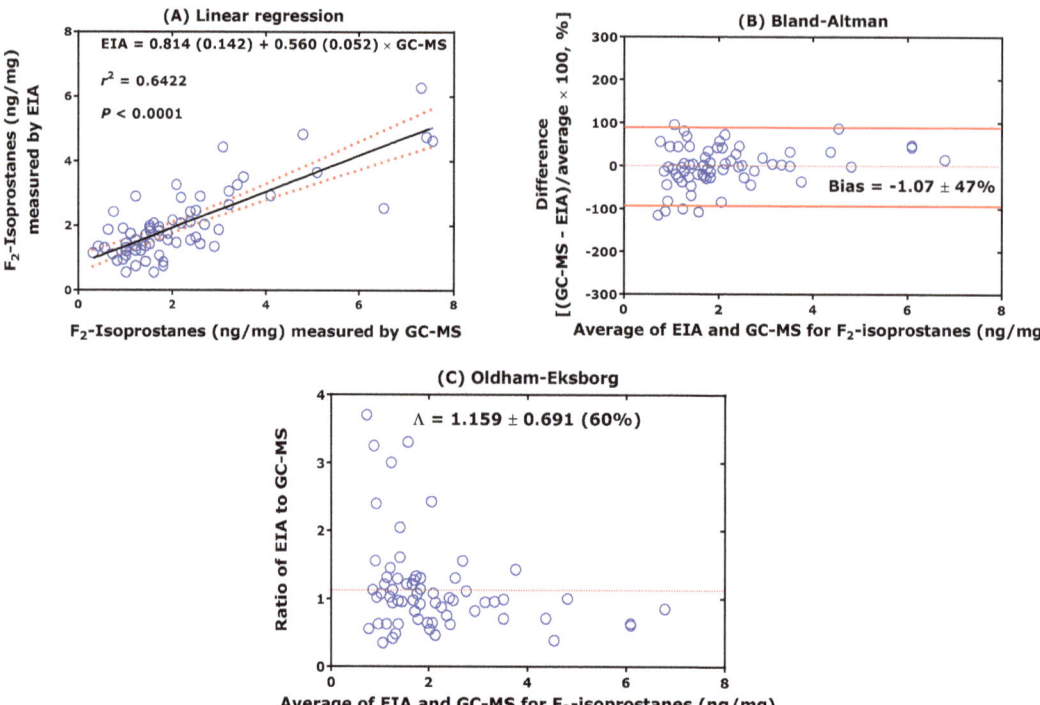

Figure 6. This Figure was constructed by using the data of Figure 1 of the article by Devaraj et al. [73]: comparison of a commercially available EIA method with a GC-MS method [72] for F_2-isoprostanes in urine; data are reported as ng F_2-isoprostanes per mg creatinine in this article: (**A**) linear regression analysis; (**B**) Bland–Altman; and (**C**) Oldham–Eksborg. Horizontal dotted lines indicate ±2 SD range. See Refs. [72–74,79,88–91] regarding analysis of F_2-isoprostanes.

The y-axis intercept α and the slope β values of the regression equation are often disregarded when methods are compared. The y-axis intercept value α of 0.81 of the regression equation in the whole concentration range of this example (Figure 6A) is clearly far from the zero which would indicate complete agreement. Moreover, this value may suggest that the LOQ value of the EIA method is about 0.8 ng/mg creatinine for the analyte in urine, with the lower values being most likely highly overestimated. The slope value β of 0.56 (for the whole range) suggests that the concentrations measured by this EIA method are statistically half of those measured by the GC-MS method. This finding seems to be supported by data in the literature [91] which suggest that the EIA method detects only one F2-isoprostane out of 64 potential isomers [89], i.e., 15(S)-8-iso-PGF$_{2\alpha}$, while the reported physicochemical methods, GC-MS [89] and GC-MS/MS [91], that do not use specific immunoaffinity column chromatography (IAC) extraction for 15(S)-8-iso-PGF$_{2\alpha}$, may detect an unknown number of additional F_2-isoprostanes. The contribution of those additional F_2-isoprostanes is presumably of the same extent (of about 50%) as that of 15(S)-8-iso-PGF$_{2\alpha}$ alone [91]. However, in the, by far more relevant, lower concentration range—both for controls and diabetes patients investigated in the study by Devaraj et al. [73]—there is no correlation between the EIA method and the GC-MS method (Figure 6A), i.e., $r^2 = 0.16$ for the 0–3 range ($n = 53$) and $r^2 = 0.006$ for the 0–3 range ($n = 39$). Figure 6A clearly demonstrates that the linearity between methods 1 and 2 solely is the result of very few (about 10% of the total) high concentration points (for a discussion see Ref. [24]). Thus, the value of linear regression analysis is limited, and the sole use of correlation coefficients may be misleading and even pretend correlation, because deviations in the lower concentration range are difficult to detect [8]. This comparison suggests that the correlation coefficient $r^2 = 0.64$ is obviously too low and the agreement is analytically insufficient [90].

The Bland–Altman approach revealed a moderate difference between t5e two methods (-0.147 ± 0.985 ng/mg) according to a percentage difference of $-1.07 \pm 47\%$ (bias) (Figure 6B). The percentage difference (y) correlated very weakly ($\rho^2 = 0.079$, $p = 0.022$) with the average concentration (x): $y = -22 + 10 \times x$. The difference (y) correlated very weakly ($\rho^2 = 0.279$, $p < 0.0001$) with the average concentration (x): $y = -0.68 + 0.392 \times x$. These findings argue for a weak agreement between the EIA and GC-MS methods with respect to F_2-isoprostanes measurement in human urine samples.

The Bland–Altman approach (Figure 6B), and a deeper examination, reveal considerable disagreement between the EIA method and the reference GC-MS method [72]. The disagreement applies to the vast majority of those concentration points in the relevant concentration range for these substances [89], i.e., for the lower concentration range (see also Refs. [74,90]). Interestingly, the mean difference between the two methods is 0.15 (see Formula (2)), which is considerably low and close to zero. However, the standard deviation of the mean difference is 0.98, which is high and within the range of measured concentrations.

The approach according to Oldham and Eksborg resulted in the ratio $\Lambda_{OE} = 1.159 \pm 0.691$ (CV, 60%) which is not very far from unity but is very variable (Figure 6C). This indicates a rather poor agreement between the EIA and the GC-MS methods, notably in the lower and more relevant concentration range (Figure 6C).

The ROC approach resulted in an AUC value of 0.510 ± 0.051 and a p value of 0.8413, suggesting agreement between the two methods.

2.7.2. Comparison between ELISA and LC-MS/MS

Figure 5 of the article by Yan et al. [74] shows the linear regression between "iPF$_{2\alpha}$-III by ELISA (pg/mL)" and "iPF$_{2\alpha}$-III by LC-MS/MS (pg/mL)" in urine as measured by the authors themselves by using a commercially available ELISA (i.e., method 1) and as measured by the same group by using an LC-MS/MS method (i.e., method 2). It should be noted that iPF$_{2\alpha}$-III and 8-*iso*-PGF$_{2\alpha}$ are abbreviations for the same F_2-isoprostane [88]. The largest part of the originally reported data of the study by Yan et al. [74], i.e., 67 out of 86 (78% of total) could be re-evaluated by the author of the present article and they are presented and discussed below.

The F_2-isoprostane iPF$_{2\alpha}$-III concentration (mean $\pm$ SD) was measured to be 934 ± 506 pg/mL (CV, 54%) by ELISA (method 2) and 417 ± 241 pg/mL (CV, 58%) by LC-MS/MS (method 1). The values differed between the methods ($p < 0.0001$; two-tailed Wilcoxon test). Other results of the methods-comparison are shown in Figure 7.

Linear regression analysis between the iPF$_{2\alpha}$-III concentrations measured in urine by ELISA (method 1) and those measured by LC-MS/MS (method 2) resulted in the regression equation with a y-axis intercept value $\alpha = 176$, a slope value $\beta = 1.82$, and a correlation coefficient ($r^2 = 0.7518$). These observations suggest a weak correlation between the ELISA and LC-MS/MS methods (Figure 7A).

Yan et al. reported in their article that ELISA and LC-MS/MS provided very similar results [74]. However, the ELISA method provides on average about two times higher values than the LC-MS/MS method. It is worth mentioning, that the theoretical value of the slope of the regression line should be about 0.5 [79]. This is because ELISA detects most likely only one F_2-isoprostane (S-form), while LC-MS and LC-MS/MS detect at least two F_2-isoprostanes (R- and S forms). Thus, actually the ELISA method provides values, which are on average about 4 times higher than those measured by the LC-MS/MS method.

The Bland–Altman approach revealed a considerably difference between the two methods (-517 ± 320 pg/mL) according to a percentage difference of $75 \pm 29\%$ (bias) (Figure 7B). The percentage difference did not correlate with the average concentration. However, the mean difference correlated with the average ($y = -9.2 - 0.752 \times x$, $\rho^2 = 0.7253$, $p < 0.0001$), indicating a considerable proportional error of the ELISA method. These findings argue for a weak agreement between the ELISA and LC-MS methods with respect to F_2-isoprostanes measurement in human urine samples.

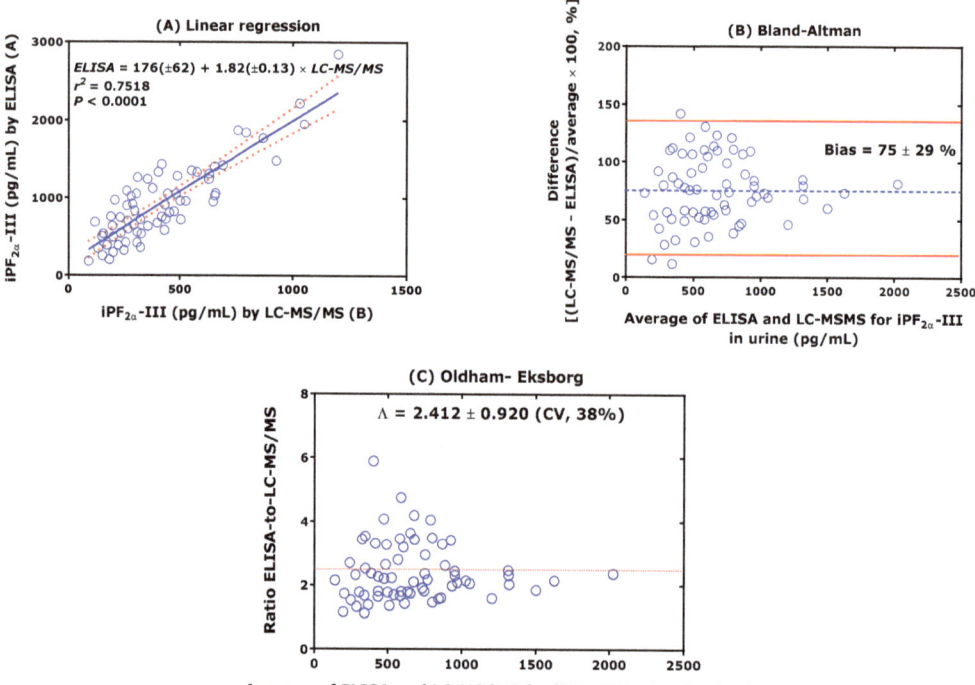

Figure 7. This Figure was constructed by using the data of Figure 5 of the article by Yan et al. [74]; due to considerable overlap of data points not all data points of the original Figure 5 from Ref. [74] could be used in the present work. Comparison of a commercially available ELISA method with a LC-MS/MS method for IPF$_{2\alpha}$-III (8-*iso*-PGF$_{2\alpha}$) in urine by: (**A**) linear regression analysis, (**B**) Bland–Altman; and (**C**) Oldham–Eksborg. Horizontal dotted lines indicate $\pm 2 \times$ SD range in (**B**). See Refs. [72–74,79,88–91] regarding analysis of F_2-isoprostanes.

The approach according to Oldham and Eksborg resulted in the ratio $\Lambda_{OE} = 2.412 \pm 0.920$ (CV, 38%) which is far from unity (Figure 7C) and supports a disagreement between ELISA and LC-MS/MS, notably in the lower concentration range (Figure 7C). This example supports the critique by Altman and Bland on the use of the correlation coefficient for evaluating method-agreement of clinical measurement [10], even when the correlation coefficient value is fairly high ($r^2 = 0.75$) as in the present example (Figure 7A).

The ROC approach yielded an AUC value of 0.845 ± 0.033 and a p value < 0.0001, suggesting a very low extent of agreement between the two methods.

2.7.3. Comparison between GC-MS and LC-MS

Sircar and Subbaiah [90] have compared their LC-MS method for 15(*S*)-8-*iso*-PGF$_{2\alpha}$ (due to the use of IAC extraction) with the GC-MS method of Morrow and Roberts [72], i.e., which measures 15(*S*)-8-*iso*-PGF$_{2\alpha}$ and additional F_2-isoprostanes in urine (Figure 8).

The F_2-isoprostane 15(*S*)-8-*iso*-PGF$_{2\alpha}$ concentration (mean $\pm$ SD) was measured to be 0.825 ± 0.349 ng/mL (CV, 42%) by LC-MS (method 1) and 3.605 ± 1.362 ng/mL (CV, 38%) by GC-MS (method 2). The values differed between the methods ($p < 0.0001$; two-tailed paired t test). The results of the methods-comparison are shown in Figure 8.

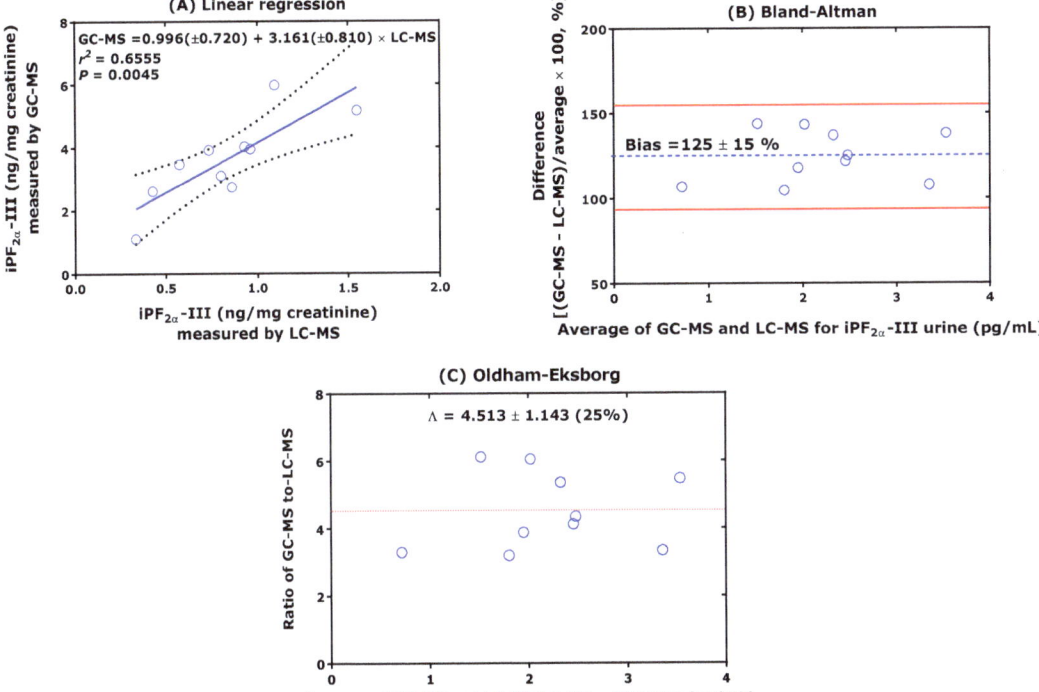

Figure 8. This Figure was constructed by using the data of Figure 4 of the article by Sircar and Subbaiah [90]. Comparison of an LC-MS method with a GC-MS method for IPF$_{2\alpha}$-III (8-*iso*-PGF$_{2\alpha}$) in urine by (**A**) linear regression analysis, (**B**) Bland–Altman, and (**C**) Oldham–Eksborg. Horizontal dotted lines indicate ± 2 SD range in (**B**). See Refs. [72–74,79,88,91] regarding analysis of F$_2$-isoprostanes.

Linear regression analysis between the iPF$_{2\alpha}$-III concentrations measured in urine by GC-MS (method 2) and those measured by LC-MS (method 1) resulted in the regression equation with a high y-axis intercept value α = 0.996, a high slope value β = 3.62, and a low correlation coefficient (r^2 = 0.6555) (Figure 8A). These observations suggest a weak correlation between GC-MS and LC-MS.

The Bland–Altman approach revealed a moderate difference between the two methods (−0.147 ± 0.985 ng/mg; CV, 335%) according to a percentage difference of 125 ± 15% (bias) (Figure 8B). The percentage difference did not correlate with the average concentration (ρ^2 = 0.028, p = 0.643). These findings argue for a weak agreement between the GC-MS and LC-MS methods. The difference correlated with the average concentration (ρ^2 = 0.906, p < 0.0001). These findings argue for a weak agreement between the GC-MS and LC-MS methods with respect to the iPF$_{2\alpha}$-III measurement in human urine samples.

The approach according to Oldham and Eksborg resulted in the very high ratio Λ_{OE} = 4.513 ± 1.143 (CV, 25%) which is far from the unity (Figure 8C). However, such a high ratio would be expectable because the GC-MS measures several F$_2$-isoprostanes in addition to 15(*S*)-8-*iso*-PGF$_{2\alpha}$ [72], while in the LC-MS method measures only 15(*S*)-8-*iso*-PGF$_{2\alpha}$ due to the use of a specific IAC extraction. The informative value of this comparison is considered low because of the small number of urine samples (n = 10).

The ROC approach resulted in an AUC value of 0.990 ± 0.016 and a p value of 0.0002, suggesting no agreement between the two methods.

2.8. An Example for Clinical Measurement-Systolic Blood Pressure [75,76]

Figure 9 shows the results from the application of the three method-comparison approaches to an example for a typical clinical measurement, i.e., for measuring systolic blood pressure (SBP) in 25 patients with essential hypertension, which was reported elsewhere [75,76]. In order to be able to evaluate the data in the same manner as in the examples discussed above, one of the methods used to measure blood pressure was chosen arbitrarily as the reference method (i.e., method 2) by the author of the present article.

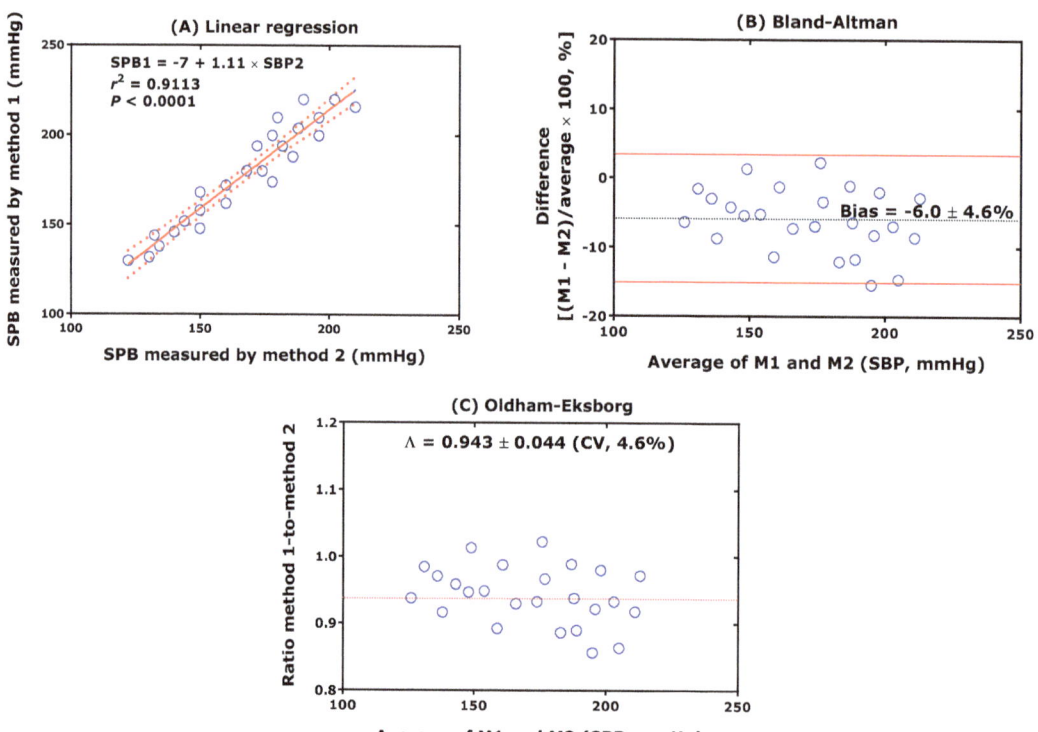

Figure 9. This Figure was constructed by using the data of Table 1 of the article by Ludbrook [75] which had been lent out from Ref. [76]. Comparison of measuring systolic blood pressure (SBP) by method 1 (M1) and method 2 (M2) in 25 patients with essential hypertension; method 2 was chosen arbitrarily as the reference method by the author of the present article. (**A**) Linear regression analysis; (**B**) Bland–Altman method. (**C**) Oldham–Eksborg method.

The SBP values were measured to be 167 ± 25 mmHg (CV, 15%) by method 2 and 178 ± 29 mmHg (CV, 16%) by method 1, and were found to differ significantly ($p < 0.0001$; two-tailed paired t test), albeit by a relatively low extent of 6%.

Linear regression analysis between the two methods resulted in the regression equation with a y-axis intercept value $\alpha = -7$, a slope value $\beta = 1.11$, and a correlation coefficient ($r^2 = 0.9113$) (Figure 9A). These observations suggest a good correlation between the two methods of blood pressure measuring, with method 2 providing on average 1.1 times higher SBP values.

The Bland–Altman approach revealed a moderate difference between the two methods according to a percentage difference of $-6 \pm 4.6\%$ (bias) (Figure 9B).

The approach according to Oldham and Eksborg resulted in the ratio $\Lambda_{OE} = 0.943 \pm 0.044$ (CV, 5%) which is close to the unity and little variable (Figure 9C).

All three approaches indicate a considerable agreement between the compared methods of SBP measurement (Figure 9). The ROC approach resulted in an AUC value of 0.609 ± 0.080 and a p value of 0.1870, also suggesting agreement between the two methods.

2.9. A Second Example for Clinical Measurement-Peak Expiratory Flow Rate [10]

Figure 10 shows the results from the application of three method-comparison approaches to a clinical measurement, i.e., for measuring peak expiratory flow rate (PEFR) by two methods in 17 subjects, originally reported by Bland and Altman [10].

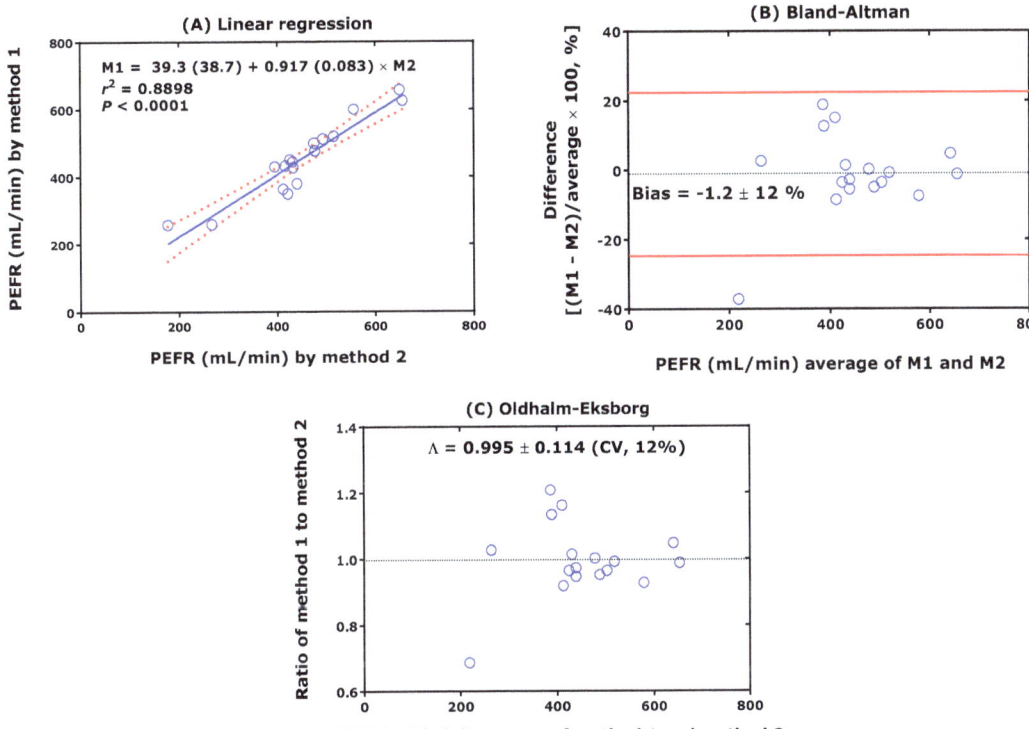

Figure 10. This Figure was constructed by using the data of the Table of the article by Bland and Altman [10]. Comparison of measuring peak expiratory flow rate (PEFR) by method 1 (M1) and method 2 (M2) in 17 subjects; method 2 was chosen arbitrarily as the reference method by the author of the present article. (**A**) Linear regression analysis; (**B**) Bland–Altman method. (**C**) Oldham–Eksborg method.

The PEFR values were measured to be 450 ± 116 mL/min (CV, 26%) by method 2 and 453 ± 113 mL/min (CV, 25%) by method 1, and differed significantly ($p < 0.0001$; two-tailed paired t test), albeit by a relatively low extent of 0.7%.

Linear regression analysis between the two methods resulted in the regression equation with a y-axis intercept value $\alpha = 39$, a slope value $\beta = 0.917$, and a correlation coefficient ($r^2 = 0.8898$) (Figure 10A). These observations suggest a good correlation between the two methods of PEFR measuring, with method 2 providing on average 0.92 times lower PEFR values.

The Bland–Altman approach revealed a small difference between the two methods (-2.1 ± 39 mL/min), yet with a considerable variability according to a percentage difference of $-1 \pm 12\%$ (bias) (Figure 10B). The difference did not correlate with the average PEFR.

The approach according to Oldham and Eksborg resulted in the ratio $\Lambda_{OE} = 0.995 \pm 0.114$ (CV, 5%) which is very close to the unity and little variable (Figure 10C).

The ROC approach resulted in an AUC value of 0.509 ± 0.1015 and a p value of 0.9314, also suggesting good agreement between the two methods of PEFR measurement.

3. Discussion

In the present work, published data from studies reporting on method-comparison were re-analyzed and re-evaluated by three method-comparison approaches, i.e., linear regression (LR) analysis and the Bland–Altman (BA) method, which are the most frequently used approaches, and the Oldham–Eksborg (OE) method, which is much less frequently used in comparison of analytical methods. The Oldham–Eksborg method is closely comparable with the Bland–Altman variant, in which the ratio of the results provided by two methods is plotted against their average. In the studies considered here, one of the applied methods was based on mass spectrometry, i.e., GC-MS, GC-MS/MS, LC-MS or LC-MS/MS. The analytes measured by the methods are all physiological low-molecular-mass substances, belong to different chemical classes and pathways, and require chemical derivatization for analysis by GC-MS-based methods (Scheme 1). Chosen biological matrices were human biological samples including plasma, serum and urine. The results of the present work are summarized in Table 1 and selected examples are illustrated in the Figures 1–10.

GC-MS/MS and LC-MS/MS methods were among the methodologies used in these comparative studies. They are generally considered best useful as reference methods, i.e., the *Gold Standards*. The tandem mass spectrometry technique, when used in combination with chromatography, e.g., GC or LC (Schemes S1 and S2) in analytical chemistry, allows to generate accurate quantitative results, i.e., the concentrations of analytes in complex biological samples, even if the analytes are isomeric (Figure S2). It is considered that they are applied properly [48]. Analytical methods based on GC-MS/MS or LC-MS/MS are superior to those based on GC-MS or LC-MS, respectively, because the tandem mass spectrometry (MS/MS) is able to exclude (entirely) potentially interfering, mostly unknown analytes (Schemes S1 and S2). Of particular importance is the unique feature of MS to enable use of stable-isotope-labelled analogs of the analytes as internal standards in quantitative analyses. In contrast to the high specificity of MS-based methods, analytical methods that utilize much less specific physicochemical properties of analytes, such as light absorbance or fluorescence detection, even in combination with GC or LC, are generally considered less accurate because of the susceptibility to interferences and a lack of analytical sensitivity, i.e., too-high limits of detection. On the basis of these facts, MS/MS methods may reasonably be considered superior to MS methods on the one hand, and MS methods superior to non-MS methods, such as GC coupled to flame ionization detection (FID) or electrochemical detection (ECD), and LC coupled to UV/vis absorbance or fluorescence detection, on the other hand. Eventually, non-MS methods may be considered superior to batch assays which are free of any chromatographic or immunologic separation. Well-documented examples for batch assays include the analysis of nitrite based on the Griess reaction [92] and of malondialdehyde (MDA) based on the use of thiobarbituric acid (TBA) [93,94]. In a comparison of analytical methods, those based on GC-MS/MS and LC-MS/MS can serve as reference methods not only for those based on GC-MS and LC-MS, but also for non-MS-based analytical methods including immunological methods. Although this thought is widely spread, few scientists practise this principle in the field of bioanalysis.

In the present work, the Bland–Altman approach, the Oldham–Eksborg approach and the standard linear regression analysis were applied to compare published analytical methods for the measurement in biological samples of a series of structurally different physiological substances. Examples were taken from the author's group and other groups who reported data from the use of two different analytical methods. Evaluations of comparability and agreement can be performed by using characteristic, preferably dimensionless, parameters of the abovementioned approaches. They include slope (β) and coefficients of correlation (r^2), the Oldham–Eksborg ratio (Λ_{OE}), the percentage difference, i.e., the bias (δ (%)), as well as the coefficient of correlation (ρ^2) obtained from linear regression analysis of the difference δ versus the average μ in the Bland–Altman approach. The relation between difference and mean in the Bland–Altman approach was addressed by Bland and Altman, who found that in some cases the difference δ may be proportional to the mean μ [10], yet it was not further considered. This issue was addressed by Ludbrook [25]. Sporadically, such as in ophthalmology and vision science, exact parametric confidence intervals for the Bland–Altman approach were proposed and reviewed [28,95]. These approaches do not consider additional methods such as those investigated in the present work. In the present study, AUC data obtained from ROC analyses were also considered, an approach that is rarely used in comparisons of analytical methods.

Statistically significant differences between the τ_1 and τ_2 values may not indicate disagreement between the methods. Values of β, r^2 and Λ_{OE} of the order of 1.00, δ(%) and ρ^2 values close to 0.00, and ROC-AUC (AUC) values close to 0.5 would indicate perfect agreement between the two methods compared. The present study indicates that perfect agreement is an exception rather than a rule. By contrast, β, r^2 and Λ_{OE} values different from 1.00, δ (%) and ρ^2 values different from 0.00, and AUC values different from 0.5 would also not decisively indicate disagreeing methods. Rather, like in statistical analyses where statistical significance is defined arbitrarily, for instance $p < 0.05$, assessing the extent of agreement or disagreement of two methods demands definition of ranges rather than discrete values for p, β, r^2, Λ_{OE}, δ (%) and possibly for AUC as well. As the definition of values and ranges is, per se, arbitrary, agreement or disagreement between two methods would also be arbitrary and relative. This resembles in many aspects the validation of analytical methods for which quantitative criteria were achieved by consensus and are widely used in the field of analytical chemistry for various types of analytical chemistry. These criteria include the precision in terms of the relative standard deviation (RSD) or the coefficient of variation (CV), the accuracy of the method in terms of recovery (%) for analytes added to biological samples at relevant concentrations, their limits of detection (LOD) and quantitation (LOQ), usually on the basis of the signal-to-noise (S/N) ratio. Such criteria have not been declared for the agreement or disagreement of methods, irrespective of the method-comparison approach. This is particularly the case with the Bland–Altman approach, which is mostly "degraded" to a simple plot.

On the basis of the results reported in the present work, each of the three currently available method-comparison approaches alone may be useful for comparing analytical methods, and for finding out which of the two compared methods is able to provide better, specifically more accurate, results. Yet, without a definition of reference methods and without a definition of quantitative criteria for the extent of agreement between two methods, no objective assessment is possible. A definition of acceptance criteria for main characteristic parameters for each method-comparison is required.

Yet, a reliable solution to this problem is likely to require the definition of a composite of all single parameters: p, β, r^2, δ(%), ρ^2, Λ_{OE}, AUC. Such a composite may provide maximum information about agreement or disagreement between the two analytical methods being compared. A way to overcome this dilemma could be to accept fully validated and published MS/MS-based methods as the reference methods. This assumption is reasonable and justified because of the inherent accuracy of the MS/MS-technique, provided it is performed correctly and errors, such as contamination, artificial formation, or the degradation of analytes during sampling, sample storage, derivatization, and analysis, are eliminated [47,48]. Strictly speaking, comparisons of two methods require the use of validated protocols for each method and the performance of comparison studies in parallel under optimum conditions for each method [96,97], and should also include the use of standardized reference compounds for an analyte and its stable-isotope-labelled analog in GC-MS/MS and LC-MS/MS methods (see Figure 3) [68,69].

Special emphasis and consideration should be given when comparing chemical and immunological methods or immunological methods such as immunoaffinity chromatography (IAC) that are used for the isolation of certain analytes from biological samples prior to chemical analysis. Without a consideration of such aspects, considerable disagreement between two methods is expected to be observed, as is shown in the present work for F_2-isoprostanes (Figures 6 and 7) [72–74,79,88–91].

Most frequently, the Bland–Altman approach plots the absolute difference of two methods versus the average, as originally proposed by Bland and Altman [10]. Plotting the percentage difference against the average of two methods is also widely used. In the present work, two examples were presented which indicate that only one of the two Bland–Altman plots may reveal additional information about the agreement/disagreement between the methods which has not been reported by Bland and Altman. Observation of a linearity between the difference of two methods and their average is often interpreted as a systematic error and is even used to identify systematic errors [70,97,98]. In contrast, the lack of linearity between the percentage difference of two methods and their average may erroneously exclude the presence of a systematic error. It is therefore advisable to test this kind of linearity.

The analysis of the results observed in the present work (Table 1), revealed that the p value from the Wilcoxon test correlated inversely with the AUC value from the ROC analysis: $r = -0.662$, $p = 0.042$. The r^2 value from the linear regression analysis correlated inversely with the $\delta(\%)$ value of the Bland–Altman assay: $r = -0.699$, $p = 0.029$, and with the CV value from the Oldham–Eksborg test: $r = -0.780$, $p = 0.010$. The $\delta(\%)$ value from the Bland–Altman test correlated directly with the correlation of coefficient from the linear regression analysis of the Bland–Altman difference vs. the average concentration ρ^2: $r = 0.729$, $p = 0.021$, as well as with CV value from the Oldham–Eksborg test: $r = 0.669$, $p = 0.039$. Such correlations may suggest that many parameters rather than a single parameter of the three methods-comparison approaches may be useful in assessing agreement and to determine its extent.

Figure 11 shows the results separately for p (Wilcoxon or paired t test), β, r^2, ρ^2, Λ, AUC, $\delta(\%)$ and CV_{OE} taken from 10 examples listed in Table 1.

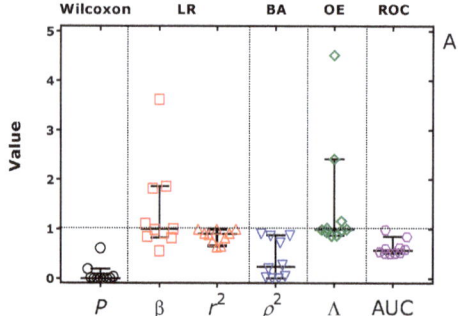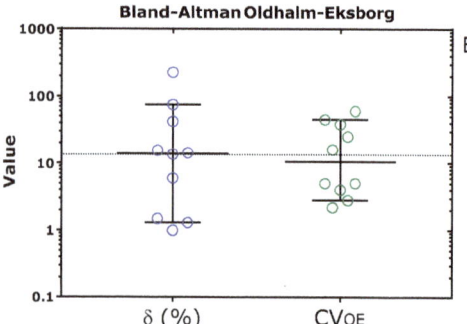

Figure 11. Presentation of values for selected statistical parameters of Table 1 obtained from the Wilcoxon or paired t test (p), the linear regression analysis (β and r^2), the coefficient of correlation ρ^2 from the linear regression analysis of the Bland–Altman difference δ vs. the average concentration of the analyte, the Oldham–Eksborg ratio Λ, the AUC value from the ROC analysis (**A**); and of the percentage difference of the Bland–Altman test δ (%), and the coefficient of variation of the Oldham–Eksborg ratio CV_{OE} (**B**). Data from 10 examples were used. Because of the greatly differing size of the values, the data are presented in two panels. Note the decadic logarithmic scale on the y axis in panel B. Data are shown as median with 95% confidence interval. LR, linear regression. See Table 1.

Figure 12 shows the results for the sum of the statistical parameters for each analyte in these examples. The analytes included ($n = 8$) were nitrate, ADMA, AEA, hArg, F_2-isoprostanes (Iso), SPB and PEFR. The numerically highest values were observed for δ (%) and CV_{OE} with respect to the statistical parameters, and for hArg (case hArg b) and the F_2-isoprostanes and CV_{OE} with respect to the analytes.

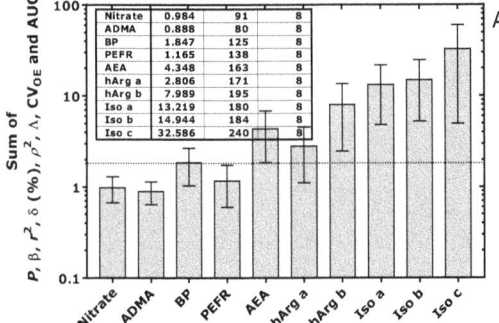
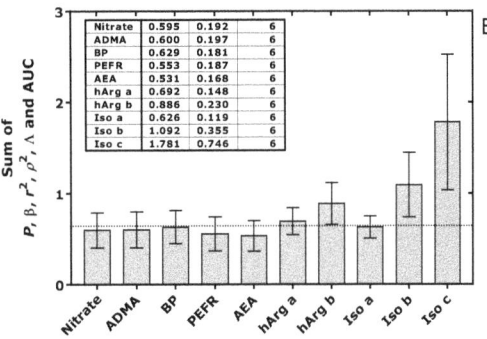

Figure 12. (**A**) Sum of values for selected statistical parameters taken from Table 1 obtained from the Wilcoxon or paired t test (p), the linear regression analysis (β and r^2), the coefficient of correlation ρ^2 from the linear regression analysis of the Bland–Altman difference δ vs. the average concentration of the analyte, the Oldham–Eksborg ratio Λ, the AUC value from the ROC analysis, the percentage difference of the Bland–Altman test $\delta(\%)$, and the coefficient of variation of the Oldham–Eksborg ratio CV_{OE}. (**B**) Like in (**A**), yet without $\delta(\%)$ and CV_{OE} in order to exclude high values. Data from 10 examples were used. Note the decadic logarithmic scale on the y axis in panel (**B**). Data are shown as median with 95% confidence interval. Iso, F_2 isoprostane. Insets indicate the tested parameters and their values. BP, systolic blood pressure; Iso, F_2-isoprostanes. See Table 1.

4. Proposal of Criteria for Method Agreement

The results of the present study suggest that reliable comparison of two analytical methods is best performed by using a combination of different statistical methods. Statistical difference between the concentrations of an analyte measured in a biological sample is not useful in assessing agreement. We can dispense with the use of the paired t test or Wilcoxon test. Linear regression analysis is useful, but we need not only the coefficient of correlation (r or r^2), but also the slope β of the regression line. The closer r and β to a value of 1.0, the higher the extent of agreement. However, linear regression does not reveal potentially important differences between the methods. This can be observed by the Bland–Altman approach, when the difference δ_{AB} is proportional to the mean μ_{AB}. The coefficient of correlation ρ^2 in the Bland–Altman approach is a useful parameter in assessing agreement. This is clearly visible in Figure 13A, notably in the measurement of hArg in plasma by GC-MS and GC-MS/MS. The Oldham–Eksborg approach provides values of the ratio Λ which correlate with β and r^2. Thus, the closer Λ to the value of 1.0, the higher the extent of the agreement. The bias, i.e., the percentage difference between the methods $\delta(\%)$ in the Bland–Altman approach correlates with the coefficient of variation CV_{OE} of the Oldham–Eksborg ratio Λ (Figure 13B).

The absolute value of the difference δ is less informative. It can therefore be concluded that β, r^2, ρ^2, Λ_{OE}, $\delta(\%)$ and CV_{OE} are useful in evaluating method agreements and in determining the extent of agreement between two analytical methods. Figure 13 suggests that good agreement between two methods exists when β, r^2, Λ_{OE} do not differ from 1.00 by 7 to 11%, and $\delta(\%)$ and CV_{OE} are below 12%.

In an analogy to currently available guidelines for chemical analytical methods with respect to method validation [52–57,99], acceptance criteria for agreement could be defined as ±15% from 1.00 for β, r^2, Λ_{OE} and ±15% for $\delta(\%)$.

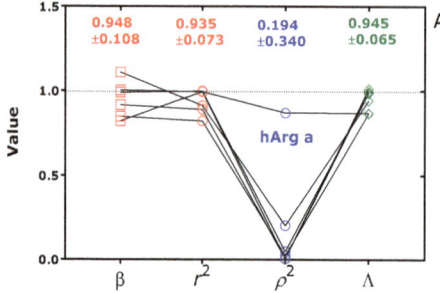
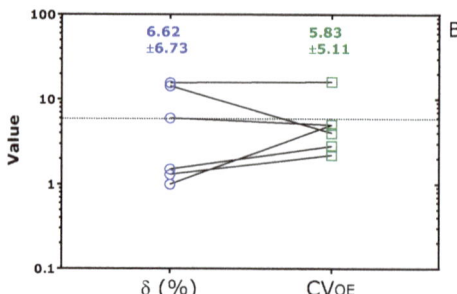

Figure 13. Presentation of values for selected statistical parameters obtained from the linear regression analysis (β and r^2), the coefficient of correlation ρ^2 from the linear regression analysis of the Bland–Altman difference δ vs. the average concentration of the analyte, and the Oldham–Eksborg ratio in (**A**); and for δ (%) and the CV_{OE} in (**B**). Lines combine symbols of the same example. Insets indicate the mean ± standard deviation of the statistical parameters from six examples (i.e., Nitrate, ADMA, AEA, hArg, BP, PEFR). Note the decadic logarithmic scale on the y axis in panel (**B**). See Figure 11.

Supplementary Materials: The following supporting information can be downloaded at: https://www.mdpi.com/article/10.3390/molecules28134905/s1. Figure S1. Number of yearly citations of the paper by (A) J.M. Bland and D.G. Altman [10] and by (B) M.M. Bradford [100] according to Scopus (Elsevier) from 1976 to 11 January 2023. Bland and Altman reported an approach on method comparison, which is widely known as the Bland–Altman plot [10]. Bradford reported in her paper a method for the measurement of protein concentration utilizing the principle of protein-dye binding [100]. The paper by M.M. Bradford is thematically not related to the present work but is suitable for a better understanding of the value of the paper by J.M. Bland and D.G. Altman in science. Scheme S1. Schematic of the principles of the mass spectrometry (MS) and tandem mass spectrometry (MS/MS) based on the quadrupole (Q) technology, exemplified for two structurally closely related analytes A and B which co-elute (same retention time, t_R) and ionize to form two isobaric ions (same mass-to-charge, m/z, ratio). (Upper left) Analytes A and B cannot be discriminated by single-stage quadrupole (SSQ) MS spectrometers. (Upper right, lower left) In the collision chamber (i.e., the second quadrupole Q2) of triple-stage quadrupole (TSQ) MS/MS spectrometers, collision induced dissociation (CID) of the precursor ions A and B with argon atoms produces several common and two distinctly different products ions (indicated by dotted arrows). (Lower, right) The third quadrupole Q3 of TSQ MS spectrometers separates the different product ions (set in dotted circles) formed in Q2. Thus, unlike SSQ MS spectrometers, TSQ MS/MS spectrometers can discriminate between analytes that co-elute and ionize in the ion source to form isobaric ions (same m/z). CID in Q2 and subsequent second mass separation in Q3 in TSQ instruments and related MS/MS instruments guarantee unique specificity. This feature makes MS/MS-based analytical methods the most qualified candidates to serve as reference methods, as the Gold Standard, for numerous analytes. See also Figure S2 and Scheme S2. Scheme S2. Schematic of the most frequently used modes in quantitative analyses of a target analyte A by using its stable-isotope-labelled analogue serving as the internal standard on quadrupole instruments. (A) Selected-ion monitoring (SIM) by mass spectrometry (MS) and (B) Selected-reaction monitoring (SRM) by tandem mass spectrometry (MS/MS). For more, details see the text. Figure S2. GC-MS (A,B) and GC-MS/MS (C,D) spectra of the pentafluorobenzyl (PFB) esters of 9-nitro-oleic acid (9-NO_2-OA) and 10-nitro-oleic acid (10-NO_2-OA). Electron-capture negative-ion chemical ionization (ECNICI) of the PFB esters of 9-NO_2-OA (A) and 10-NO_2-OA (B) leads to almost identical mass spectra, with the most intense ion being [M–PFB]$^-$ with m/z 326. In GC-MS/MS, the isobaric (m/z 326) parent (P) ions ([M–PFB]$^-$, [P]$^-$) are separated by Q1, subjected in Q2 to collision induced dissociation (CID), and the product ions formed in Q2 are separated by Q3. The product ion mass spectra of 9-NO_2-OA (C) and 10-NO_2-OA (D) are different. The product ions m/z 195 and m/z 197 are produced from 9-NO_2-OA, but not from 10-NO_2-OA. Thus, 9-NO_2-OA and 10-NO_2-OA can be discriminated by GC-MS/MS even if their PFB esters would co-elute. See also Scheme S1 and Ref. [101].

Funding: This research received no external funding.

Institutional Review Board Statement: This study did not use animal or human materials.

Informed Consent Statement: Not applicable.

Data Availability Statement: The study did not report any data.

Acknowledgments: The author is grateful to previous and current members of his group for their contributions over the last four decades. The author sincerely thanks Pedro Araujo (Norwegian Institute of Marine Research, Feed and Nutrition Group, N-5817 Bergen, Norway) and Stefanos A. Tsikas (Hannover Medical School, Academic Controlling, Hannover, Germany) for fruitful discussions.

Conflicts of Interest: The author declares no conflict of interest.

Sample Availability: Not available.

References

1. Deming, W.E. *Statistical Adjustment of Data*; John Wiley and Sons: New York, NY, USA, 1943; p. 184.
2. Oldham, P.D. *Measurement in Medicine: The Interpretation of Numerical Data*; English Universities Press: London, UK, 1968.
3. Westgard, J.O.; Hunt, M.R. Use and interpretation of common statistical tests in method-comparison studies. *Clin. Chem.* **1973**, *19*, 49–57. [CrossRef]
4. Wakkers, P.J.M.; Hellendoorn, H.B.A.; Op De Weegh, G.J.; Heerspink, W. Applications of statistics in clinical chemistry. A critical evaluation of regression lines. *Clin. Chim. Acta* **1975**, *64*, 173–184. [CrossRef]
5. Brace, R.A. Fitting straight lines to experimental data. *Am. J. Physiol.* **1977**, *233*, R94–R99. [CrossRef]
6. Cornbleet, P.J.; Gochman, N. Incorrect least-squares regression coefficients in method-comparison analysis. *Clin. Chem.* **1979**, *25*, 432–438. [CrossRef]
7. Smith, D.S.; Pourfarzaneh, M.; Kamel, R.S. Linear regression analysis by Deming's method. *Clin. Chem.* **1980**, *26*, 1105–1106. [CrossRef]
8. Eksborg, S. Evaluation of method-comparison data. *Clin. Chem.* **1981**, *27*, 1311–1312. [CrossRef] [PubMed]
9. Altman, D.G.; Bland, J.M. Measurement in Medicine: The analysis of method comparison studies. *Statistician* **1983**, *32*, 307–317. [CrossRef]
10. Bland, J.M.; Altman, D.G. Statistical methods for assessing agreement between two methods of clinical measurement. *Lancet* **1986**, *1*, 307–310. [CrossRef] [PubMed]
11. Linnet, K.; Bruunshuus, I. HPLC with enzymatic detection as a candidate reference method for serum creatinine. *Clin. Chem.* **1991**, *37*, 1669–1675. [CrossRef]
12. Pollock, M.A.; Jefferson, S.G.; Kane, J.W.; Lomax, K.; MacKinnon, G.; Winnard, C.B. Method comparison—A different approach. *Ann. Clin. Biochem.* **1992**, *29*, 556–560. [CrossRef]
13. Bland, J.M.; Altman, D.G. Comparing methods of measurement: Why plotting difference against standard method is misleading. *Lancet* **1995**, *346*, 1085–1087. [CrossRef] [PubMed]
14. National Committee for Clinical Laboratory Standards. *Method Comparison and Bias Estimation Using Patient Samples, Approved Guideline*; NCCLS publication EP9-A; NCCLS: Villanova, PA, USA, 1995.
15. Hollis, S. Analysis of method comparison studies. *Ann. Clin. Biochem.* **1996**, *33*, 1–4. [CrossRef] [PubMed]
16. Petersen, P.H.; Stöckl, D.; Blaabjerg, O.; Pedersen, B.; Birkemose, E.; Thienpont, L.; Lassen, J.F.; Kjeldsen, J. Graphical interpretation of analytical data from comparison of a field method with reference method by use of difference plots. *Clin. Chem.* **1997**, *43*, 2039–2046. [CrossRef] [PubMed]
17. Thienpont, L.M.; Van Nuwenborg, J.E.; Stöckl, D. Intrinsic and routine quality of serum total potassium measurement as investigated by split-sample measurement with an ion chromatography candidate reference method. *Clin. Chem.* **1998**, *44*, 849–857. [CrossRef]
18. Bland, J.M.; Altman, D.G. Measuring agreement in method comparison studies. *Stat. Methods Med. Res.* **1999**, *8*, 135–160. [CrossRef]
19. Mantha, S.; Roizen, M.F.; Fleisher, L.A.; Thisted, R.; Foss, J. Comparing methods of clinical measurement: Reporting standards for bland and altman analysis. *Anesth. Analg.* **2000**, *90*, 593–602. [CrossRef]
20. Dewitte, K.; Fierens, C.; Stöckl, D.; Thienpont, L.M. Application of the Bland-Altman plot for interpretation of method-comparison studies: A critical investigation of its practice. *Clin. Chem.* **2002**, *48*, 801–802. [CrossRef]
21. Altman, D.G.; Bland, J.M. Commentary on quantifying agreement between two methods of measurement. *Clin. Chem.* **2002**, *48*, 801–802. [CrossRef]
22. Lin, L.I.-K. A concordance correlation coefficient to evaluate reproducibility. *Biometrics* **1989**, *45*, 255–268. [CrossRef]
23. Eastwood, B.J.; Farmen, M.W.; Iversen, P.W.; Craft, T.J.; Smallwood, J.K.; Garbison, K.E.; Delapp, N.W.; Smith, G.F. The minimum significant ratio: A statistical parameter to characterize the reproducibility of potency estimates from concentration-response assays and estimation by replicate-experiment studies. *J. Biomol. Screen.* **2006**, *11*, 253–261. [CrossRef]
24. Dewé, W. Review of statistical methodologies used to compare (bio)assays. *J. Chromatogr. B* **2009**, *877*, 2208–2213. [CrossRef]
25. Ludbrook, J. Confidence in Altman-Bland plots: A critical review of the method of differences. *Clin. Exp. Pharmacol. Physiol.* **2010**, *37*, 143–149. [CrossRef]

26. Woodman, R.J. Bland-Altman beyond the basics: Creating confidence with badly behaved data. *Clin. Exp. Pharmacol. Physiol.* **2020**, *37*, 141–142. [CrossRef]
27. Giavarina, D. Understanding Bland Altman analysis. *Biochem. Med.* **2015**, *25*, 141–151. [CrossRef]
28. Carkeet, A. Exact parametric confidence intervals for Bland-Altman limits of agreement. *Optom. Vis. Sci.* **2015**, *92*, e71–e80. [CrossRef]
29. Francq, B.G.; Govaerts, B. How to regress and predict in a Bland-Altman plot? Review and contribution based on tolerance intervals and correlated-errors-in-variables models. *Stat. Med.* **2016**, *35*, 2328–2358. [CrossRef]
30. Hofman, C.S.; Melis, R.J.; Donders, A.R. Adapted Bland-Altman method was used to compare measurement methods with unequal observations per case. *J. Clin. Epidemiol.* **2015**, *68*, 939–943. [CrossRef] [PubMed]
31. Lu, M.J.; Zhong, W.H.; Liu, Y.X.; Miao, H.Z.; Li, Y.C.; Ji, M.H. Sample size for assessing agreement between two methods of measurement by Bland-Altman method. *Int. J. Biostat.* **2016**, *12*, 20150039. [CrossRef] [PubMed]
32. Tipton, E.; Shuster, J. A framework for the meta-analysis of Bland-Altman studies based on a limits of agreement approach. *Stat. Med.* **2017**, *36*, 3621–3635. [CrossRef] [PubMed]
33. Carkeet, A.; Goh, Y.T. Confidence and coverage for Bland-Altman limits of agreement and their approximate confidence intervals. *Stat. Methods Med. Res.* **2018**, *27*, 1559–1574. [CrossRef] [PubMed]
34. Misyura, M.; Sukhai, M.A.; Kulasignam, V.; Zhang, T.; Kamel-Reid, S.; Stockley, T.L. Improving validation methods for molecular diagnostics: Application of Bland-Altman, Deming and simple linear regression analyses in assay comparison and evaluation for next-generation sequencing. *J. Clin. Pathol.* **2018**, *71*, 117–124. [CrossRef]
35. Sadler, W. ANNALS EXPRESS: Using the variance function to generalise Bland-Altman analysis. *Ann. Clin. Biochem.* **2018**, *56*, 198–203. [CrossRef] [PubMed]
36. Jan, S.L.; Shieh, G. The Bland-Altman range of agreement: Exact interval procedure and sample size determination. *Comput. Biol. Med.* **2018**, *100*, 247–252. [CrossRef] [PubMed]
37. Hill, R.E.; Whelan, D.T. Mass spectrometry and clinical chemistry. *Clin. Chim. Acta* **1984**, *139*, 231–294. [CrossRef]
38. Lawson, A.M.; Gaskell, S.J.; Hjelm, M. International Federation of Clinical Chemistry (IFCC), Office for Reference Methods and Materials (ORMM). Methodological aspects on quantitative mass spectrometry used for accuracy control in clinical chemistry. *J. Clin. Chem. Clin. Biochem.* **1985**, *23*, 433–441.
39. Tsikas, D. Application of gas chromatography-mass spectrometry and gas chromatography-tandem mass spectrometry to assess in vivo synthesis of prostaglandins, thromboxane, leukotrienes, isoprostanes and related compounds in humans. *J. Chromatogr. B* **1988**, *717*, 201–245. [CrossRef]
40. Wilcken, B.; Wiley, V.; Hammond, J.; Carpenter, K. Screening newborns for inborn errors of metabolism by tandem mass spectrometry. *N. Engl. J. Med.* **2003**, *348*, 2304–2312. [CrossRef]
41. Kuksis, A.; Myher, J.J. Application of tandem mass spectrometry for the analysis of long-chain carboxylic acids. *J. Chromatogr. B* **1995**, *671*, 35–70. [CrossRef]
42. Oehme, M. *Praktische Einführung in die GC-MS-Analytik mit Quadrupolen*; Hüthig: Heidelberg, Germany, 1996.
43. Busch, K.L.; Glish, G.L.; McLuckey, S.A. *Mass Spectrometry/Mass Spectrometry*; VCH Publishers: New York, NY, USA; Weinheim, Germany, 1998.
44. Gross, J.H. *Mass Spectrometry*; Springer: Berlin/Heidelberg, Germany, 2004.
45. Watson, J.T.; Sparkman, O.D. *Introduction in Mass Spectrometry*, 4th ed.; Wiley: Chichester, UK, 2007.
46. Jennings, K.R. (Ed.) *A History of European Mass Spectrometry*; IM Publications LLP: Chichester, UK, 2012.
47. Tsikas, D.; Duncan, M.W. Mass spectrometry and 3-nitrotyrosine: Strategies, controversies, and our current perspective. *Mass Spectrom. Rev.* **2014**, *33*, 237–276. [CrossRef]
48. Duncan, M.W. Good mass spectrometry and its place in good science. *J. Mass Spectrom.* **2012**, *47*, 795–809. [CrossRef] [PubMed]
49. Araujo, P. Key aspects of analytical method validation and linearity evaluation. *J. Chromatogr. B* **2009**, *877*, 2224–2234. [CrossRef]
50. Shah, V.P.; Midha, K.K.; Dighe, S.; McGilveray, I.J.; Skelly, J.P.; Yacobi, A.; Layloff, T.; Viswanathan, C.T.; Cook, C.E.; Mcdowall, R.D.; et al. Analytical methods validation: Bioavailability, bioequivalence, and pharmacokinetic studies. *Pharmac. Sci.* **1992**, *81*, 309. [CrossRef]
51. Bansal, S.; DeStefano, S.A. Key elements of bioanalytical method validation for small molecules. *AAPS J.* **2007**, *9*, E109–E114. [CrossRef] [PubMed]
52. Kaza, M.; Karaźniewicz-Łada, M.; Kosicka, K.; Siemiątkowska, A.; Rudzki, P.J. Bioanalytical method validation: New FDA guidance vs. EMA guideline. Better or worse? *J. Pharm. Biomed. Anal.* **2019**, *165*, 381–385. [CrossRef]
53. Lindner, W.; Wainer, I.W. Requirements for initial assay validation and publication in J. Chromatography B. *J. Chromatogr. B* **1998**, *707*, 1–2.
54. Bischoff, R.; Hopfgartner, G.; Karnes, H.T.; Lloyd, D.; Phillips, T.M.; Tsikas, D.; Xu, G. Summary of a recent workshop/conference report on validation and implementation of bioanalytical methods: Implications on manuscript review in the Journal of Chromatography B. *J. Chromatogr. B* **2007**, *860*, 1–3. [CrossRef] [PubMed]
55. Booth, B.; Stevenson, L.; Pillutla, R.; Buonarati, M.; Beaver, C.; Fraier, D.; Garofolo, F.; Haidar, S.; Islam, R.; James, C.; et al. 2019 White Paper On Recent Issues in Bioanalysis: FDA BMV Guidance, ICH M10 BMV Guideline and Regulatory Inputs (Part 2—Recommendations on 2018 FDA BMV Guidance, 2019 ICH M10 BMV Draft Guideline and Regulatory Agencies' Input on Bioanalysis, Biomarkers and Immunogenicity). *Bioanalysis* **2019**, *11*, 2099–2132. [CrossRef]

56. Tsikas, D. Bioanalytical method validation of endogenous substances according to guidelines by the FDA and other organizations: Basic need to specify concentration ranges. *J. Chromatogr. B Analyt. Technol. Biomed. Life Sci.* **2018**, *1093–1094*, 80–81. [CrossRef]
57. Tsikas, D. A proposal for comparing methods of quantitative analysis of endogenous compounds in biological systems by using the relative lower limit of quantification (rLLOQ). *Chromatogr. B Analyt. Technol. Biomed. Life Sci.* **2009**, *877*, 2244–2251. [CrossRef]
58. Tsikas, D.; Gutzki, F.M.; Sandmann, J.; Schwedhelm, E.; Frölich, J.C. Gas chromatographic-tandem mass spectrometric quantification of human plasma and urinary nitrate after its reduction to nitrite and derivatization to the pentafluorobenzyl derivative. *J. Chromatogr. B Biomed. Sci. Appl.* **1999**, *731*, 285–291. [CrossRef]
59. Tsikas, D. Mass spectrometry-validated HPLC method for urinary nitrate. *Clin. Chem.* **2004**, *50*, 1259–1261. [CrossRef] [PubMed]
60. Tsikas, D. Simultaneous derivatization and quantification of the nitric oxide metabolites nitrite and nitrate in biological fluids by gas chromatography/mass spectrometry. *Anal. Chem.* **2000**, *72*, 4064–4072. [CrossRef]
61. Becker, A.J.; Ückert, S.; Tsikas, D.; Noack, H.; Stief, C.G.; Frölich, J.C.; Wolf, G.; Jonas, U. Determination of nitric oxide metabolites by means of the Griess assay and gas chromatography-mass spectrometry in the cavernous and systemic blood of healthy males and patients with erectile dysfunction during different functional conditions of the penis. *Urol. Res.* **2000**, *28*, 364–369. [CrossRef] [PubMed]
62. Sitar, M.E.; Kayacelebi, A.A.; Beckmann, B.; Kielstein, J.T.; Tsikas, D. Asymmetric dimethylarginine (ADMA) in human blood: Effects of extended haemodialysis in the critically ill patient with acute kidney injury, protein binding to human serum albumin and proteolysis by thermolysin. *Amino Acids* **2015**, *47*, 1983–1993. [CrossRef] [PubMed]
63. Tsikas, D.; Schubert, B.; Gutzki, F.M.; Sandmann, J.; Frölich, J.C. Quantitative determination of circulating and urinary asymmetric dimethylarginine (ADMA) in humans by gas chromatography-tandem mass spectrometry as methyl ester tri(N-pentafluoropropionyl) derivative. *J. Chromatogr. B* **2003**, *798*, 87–99. [CrossRef]
64. Schulze, F.; Wesemann, R.; Schwedhelm, E.; Sydow, K.; Albsmeier, J.; Cooke, J.P.; Böger, R.H. Determination of asymmetric dimethylarginine (ADMA) using a novel ELISA assay. *Clin. Chem. Lab. Med.* **2004**, *42*, 1377–1383. [CrossRef]
65. Martens-Lobenhoffer, J.; Westphal, S.; Awiszus, F.; Bode-Böger, S.M.; Luley, C. Determination of asymmetric dimethylarginine: Liquid chromatography-mass spectrometry or ELISA? *Clin. Chem.* **2005**, *51*, 2188–2189. [CrossRef]
66. Valtonen, P.; Karppi, J.; Nyyssönen, K.; Vakonen, V.P.; Halonen, T.; Punnonen, K. Comparison of HPLC method and commercial ELISA assay for asymmetric dimethylarginine (ADMA) determination in human serum. *J. Chromatogr. B* **2005**, *828*, 97–102. [CrossRef]
67. Široká, R.; Trefil, L.; Rajdl, D.; Racek, J.; Cibulka, R. Asymmetric dimethylarginine-comparison of HPLC and ELISA methods. *J. Chromatogr. B* **2007**, *850*, 586–587. [CrossRef]
68. Zoerner, A.A.; Gutzki, F.M.; Suchy, M.T.; Beckmann, B.; Engeli, S.; Jordan, J.; Tsikas, D. Targeted stable-isotope dilution GC-MS/MS analysis of the endocannabinoid anandamide and other fatty acid ethanol amides in human plasma. *J. Chromatogr. B Analyt. Technol. Biomed. Life Sci.* **2009**, *877*, 2909–2923. [CrossRef]
69. Zoerner, A.A.; Batkai, S.; Suchy, M.T.; Gutzki, F.M.; Engeli, S.; Jordan, J.; Tsikas, D. Simultaneous UPLC-MS/MS quantification of the endocannabinoids 2-arachidonoyl glycerol (2AG), 1-arachidonoyl glycerol (1AG), and anandamide in human plasma: Minimization of matrix-effects, 2AG/1AG isomerization and degradation by toluene solvent extraction. *J. Chromatogr. B Analyt. Technol. Biomed. Life Sci.* **2012**, *883–884*, 161–171. [CrossRef]
70. Kayacelebi, A.A.; Beckmann, B.; Gutzki, F.M.; Jordan, J.; Tsikas, D. GC-MS and GC-MS/MS measurement of the cardiovascular risk factor homoarginine in biological samples. *Amino Acids* **2014**, *46*, 2205–2217. [CrossRef] [PubMed]
71. Cordts, K.; Atzler, D.; Qaderi, V.; Sydow, K.; Böger, R.H.; Choe, C.U.; Schwedhelm, E. Measurement of homoarginine in human and mouse plasma by LC-MS/MS and ELISA: A comparison and a biological application. *Amino Acids* **2015**, *47*, 2015–2022. [CrossRef]
72. Morrow, J.D.; Roberts, J.L. Mass spectrometric quantification of F2-isoprostanes in biological fluids and tissues as measure of oxidant stress. *Methods Enzymol.* **1998**, *300*, 3–12.
73. Devaraj, S.; Hirany, S.V.; Burk, R.F.; Jialal, I. Divergence between LDL oxidative susceptibility and urinary F(2)-isoprostanes as measures of oxidative stress in type 2 diabetes. *Clin. Chem.* **2001**, *47*, 1974–1979. [CrossRef]
74. Yan, W.; Byrd, G.D.; Ogden, M.W. Quantitation of isoprostane isomers in human urine from smokers and nonsmokers by LC-MS/MS. *J. Lip. Res.* **2007**, *48*, 1607–1617. [CrossRef] [PubMed]
75. Ludbrook, J. Comparing methods of measurements. *Clin. Exp. Pharmacol. Physiol.* **1997**, *24*, 193–203. [CrossRef]
76. Daniel, W.W. *W.W. Biostatistics: A Foundation for Analysis in the Health Sciences*, 2nd ed.; John Wiley & Sons: New York, NY, USA, 1978.
77. Erpenbeck, V.; Jorres, R.A.; Discher, M.; Krentel, K.; Tsikas, D.; Luettig, B.; Krug, N.; Hohlfeld, J.M. Local nitric oxide levels reflect the degree of allergic airway inflammation after segmental allergen challenge in asthmatics. *Nitric Oxide* **2005**, *13*, 125–133. [CrossRef]
78. Tsikas, D.; Schwedhelm, E.; Stutzer, F.K.; Gutzki, F.M.; Rode, I.; Mehls, C.; Frölich, J.C. Accurate quantification of basal plasma levels of 3-nitrotyrosine and 3-nitrotyrosinoalbumin by gas chromatography-tandem mass spectrometry. *J. Chromatogr. B* **2003**, *784*, 77–90. [CrossRef]
79. Tsikas, D.; Schwedhelm, E.; Suchy, M.T.; Niemann, J.; Gutzki, F.M.; Erpenbeck, V.J.; Hohlfeld, J.M.; Surdacki, A.; Frölich, J.C. Divergence in urinary 8-iso-PGF(2alpha) (iPF(2alpha)-III, 15-F(2t)-IsoP) levels from gas chromatography-tandem mass spectrometry quantification after thin-layer chromatography and immunoaffinity column chromatography reveals heterogeneity of 8-iso-PGF(2alpha). Possible methodological, mechanistic and clinical implications. *J. Chromatogr. B* **2003**, *794*, 237–255.

80. Tsikas, D.; Wolf, A.; Frölich, J.C. Simplified HPLC method for urinary and circulating creatinine. *Clin. Chem.* **2004**, *50*, 201–203. [CrossRef]
81. Tsikas, D. A critical review and discussion of analytical methods in the L-arginine/nitric oxide area of basic and clinical research. *Anal. Biochem.* **2008**, *379*, 139–163. [CrossRef]
82. Teerlink, T. HPLC analysis of ADMA and other methylated L-arginine analogs in biological fluids. *J. Chromatogr. B* **2007**, *851*, 21–29. [CrossRef] [PubMed]
83. Horowitz, J.D.; Heresztyn, T. An overview of plasma concentrations of asymmetric dimethylarginine (ADMA) in health and disease and in clinical studies: Methodological considerations. *J. Chromatogr. B* **2007**, *851*, 42–50. [CrossRef] [PubMed]
84. Martens-Lobenhoffer, J.; Bode-Böger, S.M. Chromatographic-mass spectrometric methods for the quantification of L-arginine and its methylated metabolites in biological fluids. *J. Chromatogr. B* **2007**, *851*, 30–41. [CrossRef]
85. Zoerner, A.A.; Gutzki, F.M.; Batkai, S.; May, M.; Rakers, C.; Engeli, S.; Jordan, J.; Tsikas, D. Quantification of endocannabinoids in biological systems by chromatography and mass spectrometry: A comprehensive review from an analytical and biological perspective. *Biochim. Biophys. Acta* **2011**, *1811*, 706–723. [CrossRef]
86. Tsikas, D.; Zoerner, A.A. Analysis of eicosanoids by LC-MS/MS and GC-MS/MS: A historical retrospect and a discussion. *J. Chromatogr. B Analyt. Technol. Biomed. Life Sci.* **2014**, *964*, 79–88. [CrossRef] [PubMed]
87. Tsikas, D. Homoarginine in health and disease. *Curr. Opin. Clin. Nutr. Metab. Care* **2023**, *26*, 42–49. [CrossRef]
88. Schwedhelm, E.; Benndorf, R.A.; Böger, R.H.; Tsikas, D. Mass spectrometric analysis of F2-isoprostanes: Markers and mediators in human disease. *Curr. Pharm. Anal.* **2007**, *3*, 39–51. [CrossRef]
89. Vassalle, C.; Andreassi, M.G. 8-Iso-Prostaglandin F2 as a Risk Marker in Patients With Coronary Heart Disease. *Circulation* **2004**, *109*, e49–e50.
90. Sircar, D.; Subbaiah, P.V. Isoprostane measurement in plasma and urine by liquid chromatography-mass spectrometry with one-step sample preparation. *Clin. Chem.* **2007**, *53*, 251–258. [CrossRef]
91. Schwedhelm, E.; Tsikas, D.; Durand, T.; Gutzki, F.M.; Guy, A.; Rossi, J.-C.; Frölich, J.C. Tandem mass spectrometric quantification of 8-iso-prostaglandin F2alpha and its metabolite 2,3-dinor-5,6-dihydro-8-iso-prostaglandin F2alpha in human urine. *J. Chromatogr. B* **2000**, *744*, 99–112. [CrossRef]
92. Tsikas, D. Analysis of nitrite and nitrate in biological fluids by assays based on the Griess reaction: Appraisal of the Griess reaction in the L-arginine/nitric oxide area of research. *J. Chromatogr. B Analyt. Technol. Biomed. Life Sci.* **2007**, *851*, 51–70. [CrossRef] [PubMed]
93. Tsikas, D. Assessment of lipid peroxidation by measuring malondialdehyde (MDA) and relatives in biological samples: Analytical and biological challenges. *Anal. Biochem.* **2017**, *524*, 13–30. [CrossRef]
94. Malaei, R.; Ramezani, A.M.; Absalan, G. Analysis of malondialdehyde in human plasma samples through derivatization with 2,4-dinitrophenylhydrazine by ultrasound-assisted dispersive liquid-liquid microextraction-GC-FID approach. *J. Chromatogr. B* **2018**, *1089*, 60–69. [CrossRef]
95. Carkeet, A. A Review of the Use of Confidence Intervals for Bland-Altman Limits of Agreement in Optometry and Vision Science. *Optom. Vis. Sci.* **2020**, *97*, 3–8. [CrossRef] [PubMed]
96. Schwedhelm, E. Quantification of ADMA: Analytical approaches. *Vasc. Med.* **2005**, *10* (Suppl. 1), S89–S95. [CrossRef]
97. Tsikas, D.; Wolf, A.; Mitschke, A.; Gutzki, F.M.; Will, W.; Bader, M. GC-MS determination of creatinine in human biological fluids as pentafluorobenzyl derivative in clinical studies and biomonitoring: Inter-laboratory comparison in urine with Jaffé, HPLC and enzymatic assays. *J. Chromatogr. B Analyt. Technol. Biomed. Life Sci.* **2010**, *878*, 2582–2592. [CrossRef] [PubMed]
98. Pannekeet, M.M.; Imholz, A.L.; Struijk, D.G.; Koomen, G.C.; Langedijk, M.J.; Schouten, N.; de Waart, R.; Hiralall, J.; Krediet, R.T. The standard peritoneal permeability analysis: A tool for the assessment of peritoneal permeability characteristics in CAPD patients. *Kidney Int.* **1995**, *48*, 866–875. [CrossRef]
99. Bioanalytical Method Validation—Scientific Guideline. Available online: https://www.ema.europa.eu/en/bioanalytical-method-validation-scientific-guideline (accessed on 1 March 2023).
100. Bradford, M.M. A rapid and sensitive method for the quantitation of microgram quantities of protein utilizing the principle of protein-dye binding. *Anal. Biochem.* **1976**, *72*, 248–254. [CrossRef]
101. Tsikas, D.; Zoerner, A.; Mitschke, A.; Homsi, Y.; Gutzki, F.M.; Jordan, J. Specific GC-MS/MS stable-isotope dilution methodology for free 9- and 10-nitro-oleic acid in human plasma challenges previous LC-MS/MS reports. *J. Chromatogr. B Analyt. Technol. Biomed. Life Sci.* **2009**, *877*, 2895–2908. [CrossRef] [PubMed]

Disclaimer/Publisher's Note: The statements, opinions and data contained in all publications are solely those of the individual author(s) and contributor(s) and not of MDPI and/or the editor(s). MDPI and/or the editor(s) disclaim responsibility for any injury to people or property resulting from any ideas, methods, instructions or products referred to in the content.

Article

GC-MS Studies on Nitric Oxide Autoxidation and *S*-Nitrosothiol Hydrolysis to Nitrite in pH-Neutral Aqueous Buffers: Definite Results Using ^{15}N and ^{18}O Isotopes

Dimitrios Tsikas

Core Unit Proteomics, Institute of Toxicology, Hannover Medical School, 30625 Hannover, Germany; tsikas.dimitros@mh-hannover.de

Abstract: Nitrite (O=N-O$^-$, NO$_2{}^-$) and nitrate (O=N(O)-O$^-$, NO$_3{}^-$) are ubiquitous in nature. In aerated aqueous solutions, nitrite is considered the major autoxidation product of nitric oxide (•NO). •NO is an environmental gas but is also endogenously produced from the amino acid L-arginine by the catalytic action of •NO synthases. It is considered that the autoxidation of •NO in aqueous solutions and in O$_2$-containing gas phase proceeds via different neutral (e.g., O=N-O-N=O) and radical (e.g., ONOO•) intermediates. In aqueous buffers, endogenous *S*-nitrosothiols (thionitrites, RSNO) from thiols (RSH) such as L-cysteine (i.e., *S*-nitroso-L-cysteine, CysSNO) and cysteine-containing peptides such as glutathione (GSH) (i.e., *S*-nitrosoglutathione, GSNO) may be formed during the autoxidation of •NO in the presence of thiols and dioxygen (e.g., GSH + O=N-O-N=O → GSNO + O=N-O$^-$ + H$^+$; p$K_a{}^{HONO}$, 3.24). The reaction products of thionitrites in aerated aqueous solutions may be different from those of •NO. This work describes in vitro GC-MS studies on the reactions of unlabeled (^{14}NO$_2{}^-$) and labeled nitrite (^{15}NO$_2{}^-$) and RSNO (RS^{15}NO, RS^{15}N^{18}O) performed in pH-neutral aqueous buffers of phosphate or *tris*(hydroxyethylamine) prepared in unlabeled (H$_2{}^{16}$O) or labeled H$_2$O (H$_2{}^{18}$O). Unlabeled and stable-isotope-labeled nitrite and nitrate species were measured by gas chromatography–mass spectrometry (GC-MS) after derivatization with pentafluorobenzyl bromide and negative-ion chemical ionization. The study provides strong indication for the formation of O=N-O-N=O as an intermediate of •NO autoxidation in pH-neutral aqueous buffers. In high molar excess, HgCl$_2$ accelerates and increases RSNO hydrolysis to nitrite, thereby incorporating ^{18}O from H$_2{}^{18}$O into the SNO group. In aqueous buffers prepared in H$_2{}^{18}$O, synthetic peroxynitrite (ONOO$^-$) decomposes to nitrite without ^{18}O incorporation, indicating water-independent decomposition of peroxynitrite to nitrite. Use of RS^{15}NO and H$_2{}^{18}$O in combination with GC-MS allows generation of definite results and elucidation of reaction mechanisms of oxidation of •NO and hydrolysis of RSNO.

Keywords: autoxidation; derivatization; GC-MS; nitric oxide; pentafluorobenzyl bromide; stable isotopes

Citation: Tsikas, D. GC-MS Studies on Nitric Oxide Autoxidation and *S*-Nitrosothiol Hydrolysis to Nitrite in pH-Neutral Aqueous Buffers: Definite Results Using ^{15}N and ^{18}O Isotopes. *Molecules* **2023**, *28*, 4281. https://doi.org/10.3390/molecules28114281

Academic Editors: Paraskevas D. Tzanavaras and Victoria Samanidou

Received: 27 April 2023
Revised: 19 May 2023
Accepted: 20 May 2023
Published: 23 May 2023

Copyright: © 2023 by the author. Licensee MDPI, Basel, Switzerland. This article is an open access article distributed under the terms and conditions of the Creative Commons Attribution (CC BY) license (https://creativecommons.org/licenses/by/4.0/).

1. Introduction

Nitric oxide (•NO) is an environmental gas originating from many sources including combustion and thunderstorms. In living organisms, nitric oxide synthases (NOSs) are expressed virtually in all types of cell and convert L-arginine to L-citrulline and •NO using molecular oxygen (•O•$_2$) as the second substrate and many cofactors [1]. •NO produced in cells such as endothelial cells needs to reach the soluble guanylyl cyclase in other cells such as the smooth muscle cells or platelets in order to exert biological effects. •NO is a potent vasodilator and inhibitor of platelet aggregation and functions as a neurotransmitter [1]. •NO may react with numerous intra- and extra-cellular biomolecules. Autoxidation of •NO, i.e., its reaction with •O•$_2$, occurs immediately at the site of its generation, and this decreases the concentration of •NO. NOS and many other enzymes generate reactive oxygen species (ROS) such as the superoxide anion (O$_2$•$^-$) and hydrogen peroxide (H$_2$O$_2$).

$O_2^{\bullet-}$ and H_2O_2 can react with $^{\bullet}NO$ before it can leave the cell. These reactions do not only decrease the concentration of $^{\bullet}NO$, but they moreover produce reactive nitrogen species (RNS) such as peroxynitrite (ONOO$^-$). Peroxynitrite is a strong oxidant and reacts with sulfhydryl (SH) groups in numerous low- and high-molecular-mass biomolecules.

Prior to the recognition of $^{\bullet}NO$ as an endogenous biomolecule, the oxidation of $^{\bullet}NO$ has been investigated in the gaseous (g) phase. The gas phase autoxidation of $^{\bullet}NO$ has the stoichiometry shown by Reaction (1) and Rate Law (2). Upon the recognition of $^{\bullet}NO$ as an endogenous signaling molecule about 35 years ago, the autoxidation of $^{\bullet}NO$ has been investigated in aqueous (aq) solutions [2–9]. The stoichiometry of the $^{\bullet}NO$ autoxidation in aqueous phases is given by Reaction (3) and its rate law by Expression (4) with $4k_{aq} = 9 \times 10^6$ M^{-2}s^{-1} at 25 °C [3]. Despite similar kinetics of the autoxidation of $^{\bullet}NO$ in the gas phase and in aqueous solutions, the reaction products differ: in aqueous solutions, nitrite is the sole autoxidation product of $^{\bullet}NO$, whereas the reaction product formed in the gas phase is most likely $^{\bullet}NO_2$, which disproportionates upon dilution in aqueous solutions to nitrite and nitrate (5) [3]. In the presence of thiols, such as glutathione (GSH), in aqueous buffered solutions of $^{\bullet}NO$ and $^{\bullet}O^{\bullet}_2$, additional reaction products are formed. They include S-nitrosothiols or thionitrites (RSNOs) such as S-nitrosoglutathione (GSNO) and disulfides such as GSSG [10]. In the absence of $^{\bullet}O^{\bullet}_2$, neither GSNO nor GSSG formation has been observed. It has been hypothesized that not $^{\bullet}NO$ itself, but a $^{\bullet}NO$-derived nitrosating intermediate, is formed, which reacts with GSH to form GSNO. This species has been proposed to be nitrous anhydride (N_2O_3) [10] (6), yet the structure of N_2O_3 has not been identified thus far [7], and the mechanisms of its formation in aqueous solutions are elusive. For N_2O_3, four isomeric structures have been suggested, including O=N-O-N=O and O=N-N$^+$(=O)O$^-$ [11]. Further proposed intermediates occurring during the autoxidation of $^{\bullet}NO$ include O=N-O-O$^{\bullet}$ and O=N-O-O-N=O [6,8,9]. Experiments performed at very low temperatures in non-aqueous systems, such as in glass-like matrices of 2-methylbutane, suggested formation of yellow-colored (at 90 K, O=N-O-O$^{\bullet}$/$^{\bullet}$N=O and/or O=N-O-O$^{\bullet}$/O=N-O-O$^{\bullet}$) and red-colored (at 110 K, O=N-O-O-N=O or O=N-O-N(=O)O) intermediates [8,9]. Such species have not been detected in aqueous buffered solutions to date.

In the laboratory, RSNOs are prepared in aqueous solutions by mixing stoichiometric amounts of RSH and nitrite salts and by acidifying them with diluted acids such as HCl acid (Scheme 1) (7). Treatment of aqueous solutions of RSNO with a molar excess of an aqueous solution of HgCl$_2$ leads to formation of nitrite (Scheme 1) (8). Both reactions are performed at room temperature. In cases of labile RSNO such as CysSNO, synthesis is preferably performed in an ice-bath (Scheme 1).

Generally, RSNOs are considered to be $^{\bullet}NO$ donors, yet this does not apply to every thionitrite. Thus, S-nitroso-L-cysteine (CysSNO) is an abundant "spontaneous" $^{\bullet}NO$ donor, whereas GSNO is not a $^{\bullet}NO$ donor. In phosphate buffer of neutral pH, as much as 50% of CysSNO may release $^{\bullet}NO$ [12], as can be specifically measured by NO-specific electrodes. The underlying mechanisms of $^{\bullet}NO$ release by RSNO are still incompletely resolved. Redox-active metal ions, most notably Cu^{2+}/Cu^{1+}, are extremely potent catalysts of the release of $^{\bullet}NO$ from CysSNO (9). Cu^{2+}/Cu^{1+} are required in catalytic amounts and can be produced by small amounts of CysSH (10). It can, therefore, be assumed that the reaction products of S-nitrosothiols and possibly their intermediates in aqueous solutions may be different from those formed during the autoxidation of authentic $^{\bullet}NO$.

$$2\ ^{\bullet}N^{(II)}{=}O_{(g)} + {}^{\bullet}O^{(\pm 0)\bullet}{}_2 \to 2\ ^{\bullet}N^{(+IV)}O_{2(g)} \tag{1}$$

$$-d[^{\bullet}NO]/dt = 2k_g \times [^{\bullet}NO]^2 \times [O_2] \tag{2}$$

$$4\ ^{\bullet}N^{(II)}{=}O_{(aq)} + {}^{\bullet}O^{(\pm 0)\bullet}{}_2 + 2\ H_2O \to 4\ O{=}N^{(III)}{-}O^- + 4\ H^+ \tag{3}$$

$$-d[^\bullet NO]/dt = 4k_{aq} \times [^\bullet NO]^2 \times [O_2] \qquad (4)$$

$$2\ ^\bullet N^{(II)}O_{2(aq)} + H_2O \rightarrow O=N^{(III)}\text{-}O^- + [O=N^{(V)}(\text{-}O)\text{-}O]^- + 2\ H^+ \qquad (5)$$

$$GSH + N_2O_3 \rightarrow GSNO + O=N\text{-}O^- + H^+ \qquad (6)$$

$$RSH + O=N\text{-}O^- + H^+ \rightarrow RSNO + H_2O \qquad (7)$$

$$2\ RSNO + 2\ H_2O + HgCl_{2(aq)}^+ \rightarrow Hg(RS)_2 + 2\ O=N\text{-}O^- + 2\ Cl^- + 4\ H^+ \qquad (8)$$

$$CysSN^{(III)}O + Cu^{1+} + H_2O^+ \rightarrow\ ^\bullet N^{(II)}O + CysSH + Cu^{2+} + OH \qquad (9)$$

$$CysS^{(-II)}H + Cu^{2+} \longleftrightarrow CysS^{(-I)\bullet} + Cu^{1+} + H^+ \qquad (10)$$

In the present work, we investigated the reactions of L-cysteine-based RSNO and nitrite in aqueous buffers of neutral pH value by gas chromatography–mass spectrometry (GC-MS) in combination with the use of stable isotopes of O (natural abundance, 0.2% ^{18}O) and N (natural abundance, 0.37% ^{15}N) in RSNO, nitrite, and water (Scheme 1). The main analytes were unlabeled nitrite ([^{14}N]nitrite), nitrite labeled with ^{15}N ([^{15}N]nitrite), and nitrite labeled with ^{15}N and ^{18}O (i.e., [^{15}N, ^{18}O]nitrite) (Scheme 2). The study provides strong indication for the formation of O=N-O-N=O as an intermediate during the autoxidation of $^\bullet NO$ derived from CysSNO in aqueous buffers of neutral pH value. In $H_2{}^{18}O$, $^\bullet NO$ autoxidizes to ^{18}O-nitrite and ^{16}O-nitrite. In aqueous solutions, $HgCl_2$ mediates the hydrolysis of the SNO groups of CysSNO and GSNO to ^{18}O-nitrite and ^{16}O-nitrite.

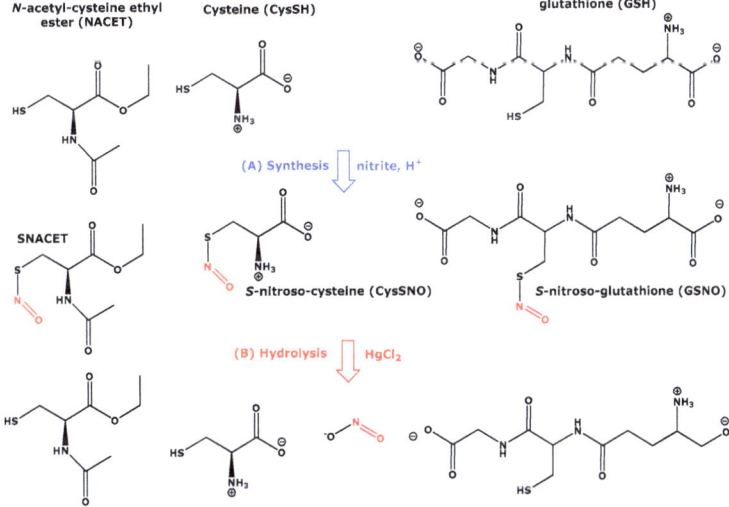

Scheme 1. Simplified schematic presentation of (A) the chemical synthesis of S-nitroso-cysteinyl thiols (thionitrites, RSNO) from their thiols (RSH) and nitrite in aqueous solutions in the presence of diluted HCl acid and of (B) the $HgCl_2$-induced hydrolysis of RSNO to the thiols and nitrite. CysSH, cysteine; GSH, glutathione; NACET, N-acetyl-cysteine ethyl ester. The corresponding S-nitrosothiols are S-nitroso-cysteine (CysSNO), S-nitroso-glutathione (GSNO), and S-nitroso-N-acetylcysteine ethyl ester (SNACET).

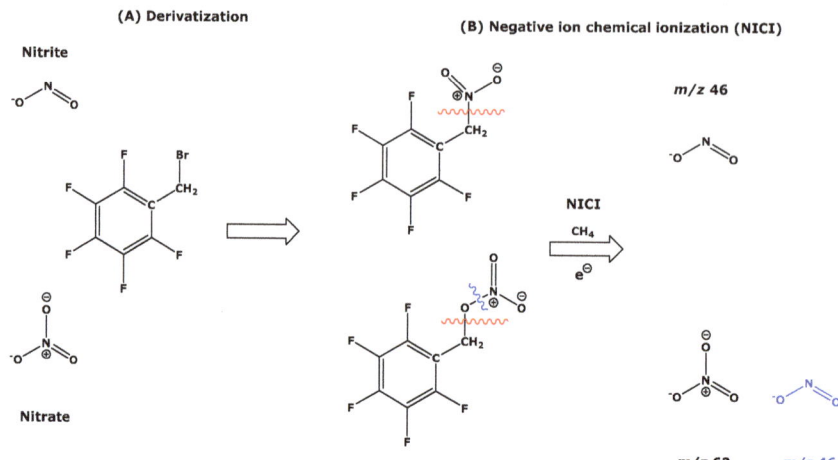

Scheme 2. Simplified schematic of the (**A**) derivatization of nitrite and nitrate with pentafluorobenzyl (PFB) bromide to their nitro and nitric acid ester derivatives in aqueous solution, respectively, and (**B**) their negative-ion chemical ionization (NICI) in gas chromatography–mass spectrometry (GC-MS) to generate nitrite and nitrate, respectively. The PFB derivative of nitrate (PFB-ONO$_2$) elutes before the PFB derivative of nitrite (PFB-NO$_2$). Under NICI conditions, PFB-NO$_2$ ionizes to form nitrite, whereas PFB-ONO$_2$ ionizes to form nitrate (99.8%) and nitrite (0.2%) [13]. Methane is used as the reagent gas. m/z, mass-to-charge ratio.

2. Materials and Methods

2.1. Chemicals and Materials

Na^{15}NO$_2$ (98.5 atom% ^{15}N) was from Cambridge Isotope Laboratories (Andover, MA, USA). Na^{15}NO$_3$ (98.5 atom% ^{15}N) was from Sigma (Munich, Germany). ^{18}O-Labeled water (95.5 atom% at ^{18}O) was purchased from Campro-Scientific (Berlin, Germany). Tetramethylammonium peroxynitrite, [Me$_4$N]$^+$[ONOO]$^-$, supplied as 1 mL portions of a 13.5 mM solution in 10 mM KOH (based on ε = 1700 M^{-1}cm^{-1} at 302 nm in 10 mM KOH), was from Alexis (Grünberg, Germany). The stock solution of Me$_4$N]$^+$[ONOO]$^-$ was stored at −80 °C. All peroxynitrite-containing solutions were kept on ice in the dark (aluminum foil). Peroxynitrite solutions were used immediately after thawing [Me$_4$N]$^+$[ONOO]$^-$ without renewed refrigerating of the remaining sample. CysSH, GSH, GSSG, HgCl$_2$, and pentafluorobenzyl (PFB) bromide were from Sigma-Aldrich (Munich, Germany). N-Acetylcysteine ethyl ester (NACET) was prepared as reported elsewhere [14]. K$_2$HPO$_4$, *tris*(hydroxymethyl)amino methane (Tris) and concentrated hydrochloric acid were obtained from Merck (Darmstadt, Germany). These salts were used to prepare 100 mM and 200 mM buffers of pH 7.4, respectively. Stock solutions of S-nitrosothiols were freshly prepared by combining equal volumes of ice-cold 10 mM solutions of nitrite and the thiols in distilled water and acidifying the samples by adding 10 μL aliquots of ice-cold 5 M HCl solutions followed by brief vortex mixing [15]. These samples were stored in an ice-bath in aluminum foil to avoid light-induced decomposition of the S-nitrosothiols and were used on the same day to prepare dilutions in the buffers. Li^{18}OH was prepared by adding a weighed amount of elemental Li (stored in paraffin) to a small volume of H$_2$^{18}O.

2.2. Experimental Conditions

All experiments were performed either in 100 mM K$_2$HPO$_4$ buffer or in 200 mM Tris buffer, both of pH 7.4, at room temperature (about 20–23 °C). For the sake of simplicity and comprehensibility, the experiments are described in detail in the Section 3.

2.3. Derivatization Procedure for Nitrite and Nitrate

Unlabeled and labeled nitrite and nitrate species were derivatized simultaneously with PFB bromide as described elsewhere [13] (Scheme 2), except for the sample volumes which varied (see Section 3). A constant sample–acetone volume ratio of 1:4 and a constant volume of toluene (1 mL) were used for the extraction of the PFB derivatives of the nitrite (PFB-NO$_2$) and nitrate (PFB-ONO$_2$) species.

2.4. GC-MS Analyses

Derivatized unlabeled and labeled nitrite and nitrate species were measured by GC-MS on an Agilent system model 5980 based on the quadrupole technology. An Optima 17 (15 m × 0.25 mm i.d., 0.25 µm film thickness) from Macherey-Nagel was used. Helium (70 kPa) and methane (200 Pa) were used as carrier and reactand gas, respectively. Aliquots (1 µL) of toluene extracts were injected in the splitless mode. Oven temperature was held at 70 °C for 1 min and then increased to 280 °C at a rate of 30 °C/min. Constant temperatures were kept at the ion source (180 °C), interface (280 °C), and injector (200 °C). Negative-ion chemical ionization (NICI) was used at an electron energy of 230 eV and an emission current of 300 µA (Scheme 2). Nitrite and nitrate species were analyzed in the selected-ion monitoring (SIM) mode using a dwell time of 50 ms for each ion (Table 1). The sum of peak area values of all ions monitored was set to 100%. Peak area values of selected ions were used to calculate their peak area ratio (PAR).

Table 1. Pentafluorobenzyl derivatives of nitrite and nitrate species quantitated by selected-ion monitoring of specific mass-to-charge (*m/z*) ions. The ^{15}N isotope (natural abundance, 0.37%); the ^{18}O isotope (natural abundance, 0.2%).

Species	*m/z*	Structure of the Anion
Nitrite	46	O=N-O$^-$
	47	O=^{15}N-O$^-$
	48	^{18}O=N-O$^-$ or O=N-^{18}O$^-$
	49	^{18}O=^{15}N-O$^-$ or O=^{15}N-^{18}O$^-$
	50	^{18}O=N-^{18}O$^-$ or ^{18}O=N-^{18}O$^-$
	51	^{18}O=^{15}N-^{18}O$^-$ or ^{18}O=^{15}N-^{18}O$^-$
Nitrate	62	O=N(-O)-O$^-$
	63	O=^{15}N(-O)-O$^-$
	64	^{18}O=N(-O)-O$^-$ or O=N(-O)-^{18}O$^-$
	65	^{18}O=^{15}N(-O)-O$^-$ or O=^{15}N(-O)-^{18}O$^-$
	66	^{18}O=N(-O)-^{18}O$^-$ or ^{18}O=N(-O)-^{18}O$^-$

3. Results

3.1. Hydrolysis of CysSNO and GSNO in $H_2^{18}O$

A 200 µL aliquot of 200 mM Tris buffer, pH 7.4, was extensively evaporated to dryness under a stream of nitrogen gas. The solid residue was reconstituted in 200 µL H$_2$^{18}O. After vortexing (highest stage), the sample was divided into four 50 µL aliquots. Two samples were spiked with CysSNO to reach a final added concentration of 100 µM (sample A, sample B). Yet another two samples were spiked with GSNO to reach final added concentrations of 100 µM (sample C, sample D). Subsequently, 2 µL aliquots of a 10 mM solution of HgCl$_2$ in deionized water (H$_2$^{16}O) were added to sample A and sample C at an approximate final concentration of 1 mM each. To allow complete HgCl$_2$-induced decomposition of CysSNO and GSNO, the samples were incubated for 60 min at room temperature [15]. Samples B and D were incubated at room temperature for 3 h and 24 h, respectively, to allow for spontaneous decomposition of CysSNO and GSNO. At the end of the incubation, all samples were treated with PFB bromide to convert nitrite species to their PFB nitro derivatives. GC-MS analysis was performed by SIM of m/z 46, m/z 48, and m/z 50. The results of this experiment are summarized in Table 2.

Table 2. Incorporation of ^{18}O from $H_2^{18}O$ into nitrite upon incubation of unlabeled CysSNO and GSNO in ^{18}O-prepared 200 mM Tris buffer, pH 7.4. Numbers in parentheses indicate the incubation time; the incubation time with $HgCl_2$ was 1 h.

Sample		m/z 46 (%)	m/z 48 (%)	m/z 50 (%)	m/z 48/m/z 46
A	CysSNO (0 h) + $HgCl_2$	58	41	1	0.7:1
C	GSNO (0 h) + $HgCl_2$	30	67	3	2.2:1
B	CysSNO (3 h)	45	33	22	0.7:1
D	GSNO (24 h)	95	5	0	0.05:1

In Tris buffer, the half-life for CysSNO is about 7 min [15]. Immediate treatment of the 100 μM solution of CysSNO (0 h) in ^{18}O-Tris buffer with $HgCl_2$ (sample A) resulted in the formation of $^{18}O=N-O^-/O=N-^{18}O^-$ (m/z 48) and $^{16}O=N-O^-/O=N-^{16}O^-$ (m/z 46) with a peak area ratio (PAR) of 1:1.4. In this experiment, the formation of $^{18}O=N-^{18}O/^{18}O=N-^{18}O^-$ (m/z 50) amounted to only 1%. This observation suggests that ^{18}O from ^{18}O-Tris buffer is incorporated into the SNO group of CysSNO induced by $HgCl_2$.

In the case of sample B, i.e., in the absence of $HgCl_2$, incubation resulted in the formation of $^{18}O=N-O^-/O=N-^{18}O^-$ (m/z 48) and $^{16}O=N-^{16}O$ (m/z 46) with a PAR of 1:1.4. The formation of $^{18}O=N-^{18}O^-$ (m/z 50) amounted to 22%. This difference is likely to be due to the longer incubation time of 3 h and the absence of $HgCl_2$. The incubation time of 3 h allows for complete decomposition of CysSNO to NO. The formation of m/z 46, m/z 48, and m/z 50 is likely to result in part by hydrolysis of the SNO group of intact CysSNO and in part due to autoxidation of CysSNO-derived to NO.

Incubation of GSNO in ^{18}O-Tris buffer in the absence of $HgCl_2$ (sample D) did not result in formation of $^{18}O=N-O^-/O=N-^{18}O^-$ (m/z 48) to an appreciable extent. This observation suggests that GSNO does not release NO nor hydrolyzes to $^{18}O=N-O^-/O=N-^{18}O^-$ in ^{18}O-Tris buffer. In contrast, GSNO immediately treated with $HgCl_2$ (sample C) resulted in the formation of $^{18}O=N-O^-/O=N-^{18}O^-$ (m/z 48) to an even greater extent compared to unlabeled nitrite (67% vs. 30%). Obviously, $HgCl_2$ is required for the hydrolysis of the SNO group of GSNO to nitrite.

3.2. Hydrolysis of CysS$^{15}N^{18}O$ and GS$^{15}N^{18}O$ in $H_2^{16}O$

The experiment described above was repeated with some modifications. Separate solutions (100 μL) of [^{15}N]nitrite, CysSH, and GSH were prepared in 200 mM Tris buffer, pH 7.4. Then, the solvent was evaporated thoroughly under a stream of nitrogen. Subsequently, the [^{15}N]nitrite samples were reconstituted in 100 μL aliquots of $H_2^{18}O$, and these solutions were used to reconstitute the CysSH and GSH residues. CysS$^{15}N^{18}O$ and GS$^{15}N^{18}O$ were synthesized separately in these samples by adding 2.5 μL aliquots of 5 M HCl. After incubation for 5 min to complete CysS$^{15}N^{18}O$ and GS$^{15}N^{18}O$, the samples were neutralized (pH 7 to 8) by adding 2.5 μL aliquots of 5 M Li^{18}OH. Then, the CysS$^{15}N^{18}O$ and GS$^{15}N^{18}O$ samples were each divided into two 50 μL aliquots. Immediately thereafter, one CysS$^{15}N^{18}O$ sample (sample A) and one GS$^{15}N^{18}O$ sample (sample C) were each spiked with 10 μL of a 10 mM solution of $HgCl_2$ prepared in $H_2^{18}O$. The second CysS$^{15}N^{18}O$ sample (sample B) and the second GS$^{15}N^{18}O$ sample (sample D) were incubated at room temperature for 3 h and 24 h, respectively, to allow complete decomposition of these RSNOs. In addition, two [^{15}N]nitrite samples were used as controls. One [^{15}N]nitrite sample (sample E) was incubated for 5 min at room temperature in 200 mM ^{18}O-Tris buffer, pH 7.4. The other [^{15}N]nitrite sample (sample F) was incubated at room temperature in acidified (about pH 2) 200 mM ^{18}O-Tris buffer. After PFB bromide derivatization, GC-MS analysis was performed for nitrite species by SIM. The results of this experiment are summarized in Table 3.

Table 3. Incorporation of ^{18}O from $H_2^{18}O$ into ^{15}N-nitrite formed from $CysS^{15}N^{18}O$ and $GS^{15}N^{18}O$ in 200 mM Tris buffer, pH 7.4, prepared in $H_2^{18}O$.

Sample		m/z 47 (%)	m/z 49 (%)	m/z 51 (%)	m/z 51/m/z 49
A	$CysS^{15}N^{18}O$ (0 h) + $HgCl_2$ ($H_2^{18}O$)	4	48	48	1:1
C	$GS^{15}N^{18}O$ (0 h) + $HgCl_2$ ($H_2^{18}O$)	4	48	48	1:1
B	$CysS^{15}N^{18}O$ (3 h)	8	44	48	1.2:1
D	$GS^{15}N^{18}O$ (24 h)	2	30	68	2.3:1
E	[^{15}N]Nitrite (5 min), pH 7.4	78	19	3	0.16:1
F	[^{15}N]Nitrite (5 min), pH 2.0	8	37	56	1.5:1

Immediate treatment of $CysS^{15}N^{18}O$ (sample A) and $GS^{15}N^{18}O$ (sample C) with $HgCl_2$ ($H_2^{18}O$) resulted in almost complete hydrolysis of the RSNO and formation of $^{18}O=^{15}N-^{16}O^-/^{16}O=^{15}N-^{18}O^-$ (m/z 49) and of $^{18}O=^{15}N-^{18}O^-$ (m/z 51) with a PAR m/z 51 to m/z 49 of 1:1 for both RSNOs. Comparable results were obtained from the incubation of $CysS^{15}N^{18}O$ in the absence of $HgCl_2$ (sample B). In the case of sample D, $GS^{15}N^{18}O$ (24 h) resulted in the formation of $^{18}O=^{15}N-^{16}O^-/^{16}O=^{15}N-^{18}O^-$ (m/z 49) and of $^{18}O=^{15}N-^{18}O^-$ (m/z 51) with a PAR of 2.3:1, suggesting higher incorporation of ^{18}O into the $S^{15}N^{18}O$ group of $GS^{15}N^{18}O$ compared to $CysS^{15}N^{18}O$.

For comparison, [^{15}N]nitrite was incubated in non-acidified ^{18}O-Tris buffer (pH 7.4, sample E) and in acidified ^{18}O-Tris buffer (pH 2.0, sample F). In non-acidified ^{18}O-Tris buffer, there was little incorporation of ^{18}O into [^{15}N]nitrite, whereas the incorporation of ^{18}O into [^{15}N]nitrite in acidified ^{18}O-Tris buffer (pH 2, sample F) was almost complete. Thus, in ^{18}O-Tris buffer (pH 7.4) a small incorporation of ^{18}O from Tris buffer into [^{15}N]nitrite is possible, yet it is lower than in RSNO. Nitrite is the conjugate base of nitrous acid (HONO; pK_a, 3.2), and HONO and/or its anhydride seems to be more easily accessible for hydrolysis than nitrite and RSNO.

3.3. $HgCl_2$-Induced Hydrolysis of $NACCysS^{14}NO$ and $NACCysS^{15}NO$ in $H_2^{16}O/H_2^{18}O$ Mixtures

In a further experiment, the $HgCl_2$-induced hydrolysis of $NACCysS^{14}NO$ and $NACCysS^{15}NO$ was investigated in phosphate buffer of pH 7.4 using $HgCl_2$ prepared in $H_2^{16}O/H_2^{18}O$ mixtures.

An equimolar mixture of $NACCysS^{14}NO$ and $NACCysS^{15}NO$ was diluted in 100 mM K_2HPO_4 buffer, pH 7.4, to reach a final concentration of 238 µM. Aliquots (12.5 µL) of this solution were treated with 12.5 µL aliquots of $HgCl_2$ solutions prepared in $H_2^{16}O/H_2^{18}O$ mixtures. The $H_2^{16}O/H_2^{18}O$ mixtures varied (v/v) as follows: sample A: 100:0; sample B, 100:5; sample C, 100:25; sample D, 100:100; sample E, 100:75; and sample F, 0:100. The final concentration of $HgCl_2$ was constant at 3.33 mM. All samples were incubated for 10 min at room temperature and then derivatized with PFB bromide. GC-MS analysis was performed in the SIM mode. The results of this experiment are summarized in Table 4.

The ^{15}N- to ^{14}N-nitrite molar ratio (m/z 47 to m/z 46) was independent of the $H_2^{16}O/H_2^{18}O$ final volume ratio in the samples and was determined to be 1.070 ± 0.045 (mean ± SD, n = 5). The molar ratios of m/z 48 to m/z 46 and of m/z 49 to m/z 47 increased with an increasing proportion of $H_2^{18}O$ in the samples. The increase was linear until a proportion of 37.5% of $H_2^{18}O$ in the sample. These observations suggest that ^{18}O from $H_2^{18}O$ is incorporated almost to the same extent into ^{15}N- to ^{14}N-nitrite released from $NACCysS^{15}NO$ and $NACCysS^{14}NO$, respectively.

Table 4. Incorporation of ^{18}O from $H_2{}^{18}O$ into ^{14}N- and ^{15}N-nitrite formed from NACCysS^{14}NO and NACCyS^{15}NO in 100 mM potassium phosphate buffer, pH 7.4, in the presence of a 3.33 mM HgCl$_2$ solution in $H_2{}^{16}O/H_2{}^{18}O$ mixtures. n.d., not detected.

Sample	$H_2{}^{18}O$ (vol%)	m/z 47/m/z 46 $O^{15}NO^-/O^{14}NO^-$	m/z 48/m/z 46 $^{18}O^{14}NO^-/O^{14}NO^-$	m/z 49/m/z 47 $^{18}O^{15}NO^-/O^{15}NO^-$
A	0	1.04	0.005	0.005
B	2.5	n.d.	0.064	0.065
C	12.5	1.03	n.d.	n.d.
D	25.0	1.05	0.432	0.435
E	37.5	1.09	0.642	0.688
F	50.0	1.14	0.676	0.744

In a further experiment, separate solutions of ^{14}N-nitrite (500 µM), GS^{14}NO (238 µM), and NACCysS^{14}NO (100 µM) were prepared in 100 mM K$_2$HPO$_4$ buffer, pH 7.4, using $H_2{}^{16}O$. From these solutions, 12.5 µL aliquots were treated with 12.5 µL aliquots of a 5 mM HgCl$_2$ solution prepared in $H_2{}^{18}O$. The final $H_2{}^{16}O:H_2{}^{18}O$ volume ratio was constant at 1:1 (v/v). Derivatization and GC-MS analyses of these samples resulted in a PAR m/z 46 to m/z 48 of 51:1 for ^{14}N-nitrite, 1.25:1 for GS^{14}NO, and 1.84:1 for NACCysS^{14}NO. These data suggest considerable incorporation of ^{18}O from $H_2{}^{18}O$ into ^{14}N-nitrite derived from GS^{14}NO and NACCysS^{14}NO, but not into authentic ^{14}N-nitrite.

3.4. Decomposition and Isomerization of Synthetic Peroxynitrite in $H_2{}^{18}O$

Similar experiments were performed with freshly prepared dilutions of commercially available peroxynitrite (i.e., tetramethylammonium peroxynitrite; 100 µM) in 0.2 M Tris buffer and in 0.1 M potassium phosphate buffer (each of pH 7.4). They resulted in formation of ^{18}O-labeled nitrite and ^{18}O-labeled nitrate to the same very low extent, closely comparable to that obtained using solutions of synthetic nitrite and nitrate (each at 100 µM) in pH-neutral 0.2 M Tris buffer and 0.1 M potassium phosphate buffer. ^{18}O incorporation was very low even in the buffers prepared in 100% $H_2{}^{18}O$ for long incubation times (up to 60 min). These results suggest that water is not involved in the decomposition of peroxynitrite to nitrite and isomerization of peroxynitrite to nitrate in aqueous buffers of neutral pH value. Synthetic peroxynitrite was found to decompose to nitrite and to isomerize to nitrate with a stoichiometry of 1:1 [16] (11). Decomposition of peroxynitrite to nitrite and dioxygen with a stoichiometry of 2:1 been reported by others [17].

$$4\,[O{=}N^{(III)}{-}O^{(-I)}{-}O^{(-I)}]^- \rightarrow 2\,[O{=}N^{(III)}{-}O^{(-II)}]^- + 2\,[O{=}N^{(V)}({-}O^{(-II)}){-}O^{(-II)}]^- + O^{(\pm 0)}{}_2 \quad (11)$$

3.5. Effects of GSH on Reaction of Synthetic Peroxynitrite in $H_2{}^{18}O$

GSH and other thiols such as CysSH react with peroxynitrite [15]. Known reaction products of peroxynitrite and GSH are GSNO, oxidized GSH, i.e., GSH disulfide (GSSG), nitrite, and nitrate. As shown above, in the absence of GSH, peroxynitrite decomposes to nitrite and isomerizes to nitrate. In the presence of GSH, peroxynitrite increasingly decomposes to nitrite at the cost of its isomerization product nitrate (11).

The reaction of peroxynitrite with GSH was investigated in aqueous buffers prepared in $H_2{}^{18}O$. At a concentration of 5 mM GSH, no incorporation of ^{18}O from $H_2{}^{18}O$ into nitrite or nitrate from decomposed peroxynitrite (100 µM) was observed. For comparison, the same experiment was performed with nitrite (100 µM) instead of peroxynitrite. Linear regression analysis between the PAR of m/z 48 to m/z 46 (y_1) or the PAR of m/z 50 to m/z 46 (y_2) and the percentage content of $H_2{}^{18}O$ (x) was performed. The regression equations were $y_1 = 2 \times 10^{-3} + 4.9 \times 10^{-4}$ x ($r^2 = 0.984$) and $y_2 = 2 \times 10^{-5} + 1.9 \times 10^{-4}$ x ($r^2 = 0.944$) for peroxynitrite. The corresponding regression equations were $y_1 = 4 \times 10^{-3} + 5.3 \times 10^{-4}$ x ($r^2 = 0.9967$) and $y_2 = 8 \times 10^{-5} + 2.6 \times 10^{-5}$ x ($r^2 = 0.958$) for nitrite. These very similar results suggest that in aqueous buffer of neutral pH, peroxynitrite is converted to nitrite (decomposition) without the participation of water.

In the presence of GSH, the peroxy group of peroxynitrite is reduced to yield nitrite and GSSG via intermediate formation of GSOH (12a). GSOH further reacts with GSH to form GSSG (12b). GSSG is the major reaction product of GSH with peroxynitrite [16]. $^{18}O=^{14}N-^{18}O^-$ (m/z 50) was formed from the reaction of GSH with peroxynitrite in $H_2^{18}O$ to a very low extent which was, however, higher than the incorporation of ^{18}O into authentic nitrite. This could be due to the occurrence of additional much less abundant reactions such as the formation of GSNO (12) and $^\bullet$NO.

$$2\ GS^{(-II)}H + [O=N^{(III)}-O^{(-I)}O^{(-I)}]^- \rightarrow GS^{(-I)}S^{(-I)}G + [O=N^{(III)}-O^{(-II)}]^- + H_2O \quad (12)$$

$$GS^{(-II)}H + [O=N^{(III)}-O^{(-I)}O^{(-I)}]^- \rightarrow GS^{(\pm 0)}OH + [O=N^{(III)}-O^{(-II)}]^- \quad (12a)$$

$$GS^{(-II)}H + GS^{(\pm 0)}OH \rightarrow GS^{(-I)}S^{(-I)}G + H_2O \quad (12b)$$

4. Discussion

The elements H, C, N, and O are mixtures of stable isotopes. The natural abundance of their heavier isotopes amounts to 0.0145% ^2H, 1.06% ^{13}C, 0.366% ^{15}N, and 0.2% ^{18}O. Analytes "labeled" with stable isotopes of these elements can be separated from their "unlabeled" analogs by mass spectrometry (MS). Stable-isotope-labeled compounds are useful as internal standards in MS-based quantitative chemical analysis, because they have almost identical physicochemical properties. The only difference is the formation of ions with different mass-to-charge (m/z) ratios, which is utilized in mass spectrometers for their separation.

Another important topic of application of stable isotopes is qualitative and quantitative physical, chemical, biochemical, and biomedical research. The present work demonstrates the unique utility of the use of stable isotopes to perform mechanistic studies on reactions of $^\bullet$NO and its metabolites S-nitrosothiols (RSNOs), peroxynitrite, nitrite, and nitrate and to obtain definite results. In these studies, $H_2^{18}O$ was used in combination with a highly specific GC-MS method [13] for the simultaneous measurement of nitrite and nitrate species that contain ^{14}N, ^{15}N, ^{16}O, or ^{18}O in their molecules. The GC-MS method uses simultaneous derivatization of nitrite and nitrate in aqueous buffers with PFB bromide, methane negative-ion chemical ionization of the PFB derivatives to nitrite and nitrate, their separation on a single quadrupole GC-MS apparatus, and detection by an electron multiplier (Scheme 2). Nitrite and nitrate are ubiquitous, i.e., they are present as contaminations in the laboratory, at concentrations lying in the lower µM range. The high specificity and sensitivity of the GC-MS method and the relatively low natural abundance of ^{15}N and ^{18}O enable performance of experiments using small amounts (volumes) of $H_2^{18}O$ and relatively small quasi-physiological concentrations of reactands. These features help overcome the ubiquity of nitrite and nitrate contaminations. $H_2^{18}O$ is a quite expensive solvent. This is an issue and may limit the number of replicates. However, the information gained by such experiments overwhelms potential limitations.

The results presented in the current study unequivocally demonstrate that $H_2^{18}O$ is involved in the generation of nitrite from RSNO, in part via autoxidation of $^\bullet$NO and hydrolysis of the unisolable intermediate N_2O_3, the anhydride of nitrous acid. This is the case in CycSNO (Scheme 3). GSNO is neither a $^\bullet$NO donor nor hydrolyzes to nitrite. In high molar excess, $HgCl_2$ in aqueous solution mediates the hydrolysis of the SNO group of CysSNO and GSNO (Scheme 3) as well as of NACCysSNO and most likely of every RSNO. There is indication that the $Hg^{(II)}$ ion in $HgCl_2$ used in the experiments forms a hydratation shell with several $H_2^{18}O$ molecules (Scheme 3). Literature reports support this observation, indicating that the first solvation shell of Hg^{2+} ions may contain 6 to 24 water molecules [18–24]. Experiments with $H_2^{18}O$ should consider possible effects of hydratation of reagents used to prepare buffers and solutions of chemicals, which may potentially form stable hydratation shells with $H_2^{16}O$ that are difficult to be completely displaced by $H_2^{18}O$.

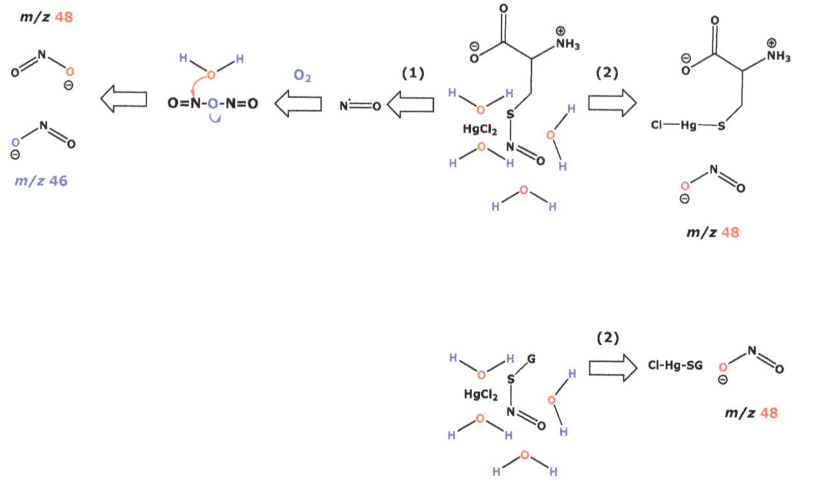

Scheme 3. Proposed mechanisms for the incorporation of ^{18}O from $H_2^{18}O$ into (1) nitrous anhydride (N_2O_3) from autoxidized NO formed from decomposed CysSNO and into (2) the SNO groups of CysSNO (upper panel) and GSNO (lower panel) mediated by aqueous $HgCl_2$. GSNO is not a NO donor. In its $H_2^{18}O$ solutions, $HgCl_2$ forms a hydration shell with $H_2^{18}O$, which attacks the SNO groups of CysSNO and GSNO to form ^{18}O-labeled nitrite.

5. Conclusions

Currently, simultaneous analysis of nitrite and nitrate is best performed by GC-MS after simultaneous derivatization with pentafluorobenzyl bromide and negative-ion chemical ionization. This is a unique technique to detect nitrite and nitrate anions as they occur in biological samples. The combination of this GC-MS approach with the use of buffers prepared in $H_2^{18}O$ enables generation of definite results in mechanistic studies allowing elucidation. In $H_2^{18}O$ buffers of neutral pH, *S*-nitroso-cysteine (CysSNO) decomposes to form ^{18}O-nitrite, indicating the involvement of water. This is not the case for *S*-nitrosoglutathione (GSNO). $HgCl_2$ mediates the hydrolysis of the SNO groups of CysSNO and GSNO. Aqueous solutions of $HgCl_2$ are likely to form Hg^{2+} ions solvated with $H_2^{16}O$ and $H_2^{18}O$, and this "isotope effect" may influence the outcome of hydrolysis studies.

Funding: This study was in part supported by the Deutsche Forschungsgemeinschaft (Grant TS 60/2-1 and TS 60/4-1).

Institutional Review Board Statement: Not applicable.

Informed Consent Statement: Not applicable.

Data Availability Statement: Not applicable.

Acknowledgments: Parts of the results reported here have been taken from previous work performed in the author's group by C.S. Rossa [25] and K. Denker [26] within the framework of their medical theses.

Conflicts of Interest: The author declares no conflict of interest.

Sample Availability: Not available.

References

1. Tsikas, D. A critical review and discussion of analytical methods in the L-arginine/nitric oxide area of basic and clinical research. *Anal. Biochem.* **2008**, *379*, 139–163. [CrossRef] [PubMed]
2. Wink, D.A.; Darbyshire, J.F.; Nims, R.W.; Saavedra, J.E.; Ford, P.C. Reactions of the bioregulatory agent nitric oxide in oxygenated aqueous media: Determination of the kinetics for oxidation and nitrosation by intermediates generated in the NO/O_2 reaction. *Chem. Res. Toxicol.* **1993**, *6*, 23–27. [CrossRef] [PubMed]
3. Ford, P.C.; Wink, D.A.; Stanbury, D.M. Autoxidation kinetics of aqueous nitric oxide. *FEBS Lett.* **1993**, *326*, 1–3. [CrossRef] [PubMed]
4. Kharitonov, V.G.; Sundquist, A.R.; Sharma, V.S. Kinetics of nitric oxide autoxidation in aqueous solution. *J. Biol. Chem.* **1994**, *269*, 5881–5883. [CrossRef] [PubMed]
5. Wink, D.A.; Nims, R.W.; Darbyshire, J.F.; Christodoulou, D.; Hanbauer, I.; Cox, G.W.; Laval, F.; Laval, J.; Cook, J.A. Reaction kinetics for nitrosation of cysteine and glutathione in aerobic nitric oxide solutions at neutral pH. Insights into the fate and physiological effects of intermediates generated in the NO/O_2 reaction. *Chem. Res. Toxicol.* **1994**, *17*, 519–525. [CrossRef]
6. Caccia, S.; Denisov, I.; Perrella, M. The kinetics of the reaction between NO and O_2 as studied by a novel approach. *Biophys. Chem.* **1999**, *76*, 63–72. [CrossRef] [PubMed]
7. Nedospasov, A.A. Is N_2O_3 the main nitrosating intermediate in aerated nitric oxide (NO) solutions in vivo? If so, where, when, and which one? *J. Biochem. Mol. Toxicol.* **2002**, *16*, 109–120. [CrossRef]
8. Galliker, B.; Kissner, R.; Nauser, T.; Koppenol, W.H. Intermediates in the autoxidation of nitrogen monoxide. *Chemistry* **2009**, *15*, 6161–6168. [CrossRef]
9. Mahmoudi, L.; Kissner, R.; Koppenol, W.H. Low-temperature trapping of intermediates in the reaction of NO• with O_2. *Inorg. Chem.* **2017**, *56*, 4846–4851. [CrossRef]
10. Kharitonov, V.G.; Sundquist, A.R.; Sharma, V.S. Kinetics of nitrosation of thiols by nitric oxide in the presence of oxygen. *J. Biol. Chem.* **1995**, *270*, 28158–28164. [CrossRef]
11. Vitturi, D.A.; Minarrieta, L.; Salvatore, S.R.; Postlethwait, E.M.; Fazzari, M.; Ferrer-Sueta, G.; Lancaster, J.R., Jr.; Freeman, B.A.; Schopfer, F.J. Convergence of biological nitration and nitrosation via symmetrical nitrous anhydride. *Nat. Chem. Biol.* **2015**, *11*, 504–510. [CrossRef] [PubMed]
12. Tsikas, D. Methods of quantitative analysis of the nitric oxide metabolites nitrite and nitrate in human biological fluids. *Free. Radic. Res.* **2005**, *39*, 797–815. [CrossRef]
13. Tsikas, D. Simultaneous derivatization and quantification of the nitric oxide metabolites nitrite and nitrate in biological fluids by gas chromatography/mass spectrometry. *Anal. Chem.* **2000**, *72*, 4064–4072. [CrossRef] [PubMed]
14. Tsikas, D.; Dehnert, S.; Urban, K.; Surdacki, A.; Meyer, H.H. GC-MS analysis of S-nitrosothiols after conversion to S-nitroso-N-acetyl cysteine ethyl ester and in-injector nitrosation of ethyl acetate. *J. Chromatogr. B Analyt Technol. Biomed. Life Sci.* **2009**, *877*, 3442–3455. [CrossRef]
15. Tsikas, D.; Sandmann, J.; Rossa, S.; Gutzki, F.M.; Frölich, J.C. Investigations of S-transnitrosylation reactions between low- and high-molecular-weight S-nitroso compounds and their thiols by high-performance liquid chromatography and gas chromatography-mass spectrometry. *Anal. Biochem.* **1999**, *270*, 231–241. [CrossRef] [PubMed]
16. Tsikas, D. GC-MS and HPLC methods for peroxynitrite (ONOO- and $O^{15}NOO$-) analysis: A study on stability, decomposition to nitrite and nitrate, laboratory synthesis, and formation of peroxynitrite from S-nitrosoglutathione (GSNO) and KO_2. *Analyst* **2011**, *136*, 979–987. [CrossRef]
17. Pfeiffer, S.; Gorren, A.C.; Schmidt, K.; Werner, E.R.; Hansert, B.; Bohle, D.S.; Mayer, B. Metabolic fate of peroxynitrite in aqueous solution. Reaction with nitric oxide and pH-dependent decomposition to nitrite and oxygen in a 2:1 stoichiometry. *J. Biol. Chem.* **1997**, *272*, 3465–3470. [CrossRef]
18. Chillemi, G.; Mancini, G.; Sanna, N.; Barone, V.; Della Longa, S.; Benfatto, M.; Pavel, N.V.; D'Angelo, P. Evidence for sevenfold coordination in the first solvation shell of Hg(II) aqua ion. *J. Am. Chem. Soc.* **2007**, *129*, 5430–5436. [CrossRef]
19. Sobolev, O.; Cuello, G.J.; Román-Ross, G.; Skipper, N.T.; Charlet, L. Hydration of Hg^{2+} in aqueous solution studied by neutron diffraction with isotopic substitution. *J. Phys. Chem. A* **2007**, *111*, 5123–5125. [CrossRef]
20. Mancini, G.; Sanna, N.; Barone, V.; Migliorati, V.; D'Angelo, P.; Chillemi, G. Structural and dynamical properties of the Hg^{2+} aqua ion: A molecular dynamics study. *J. Phys. Chem. B* **2008**, *112*, 4694–4702. [CrossRef]
21. Castro, L.; Dommergue, A.; Renard, A.; Ferrari, C.; Ramirez-Solis, A.; Maron, L. Theoretical study of the solvation of $HgCl_2$, HgClOH, $Hg(OH)_2$ and $HgCl_3(-)$: A density functional theory cluster approach. *Phys. Chem. Chem. Phys.* **2011**, *13*, 16772–16779. [CrossRef] [PubMed]
22. Amaro-Estrada, J.I.; Maron, L.; Ramírez-Solís, A. Aqueous solvation of $Hg(OH)_2$: Energetic and dynamical density functional theory studies of the $Hg(OH)_2$-$(H_2O)_n$ (n = 1–24) structures. *J. Phys. Chem. A* **2013**, *117*, 9069–9075. [CrossRef]
23. Afaneh, A.T.; Schreckenbach, G.; Wang, F. Theoretical study of the formation of mercury (Hg^{2+}) complexes in solution using an explicit solvation shell in implicit solvent calculations. *J. Phys. Chem. B* **2014**, *118*, 11271–11283. [CrossRef] [PubMed]
24. Amaro-Estrada, J.I.; Maron, L.; Ramírez-Solís, A. Aqueous solvation of HgClOH. Stepwise DFT solvation and Born-Oppenheimer molecular dynamics studies of the HgClOH-$(H_2O)24$ complex. *Phys. Chem. Chem. Phys.* **2014**, *16*, 8455–8464. [CrossRef] [PubMed]

25. Rossa, C.S. Chemistry, Biochemistry and Pharmacology of Low-Molecular-Mass S-Nitroso Compounds and S-Nitroso-Albumin. Ph.D. Thesis, Hannover Medical School, Hannover, Germany, 1997.
26. Denker, K. Chemistry and Biochemistry of Peroxynitrite. Ph.D. Thesis, Hannover Medical School, Hannover, Germany, 2006.

Disclaimer/Publisher's Note: The statements, opinions and data contained in all publications are solely those of the individual author(s) and contributor(s) and not of MDPI and/or the editor(s). MDPI and/or the editor(s) disclaim responsibility for any injury to people or property resulting from any ideas, methods, instructions or products referred to in the content.

Article

An Aptamer-Based Lateral Flow Biosensor for Low-Cost, Rapid and Instrument-Free Detection of Ochratoxin A in Food Samples

Electra Mermiga, Varvara Pagkali, Christos Kokkinos and Anastasios Economou *

Department of Chemistry, National and Kapodistrian University of Athens, 157 71 Athens, Greece; electram24@gmail.com (E.M.); bpagali@yahoo.gr (V.P.); christok@chem.uoa.gr (C.K.)
* Correspondence: aeconomo@chem.uoa.gr; Tel.: +30-210-7274328

Citation: Mermiga, E.; Pagkali, V.; Kokkinos, C.; Economou, A. An Aptamer-Based Lateral Flow Biosensor for Low-Cost, Rapid and Instrument-Free Detection of Ochratoxin A in Food Samples. *Molecules* 2023, 28, 8135. https://doi.org/10.3390/molecules28248135

Academic Editors: Paraskevas D. Tzanavaras and Victoria Samanidou

Received: 4 December 2023
Revised: 14 December 2023
Accepted: 15 December 2023
Published: 17 December 2023

Copyright: © 2023 by the authors. Licensee MDPI, Basel, Switzerland. This article is an open access article distributed under the terms and conditions of the Creative Commons Attribution (CC BY) license (https://creativecommons.org/licenses/by/4.0/).

Abstract: In this work, a simple and cost-efficient aptasensor strip is developed for the rapid detection of OTA in food samples. The biosensor is based on the lateral flow assay concept using an OTA-specific aptamer for biorecognition of the target analyte. The strip consists of a sample pad, a conjugate pad, a nitrocellulose membrane (NC) and an absorbent pad. The conjugate pad is loaded with the OTA-specific aptamer conjugated with gold nanoparticles (AuNPs). The test line of the NC membrane is loaded with a specific OTA-aptamer probe and the control line is loaded with a control probe. The assay is based on a competitive format, where the OTA present in the sample combines with the OTA aptamer-AuNP conjugate and prevents the interaction between the specific probe immobilized on the test line and the OTA aptamer-AuNP conjugates; therefore, the color intensity of the test line decreases as the concentration of OTA in the sample increases. Qualitative detection of OTA is performed visually, while quantification is performed by reflectance colorimetry using a commercial scanner and image analysis. All the parameters of the assay are investigated in detail and the analytical features are established. The visual limit of detection (LOD) of the strip is 0.05 ng mL^{-1}, while the LOD for semi-quantitative detection using reflectance colorimetry is 0.02 ng mL^{-1}. The lateral flow strip aptasensor is applied to the detection of OTA in wine, beer, apple juice and milk samples with recoveries in the range from 91 to 114%. The assay exhibits a satisfactory selectivity for OTA with respect to other mycotoxins and lasts 20 min. Therefore, the lateral flow strip aptasensor could be useful for the rapid, low-cost and fit-for-purpose on-site detection of OTA in food samples.

Keywords: ochratoxin A; aptamer; aptasensor; biosensor; gold nanoparticles; lateral flow; colorimetry

1. Introduction

Ochratoxin A (OTA) is a mycotoxin produced as a secondary metabolite by several fungal species such as *Aspergillus* and *Penicillium* [1,2]. Various studies have shown that OTA can cause several adverse health effects to animals and humans through the consumption of contaminated agricultural goods, such as cereal grains, vegetables, coffee beans, wine and beer [2,3]. OTA has been shown to be nephrotoxic, teratogenic, immunotoxic and carcinogenic and therefore the International Agency for Research on Cancer (IARC) has classified OTA as a group 2B possible human carcinogen [4]. Due to the toxicity of OTA, the European Union has set maximum limits (MLs) for OTA in foods in the range of 0.5–10 μg kg^{-1} [4].

Considering the severe toxic effects and the low permitted MLs of OTA in food, it is of great importance to develop rapid and sensitive sensing platforms for OTA monitoring to address issues of food safety and avoid or minimize the risk of OTA intake by consumers. The detection of OTA in food is mostly based on conventional chromatographic techniques such as thin-layer chromatography (TLC), high-performance liquid chromatography (HPLC) or gas chromatography (GC), which, although powerful, require expensive and bulky equipment, trained personnel and a complex sample preparation and are not suitable for on-site rapid assays [2,4,5]. On the contrary, immunochromatographic (lateral flow) assays, utilizing antibodies as biorecognition elements, are convenient, portable, cost-effective and simple-to-use analytical and diagnostic platforms, providing satisfactory sensitivity with the potential for on-site rapid screening by non-trained personnel [6,7]; however, these devices require the use of expensive antibodies which are available for only a limited spectrum of target analytes and possess low chemical stability [8–12].

Aptamers have attracted the interest of the scientific community since their discovery in 1990 [13,14]. Aptamers are short single-stranded oligonucleotides (DNA or RNA) which can interact with a wide range of target analytes, such as antibiotics, proteins or other organic compounds [15,16]. Aptamers are selected from large synthetic DNA or RNA libraries, through a process called SELEX (Systematic Evolution of Ligands by EXponential enrichment). Through the SELEX process, it is possible to identify specific DNA sequences which bind to the target analyte(s) with high specificity by exploiting conformational changes in the three-dimensional structure of the oligonucleotides [13–17]. When serving as biorecognition elements, aptamers possess several advantages over antibodies, including their low cost, facile artificial synthesis, wider potential target range, good stability, resistance to heating and long shelf-life; these advantages make them ideal for rapid on-site detection and the development of inexpensive and portable analytical devices [13–19]. Over the last few years, aptamer-based sensors with optical or electrochemical detection have been used to detect a wide range of analytes [19,20], including OTA [4,12].

The aim of this work is the development of a simple, portable, instrument-free and cost-efficient lateral flow strip aptasensor for OTA determination in foodstuffs with improved analytical features over existing sensors. The biosensor strip is based on a competitive lateral flow assay principle using a conjugate of an OTA-specific aptamer with gold nanoparticles (AuNPs) as the biorecognition element. The test line of the strip contains a specific OTA-aptamer probe. In the presence of OTA, the OTA aptamer-AuNP conjugate is bound by the target analyte and is not allowed to bind with the specific probe of the test line in the strip. Qualitative detection of OTA is performed visually while quantification is performed by reflectance colorimetry using a scanner and image analysis. All the parameters of the assay are investigated in detail to provide optimum assay performance.

2. Results and Discussion

2.1. Assembly and Working Principle of the Lateral Flow Aptasensor

A schematic of the lateral flow strip with dimensions is illustrated in Figure 1a. It consists of four successively placed segments: a sample pad; a conjugate pad; a NC membrane; and an absorbent pad, all with a width of 3 mm and lengths of 20 mm, 5 mm, 20 mm and 30 mm, respectively. The components mutually overlap by 2 mm to ensure continuous liquid flow.

The principle of the aptamer-based lateral flow strip for the detection of OTA is illustrated in Figure 1b. It is based on the competition between probe 1 immobilized on the test line and the OTA target present in the sample to combine with the OTA-specific aptamer that is conjugated with AuNPs, loaded onto the conjugate pad. If OTA is present in the sample, it will bind to the aptamer-AuNP conjugate, preventing the interaction between the aptamer-AuNP conjugate and probe 1 at the test line. If OTA is not present in the sample, the aptamer-AuNP conjugate will be available to interact with, and bind to, probe 1 at the test line. Therefore, the intensity of the test line decreases as the concentration of OTA in the sample increases. To ensure the validity of the test, the aptamer-AuNP conjugate interacts with control probe 2 at the control line to provide a visual signal regardless of whether or not OTA is present in the sample.

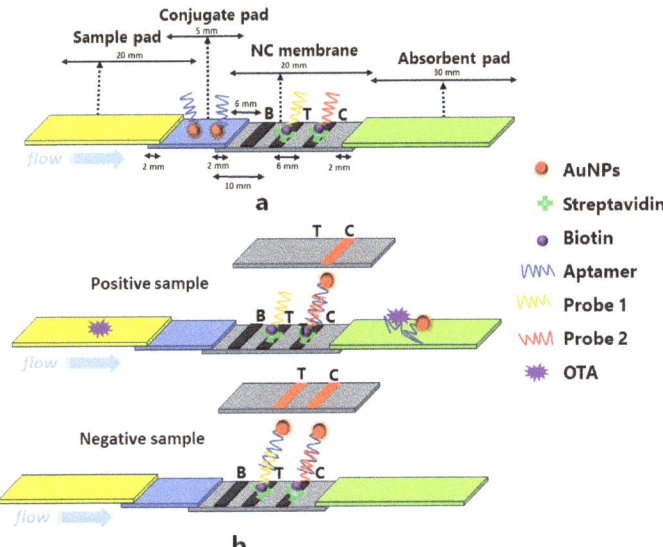

Figure 1. (a) Structure of the lateral flow strip and (b) principle of lateral flow immunoassay (dimensions are not drawn to scale). B is the PVA barrier, T is the test line and C is the control line.

2.2. Optimization of Experimental Parameters

In order to achieve the best analytical performance of the developed assay, it was necessary to study various experimental parameters such as the signal processing method used for quantification; the aptamer sequence; the preparation of the OTA aptamer-AuNP conjugate; the method of AuNP preparation; the concentrations of streptavidin and of probe 1 and probe 2 immobilized at the test and control line, respectively; the selection of the appropriate NC membrane, the conjugate pad and the sample pad; the volume of the OTA aptamer-AuNP conjugate loaded onto the conjugate pad; the volume of sample applied at the sample pad; and the composition of the running buffer. The selection of the experimental conditions was based on the blank solution (which gives the maximum signal intensity) as well as on the assay sensitivity as determined by the % color intensity inhibition values (with respect to the intensity of the blank solution) or the color intensity values corresponding to signals of standard solutions.

Initially, the influence of the color mode detection was evaluated. The color intensities at the test and control lines were converted to grayscale and RGB color modes by using Inkscape. As shown in Figure 2a, when the color intensity was measured by reading the green component, the sensitivity of detection was highest. In addition, the effect of different color filters was also examined. As demonstrated in Figure 2b, the combination of fluorescence and brilliance filters led to a greater difference between the blank and the OTA standard. Therefore, the processing of all the colorimetric scanned images was performed using the green color with the application of both fluorescence and brilliance filters.

Then, a comparison of two OTA-thiol-modified specific aptamers (aptamer 1: 5′/-thiolMC6-D/**GAT CGG GTG TGG GTG GCG TAA AGG GAG CAT CGG ACA** AAA AAA AAA AAA AAA-3′ and aptamer 2: 5′/-thiolMC6-D/AAA AAA AAA AAA AAA AAA **GAT CGG GTG TGG GTG GCG TAA AGG GAG CAT CGG ACA**-3′) was carried out. The active part of the two aptamers (denoted by the sequence in bold) is the same as those of the two aptamers only differing in the positioning of the poly-A chain which is located close to 3′ and 5′ end of the aptamers, respectively. As shown in Figure 3a, the assay sensitivity is greatly improved with aptamer 2 and, therefore, this is selected for further investigation.

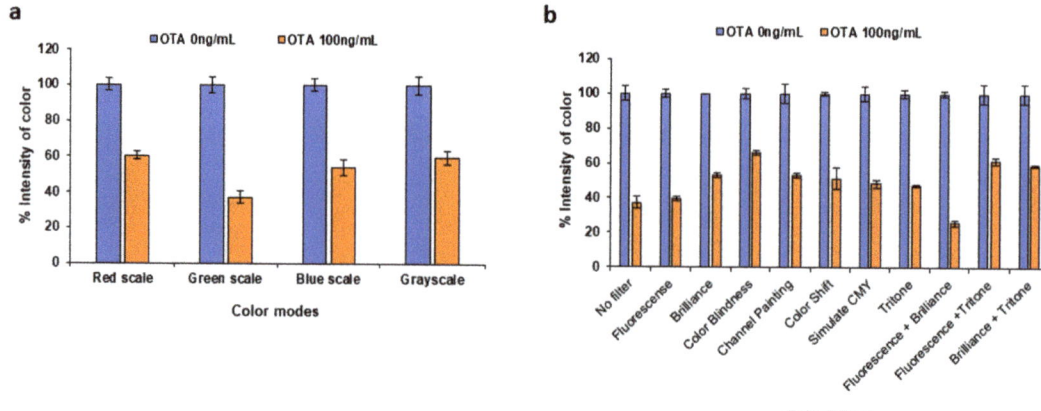

Figure 2. Effect of (**a**) the color mode (grayscale and RGB) detection and (**b**) the filters applied on the analytical signal.

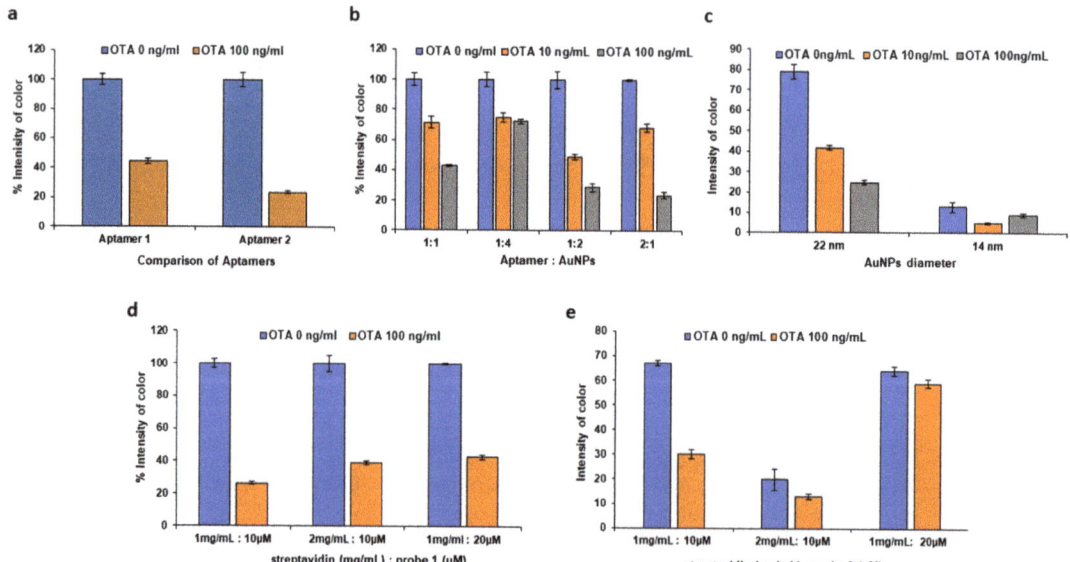

Figure 3. Effect of (**a**) the two OTA-specific aptamers (aptamer 1 and aptamer 2), (**b**) the volume ratio of the aptamer to AuNPs for the formation of OTA aptamer-AuNP conjugate, (**c**) the AuNP diameter, (**d**) the concentration of streptavidin and probe 1 for the test line and (**e**) the concentration of streptavidin and probe 2 for the control line.

The volume ratio of the aptamer to AuNPs for the formation of the aptamer-AuNP conjugate was further studied in the range from 1:4 to 2:1; the 1:2 volume ratio of aptamer to AuNPs exhibited the highest color intensity changes and better differentiation among the OTA concentrations (Figure 3b).

AuNPs were prepared according to two different methods, the modified Frens method and the Turkevich method, with particle diameters of 22 and 14 nm, respectively. The AuNPs with a diameter of 22 nm, prepared with the modified Frens method, yielded higher signal intensities (Figure 3c).

Another important parameter is the concentration of streptavidin and of probes 1 and 2 during the preparation of streptavidin–biotin–probe conjugates. When the concentration of streptavidin was 1 mg mL^{-1} and the concentration of probe 1 at the test line was 10 µmol L^{-1}, the addition of an OTA standard caused the most effective decrease in the signal of the test line, thus improving the assay sensitivity (Figure 3d). As for probe 2 at the control line, the addition of OTA should not affect the intensity of the control line; as illustrated in Figure 3e, the effect of OTA was minimal and the intensity of the control line was highest at 1 mg mL^{-1} of streptavidin and 20 µmol L^{-1} of probe 2.

Among different types of membranes, NC membranes are the most suitable for lateral flow assays. An important parameter for the selection of the NC membrane is the solution capillary flow rate. Two commonly used NC membranes were tested, Immunopore FP and Immunopore RP membrane, with flow rates of 140–200 s/4 cm and 90–150 s/4 cm, respectively. As shown in Figure S2a (Supplementary Information), when the Immunopore RP membrane was used, the addition of the OTA caused a more significant decrease in the signal of the test line.

As for the conjugate pad, two types of glass fiber pads (Whatman Standard 17, glass fiber cellulose pads) were investigated and the results (Figure S2b, Supplementary Information) indicated that Whatman Standard 17 resulted in a lower color intensity in the test line when OTA was introduced in the strip.

Four types of sample pad (Whatman Standard 17, Whatman Chromatography Paper No. 1, Whatman Chromatography Paper 3MM, Cellulose Fiber Sample Pads) were tested. The pad that proved most suitable for the strip was Whatman chromatography paper 3MM (Figure S2c, Supplementary Information), as it enhanced the sensitivity of the assay by ensuring a homogenous transportation of the sample and prevented the strip from overflowing.

Another important parameter is the volume of the AuNP-aptamer conjugate and the volume of sample loaded on the conjugate and sample pads. Varying volumes (5–20 µL) of the conjugate were loaded on the conjugate pad and the results are shown in Figure 4a. Using low volumes of the conjugate (<10 µL) resulted in lower test line intensities and the signal obtained for the OTA standard was not distinguishable from the blank signal. An amount of 15 µL of the OTA aptamer-AuNP conjugate solution provided the highest sensitivity and was selected for subsequent studies.

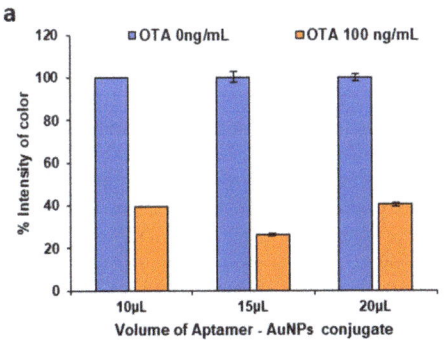 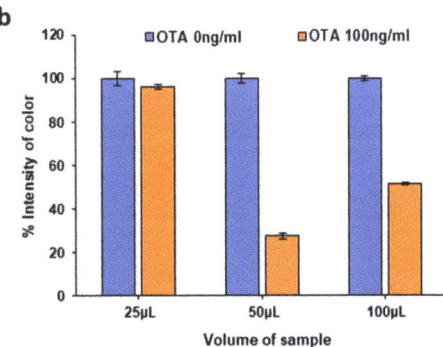

Figure 4. Selection of (**a**) the volume of the OTA aptamer/AuNP conjugate applied on the conjugate pad and (**b**) the volume of sample added on the sample pad.

Regarding the volume of the sample on the sample pad, when 25 µL of sample was applied, the flow towards the NC membrane was not uniform and continuous; thus, the signal of the OTA standard did not differ from the blank signal. Conversely, the addition of 100 µL of sample caused the sample pad to overflow, leading to a loss of analyte and consequently to the decrease in assay sensitivity. As shown in Figure 4b, the most suitable sample volume was 50 µL as this produced the maximum signal difference between the blank signal and the signal due to the OTA standard.

In order to control the sample flow rate and increase the interaction time between the target analyte and the aptamer-AuNP conjugates, a PVA flow barrier was introduced into the NC membrane upstream of the test line. The PVA barrier contributes to enhanced sensitivity by delaying the sample flow, until it gradually dissolves, as the sample passes through the NC membrane pores [21,22]. In this study, by using a PVA barrier, the flow of analyte was effectively slowed down, leading to an increased biorecognition interaction time between the OTA in the sample and the aptamer-AuNP conjugates. Different concentrations of PVA (0–2% w/w PVA) (Figure 5a) and volumes of PVA (0–2.0 µL) (Figure 5b) were tested. As shown in Figure 5a, as the PVA concentration was increased, the sensitivity improved up to 1% w/w PVA. The signal was also enhanced by increasing the volume of the PVA solution placed on the NC membrane (Figure 5b); a volume of 1.5 µL of 1% w/w PVA solution was finally selected. The results indicate that using a barrier with 1% w/w PVA compared to a strip without a PVA barrier improved the visual LOD from 1 ng mL^{-1} to 0.05 ng mL^{-1}, at the expense of a higher assay time.

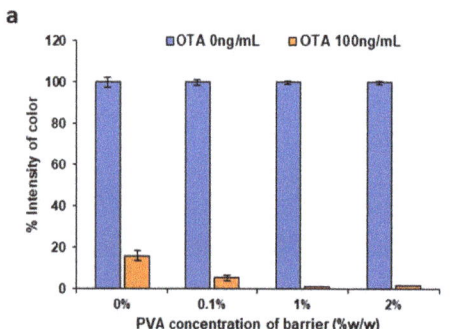 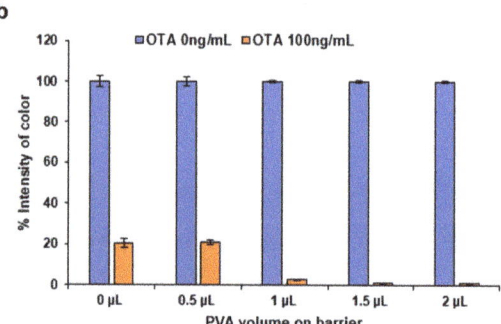

Figure 5. Effect of (**a**) the concentration and (**b**) the volume of PVA used to form PVA barrier.

To improve the sensitivity of detection, the running buffer composition is an important factor; thus, several types of running buffers were tested. Initially, four basic running buffers were tested containing 2% w/w sucrose, 1% v/v Tween 20, 0.02% w/w MgSO$_4$, 0.05% w/w (NH$_4$)$_2$SO$_4$ and 1% w/w BSA as common ingredients, with the addition of a. 50 mmol L^{-1} PBS (pH 7.4), b. 200 mmol L^{-1} tris-HCl buffer (pH 8.8), c. 50 mmol L^{-1} sodium citrate buffer (pH 6.5), and d. 50 mmol L^{-1} sodium borate buffer (pH 8.5). The use of the tris-HCl buffer and the sodium citrate buffers led to aggregation of the nanoparticles, as the color of the OTA aptamer-AuNP conjugates turned purple. Comparing the PBS and the sodium borate buffer, the results indicated that a higher sensitivity was observed with the PBS (Figure S3a, Supplementary Information).

Another important factor towards sensitivity enhancement is the addition of macromolecular aggregating agents to the running buffer. The influence of such reagents on assay sensitivity has been previously demonstrated on lateral flow immunoassays [23]. In this work, we studied how macromolecular crowding agents affect the sensitivity of the lateral flow aptasensor using PEG 20000, Ficoll F-400 and Ficoll F-70. The addition of PEG at various concentrations in the running buffer resulted in the appearance of a strong non-specific signal (Figure S3b, Supplementary Information). On the contrary, the addition of Ficoll, especially Ficoll 70, improved the assay sensitivity. Comparing F-400 and F-70, F-70 caused a more drastic decrease in signal intensity among the different OTA concentrations (Figure S3c, Supplementary Information). As shown in Figure S3d (Supplementary Information), the optimum concentration of F-70 was found to be 0.1% w/w, while concentrations higher than 0.1% w/w led to an increase in both the specific and the non-specific signal. The addition of F-70 contributed to the sensitivity enhancement, but most importantly increased the signal intensity of the control line by 45%. Another essential component of the running buffer is sucrose, in accordance with earlier reports [24]. The addition of 2% w/w sucrose to the running buffer resulted in the maximum signal intensity of the blank without increasing the non-specific signal of the OTA standard (Figure S4a, Supplementary Information). The effect of different surfactants (Tween 20, Triton X-114, SDS), proteins (BSA) and other compounds (PVP) in the running buffer was investigated. As shown in Figure S4 (Supplementary Information), the best composition for the running buffer was found to be 0.1% w/w Ficoll F-70, 2% w/w sucrose, 1% v/v Tween 20, 0.02% w/w $MgSO_4$, 0.05% w/w $(NH_4)_2SO_4$ and 1% w/w BSA in 50 mmol L^{-1} PBS (pH 7.4).

2.3. Analytical Performance

Based on the selected detection conditions, the sensitivity, the specificity and the stability of the lateral flow assay for OTA were assessed. Standard solutions of OTA at various concentrations (0 to 50 ng mL^{-1}) were prepared by diluting the OTA stock solution in running buffer and each standard was measured in triplicate. Typical scanned images of the strips using various concentrations of OTA are shown in Figure 6a. It can be seen that the blank solution produced well-defined test and control lines while the color intensity at the test lines decreased as the OTA concentration increased from 0 ng mL^{-1} to 50 ng mL^{-1}. For OTA concentration ≥ 0.05 ng mL^{-1}, distinguishable differences in the test line intensity were visually observed compared to the intensity of the blank, while the test line disappeared when the OTA concentration was >25 ng mL^{-1}. According to the definition of the visual limit of detection (LOD)—the minimum concentration of OTA leading to the color of the test line to be visually identified as weaker than the color of the control line [25]—the visual LOD of the aptamer-based strip was considered to be 0.05 ng mL^{-1}. The quantitative LOD was calculated from the calibration curve (Figure 6b) as 0.02 ng mL^{-1}. The within day assay reproducibility (expressed as the average % RSD of color intensities across the entire calibration range calculated from triplicate assays at each calibration level in a single day) was 5.5%. The between-day assay reproducibility (expressed as the average % RSD of color intensities across the entire calibration range calculated from triplicate assays at each calibration level over three different days) was 8.2%.

The specificity of the aptamer-based test strip is of great importance and was confirmed by studying potential interferences from three common mycotoxins with a similar structure to OTA: deoxynivalenol (DON), aflatoxin B1 (AFB1) and fumonisin B1 (FB1). These mycotoxins displayed a negligible influence on the test lines at equal concentrations (meaning that the average signal intensities of the control line in the presence of interferents were within ±5% of the blank signal), indicating a negligible cross-reactivity to other common mycotoxins and a good specificity of the strip for the detection of OTA (Figure 7).

In order to verify the stability of the aptasensor, strips from the same batch were prepared and stored at 4 °C for one month, three months and six months, respectively. As illustrated in Figure 8, there was no statistical difference between the signal of the test line at the strips stored for one, three and six months. Hence, the stability of the lateral flow strip remained satisfactory after a storage period of 6 months.

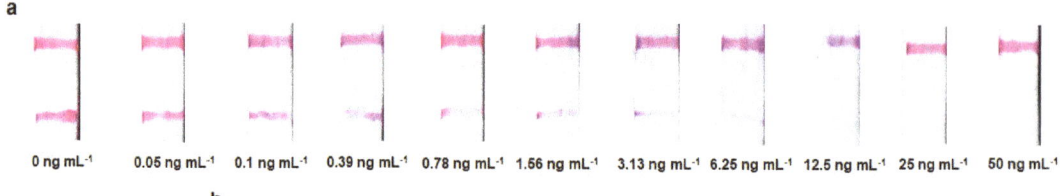

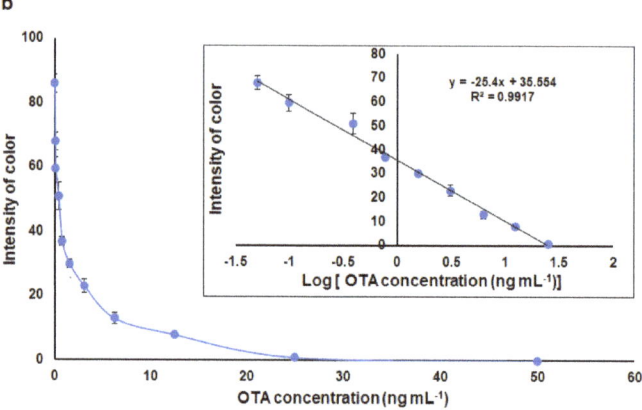

Figure 6. (**a**) Scanned images of the strips at different OTA concentrations in the range of 0–50 ng mL^{-1}, (**b**) calibration curve for OTA (the inset shows the linear relationship between the signal of the test line and the logarithm of the OTA concentration).

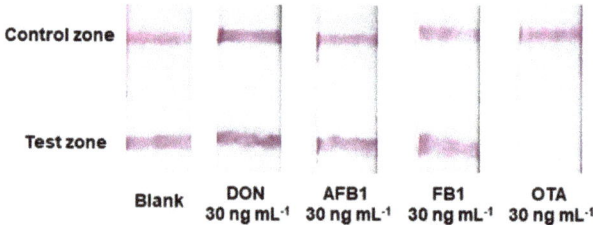

Figure 7. Selectivity study of the lateral flow assay.

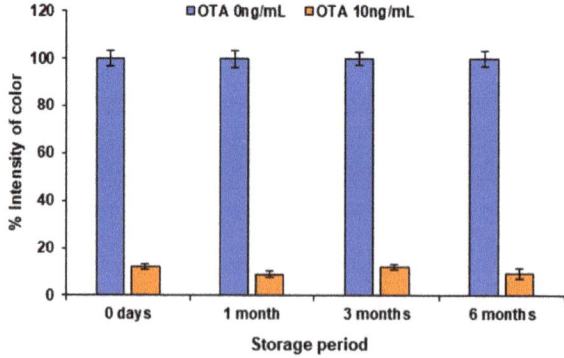

Figure 8. Stability of test of the lateral flow aptasensor.

2.4. Application of the Lateral Flow Aptasensor to Food Samples

As a way to evaluate and verify the applicability and the accuracy of the developed aptamer-based LFA, samples spiked with OTA at specific concentrations (0.1, 1, 10 ng mL^{-1}) were prepared by spiking appropriate volumes of a 2 µg mL^{-1} standard OTA solution in OTA-free beer, wine, apple juice and milk samples. The samples were analyzed in triplicate and the results are shown in Table 1. The recoveries ranged from 95 to 105%, 91 to 109%, 95 to 114% and 91 to 113% for beer, wine, apple juice and milk samples, respectively, and the % relative standard deviations (% RSDs) were in the range from 3.4 to 11.7 for beer, 3.3 to 11.8 for wine samples, 2.2 to 4.4 for apple juice and 3.3 to 13.3 for milk samples, demonstrating that the aptasensor had satisfactory accuracy and precision for OTA detection in various samples. It should be pointed out that, to the best of our knowledge, this is the first time that OTA has been effectively detected in spiked apple juice and milk samples using a lateral flow biosensor.

Table 1. Results of OTA determination in spiked samples obtained using the aptasensor developed in this work.

Sample	Amount Added (ng/mL)	Amount Determined (ng mL^{-1})	% Recovery	% RSD (n = 3)
Beer	0.1	0.104 ± 0.012	104	11.7
	1	0.95 ± 0.11	95	11.1
	10.0	10.50 ± 0.36	105	3.4
White wine	0.1	0.096 ± 0.011	96	11.8
	1	1.09 ± 0.04	109	3.7
	10.0	10.93 ± 0.85	109	7.8
Rose wine	0.1	0.104 ± 0.004	104	3.8
	1	0.937 ± 0.035	94	3.7
	10.0	9.8 ± 1.1	98	11.3
Red wine	0.1	0.090 ± 0.003	90	3.3
	1	0.977 ± 0.075	98	7.7
	10.0	9.07 ± 0.35	91	3.9
Apple Juice	0.1	0.114 ± 0.005	114	4.4
	1	1.12 ± 0.05	112	4.0
	10.0	9.45 ± 0.21	95	2.2
Milk (3.2% w/v)	0.1	0.094 ± 0.007	94	7.0
	1	0.95 ± 0.1	95	10.6
	10.0	9.07 ± 0.35	91	3.9
Milk (0% w/v)	0.1	0.105 ± 0.014	105	13.3
	1	1.13 ± 0.04	113	3.6
	10.0	9.69 ± 0.32	97	3.3

The analytical performance of the developed lateral flow strip was further confirmed by comparing the analytical and operational features with those of other validated aptamer-based lateral flow assays for OTA detection (Table 2). The aptasensor developed in this work exhibits better analytical features than existing colorimetric, visual and fluorometric lateral flow aptasensors at comparable assay times using a simple assay protocol. Only one work has reported better analytical characteristics than the current work, but this method makes use of a very complicated assay protocol, requires additional reagents, is less cost-effective and is more time consuming [26]; therefore, it is less suitable for on-site rapid screening purposes.

Table 2. Comparison of the assay proposed in this study with other reported methods for OTA detection using aptamer-based LFAs.

Detection	Labels	Visual LOD (ng/mL)	Quantitative LOD (ng/mL)	Linear Range	Total Assay Time (min)	Use of Activated Aptamer	Complex Preparation of Biorecognition Conjugate	Samples	Reference
Colorimetry	AuNPs	10^{-4}	35×10^{-6}	10^{-4}–10^2 ng/mL	200	Yes	Yes	Corn, rice, coffee bean	[26]
Colorimetry	AuNPs	1	0.18	0–2.5 ng/mL	<10	No	No	Red Wine	[27]
Visual	AuNPs	1	-	-	15	Yes	No	*Astragalus membranaceus*	[28]
Colorimetry	AuNPs	0.4	0.04	0–0.4 ng/mL	15	No	No	Red Wine	[29]
Colorimetry	AuNPs/AgNPs	0.25	1.6	0.02–25.4 ng/mL	60	No	Yes	-	[30]
Fluorescence	Cy5	-	0.4	1–1000 ng/mL	20	No	No	Corn	[31]
Fluorescence	QDs	5	1.9	0–10 ng/mL	<15	No	No	Red Wine	[32]
Fluorescence	UCNPs	-	3	0.01–50 µg/mL	30	No	Yes	Real water	[33]
Fluorescence	UCNPs	-	1.86	5–100 ng/mL	15	Yes	Yes	Wheat, Beer	[34]
Colorimetry	AuNPs	0.05	0.02	0.05–25 ng/mL	20	No	No	Beer, Wine, Apple juice, Milk	This work

AuNPs: gold nanoparticles; AgNPs: silver nanoparticles; UCNPs: upconversion nanoparticles; QDs: quantum dots; Cy5: cyanine-5.

3. Materials, Equipment and Methods

3.1. Reagents and Materials

Ochratoxin A (OTA), aflatoxin B1 (AFB1), fumonisin B1 (FB1), deoxynivalenol (DON), streptavidin, sucrose, polysorbate 20 (Tween-20), polyethylene glycol (PEG 20000), polyvinylpyrrolidone (PVP), poly vinyl alcohol (PVA), potassium carbonate (K_2CO_3), sodium hydroxide (NaOH), sodium tetraborate decahydrate ($Na_2B_4O_7 \cdot 10H_2O$), tetrachloroauric (III) acid trihydrate 99% ($HAuCl_4 \cdot 3H_2O$, ≥99.9%) and boric acid (H_3BO_3) were purchased from Sigma-Aldrich (St. Louis, MO, USA). Nitric acid (HNO_3) 65%, trisodium citrate dihydrate ($C_6H_5Na_3O_7 \cdot 2H_2O$), sodium chloride (NaCl), disodium hydrogen phosphate (Na_2HPO_4) potassium chloride (KCl) and monobasic potassium phosphate (KH_2PO_4) were obtained from Chem-Lab NV (Zedelgem, Belgium). Hydrochloric acid (HCl) 30%, magnesium sulfate heptahydrate ($MgSO_4 \cdot 7H_2O$) and ammonium sulfate ($(NH_4)_2SO_4$) were obtained from Merck (Darmstadt, Germany). Tris (hydroxymethyl) aminomethane ($C_4H_{11}NO_3$) was obtained from Duchefa Biochemie (Haarlem, The Netherlands), bovine serum albumin (BSA) was obtained from ThermoFisher Scientific (Waltham, MA, USA) and sodium dodecyl sulfate (SDS) was obtained from Fluka Biochemika (Buchs, Switzerland). Ficoll F-400 (Mr 400,000) was purchased from Sigma-Aldrich, while Ficoll F-70 (Mr 70,000) was obtained from Santa Cruz Biotechnology (Dallas, TX, USA). Water for molecular biology was supplied by PanReac AppliChem (Darmstadt, Germany). All other inorganic chemicals and organic solvents were of analytical grade. All the solutions were prepared with ultrapure water (resistivity 18.2 MΩ·cm).

Nitrocellulose (NC) membranes (Immunopore FP 5 µm, 25 mm × 50 m, 140–200 s/4 cm; Immunopore RP 8 µm, 25 mm × 50 m, 90–150 s/4 cm), Whatman STANDARD 17 bound glass fiber (34.5 s/4 cm), Whatman 3 MM Chromatography Paper and Whatman Chromatography Paper No. 1 were obtained from Whatman (Kent, UK). Glass fiber diagnostic pads and cellulose fiber sample pads were purchased from Merck Millipore (Burlington, MA, USA). HPLC-certified disposable syringe filters (pore size 0.20 µm) were obtained from MACHEREY-NAGEL (Duren, Germany)

Thiol-modified aptamers specific to OTA and biotin-modified DNA probes were obtained from Integrated DNA Technologies (Coralville, IA, USA). The detailed sequence of the aptamers and DNA probes were as follows:

Aptamer 1: 5′/-thiolMC6-D/GAT CGG GTG TGG GTG GCG TAA AGG GAG CAT CGG ACA AAA AAA AAA AAA AAA AAA-3′.

Aptamer 2: 5′/-thiolMC6-D/AAA AAA AAA AAA AAA AAA GAT CGG GTG TGG GTG GCG TAA AGG GAG CAT CGG ACA-3′.

Probe 1: 5′-/5biosg/TGT CCG ATG CTC CCT TTA CGC CAC CCA CAC CCG ATC-3′ (Test Line).

Probe 2: 5′-/5biosg/TTT TTT TTT TTT TTT TTT-3′ (Control Line).

In addition, 100 µmol L^{-1} stock solutions of the two aptamers and the complementary strands (probe 1 and probe 2) were prepared by dissolving the as-received compounds with molecular-biology-grade water and storing at -20 °C. An amount of 10 µmol L^{-1} of the aptamer solutions was prepared by adding 20 µL of the 100 µmol L^{-1} stock solutions to 180 µL of molecular-biology-grade water. A 10 µmol L^{-1} solution of probe 1 was prepared by diluting 3.5 µL of the 100 µmol L^{-1} stock solution to a final volume of 35 µL with molecular-biology-grade water. A 20 µmol L^{-1} solution of probe 2 was prepared by adding 7 µL of the 100 µmol L^{-1} stock solution to 28 µL with molecular-biology-grade water.

A 1 mg mL^{-1} stock OTA solution was prepared by dissolving 2 mg of OTA in 2 mL of methanol and stored in aliquots at -20 °C. More dilute working OTA solutions were prepared by dilution of the stock solution in the running buffer.

3.2. Equipment

A 10 µL syringe (Agilent Technologies, Santa Clara, CA, USA) was used to immobilize the reagents of the test and the control lines on the NC membrane. The Vortex mixer (VM + 10), table-top centrifuge (Eppendorf CF-10), hotplate magnetic stirrer (MSH-20D) and ultrasonic bath were supplied by Witeg (Wertheim, Germany). The analytical balance (220 g capacity, 0.1 mg readability) was obtained from Radwag (Radom, Poland), while a UV/Vis spectrophotometer (UV-1800) was obtained from Shimadzu (Kyoto, Japan)).

Optical detection on the lateral flow strips was based on reflectance colorimetry. The strips were scanned with an HP Deskjet 2720 scanner (Palo Alto, CA, USA) using 1200 dpi resolution. The further processing of images to convert the color intensity of test line to green mode was performed using the free-access Inkscape 1.3.1 vector graphics editor (https://inkscape.org/, accessed on 16 December 2023).

3.3. Preparation of the AuNPs and the AuNP-Aptamer Conjugate

AuNPs were prepared following two different modified procedures based on the Frens method [27] and the Turkevich method [35] (for details see Supplementary Material). AuNPs were characterized by using UV/vis spectrophotometry and their average size was estimated based on the absorption maximum λ_{max} = 523 nm [36]. The AuNPs prepared using the modified Frens method had an average diameter of 22 nm, while those prepared with the modified Turkevich method had an average diameter of 14 nm (Figure S1, Supplementary Information).

The formation of aptamer-AuNP conjugates depends on the reaction between AuNPs and the sulfhydryl group of the thiol-modified aptamer to form Au-S bonds. The AuNP-aptamer conjugate was prepared based on a modified published procedure [27]. More specifically, 2 mL of AuNP solution was adjusted to pH 8.5 with 0.2 mol L^{-1} K_2CO_3 and centrifuged at 10,000 rpm for 30 min, the supernatant was discarded and the residue was re-suspended to 400 µL with molecular-biology-grade water. Afterwards, 10 µmol L^{-1} of the thiol-modified aptamer was mixed with AuNP solution at a volume ratio of 1:2 and left at room temperature for 8 h. Then, the mixture was salt-aged by adding 2 mol L^{-1} NaCl solution until a final concentration of 50 mmol L^{-1} was achieved and left for another 12 h. Finally, the mixture was centrifuged at 10,000 rpm for 30 min and the supernatant was discarded to remove unbound thiol-modified aptamer and AuNPs. The precipitate was redispersed in 10 mmol L^{-1} phosphate buffer (PBS) solution (pH 7.4) and stored at 4 °C.

3.4. Preparation of Streptavidin–Biotin–DNA Probe Conjugates for Test and Control Lines

The streptavidin–biotin DNA probe conjugates were formed by exploiting streptavidin–biotin binding, as reported previously [28]. The solution added to the test line was prepared as follows: streptavidin was dissolved into 10 mmol L^{-1} PBS (pH 7.4) solution to a final concentration of 1 mg mL^{-1}. An amount of 35 µL of this solution was added to 35 µL of the 10 µmol L^{-1} probe 1 solution and incubated for 2 h at 4 °C, and then 30 µL of 10 mmol L^{-1} PBS (pH 7.4) was added into the mixture. Similarly, for the solution added to the control line, 35 µL of 1 mg mL^{-1} streptavidin solution and 35 µL of 20 µmol L^{-1} probe 2 were mixed and incubated for 2 h at 4 °C. The mixture was then added to 30 µL of 10 mmol L^{-1} PBS (pH 7.4). The streptavidin–biotin DNA probe conjugates were stored at 4 °C until their immobilization on the NC membrane.

3.5. Preparation of PVA Barrier Solution

An aqueous 1% w/w PVA solution was prepared by dissolving 0.1 g of PVA in 10 mL deionized water. The solution was placed on a heating plate and stirred at 80 °C for 2 h to obtain a homogeneous solution. Finally, the solution was stored at 4 °C.

3.6. Preparation of the Lateral Flow Strip Components

The test line (T) and control line (C) of the aptasensor were drawn on the NC membrane by using an injection needle to dispense 2 µL solution of streptavidin–biotin–probe 1 and streptavidin–biotin–probe 2, respectively. The PVA barrier (B) was loaded onto the NC membrane upstream of the test line by dispensing 1.5 µL of 1% w/w PVA solution via an injection needle. The membranes were dried at room temperature for 2 h and stored at 4 °C until use.

The sample pad was soaked with 10 mmol L^{-1} PBS solution (pH 7.4) containing 1% w/w BSA and dried at 37 °C. The conjugate pad was soaked in 10 mmol L^{-1} PBS solution (pH 7.4) containing 1% w/w BSA, 0.25% v/v Tween 20 and 2% w/w sucrose and dried at 65 °C; then, 15 µL of the synthesized AuNP-aptamer conjugate solution was loaded on the conjugate pad and dried at 37 °C before use.

The consumables for each strip cost <0.10 in terms of reagents/materials.

3.7. Sample Preparation for OTA Detection

Beer, wine samples (white, rose and red), apple juice and milk with fat contents of 3.2% w/v and 0% w/v (zero-fat) were purchased from a local supermarket. Beer samples were degassed in an ultrasonic bath for 10 min and diluted 4 times with running buffer, prior to analysis. As for wine samples, the pH was adjusted from 3.3 to 7.0 using 2 mol L^{-1} NaOH. Then, white wine samples were diluted 10 times with running buffer, while rose and red wine samples were diluted 5 times with running buffer before analysis. pH adjustment to 7.0 was also carried out for apple juice samples using 2 mol L^{-1} NaOH and the samples were filtered through 0.20 µm filters and diluted with running buffer in a volume ratio 1:4. Finally, milk samples with 3.2% w/v fat content were centrifuged at 9500 rpm for 15 min, the upper fat layer was discarded and the lower milk layer was diluted with running buffer in a volume ratio of 1:20. Zero-fat milk samples were diluted 20 times with running buffer before analysis.

3.8. Detection of OTA

The detection of OTA was carried out by applying 50 µL of the appropriate OTA standard solution or sample to the sample pad. The solution was left to move across the strip for 10 min and then the strip was washed with 50 µL of running buffer, to remove any non-specifically adsorbed conjugates and other unretained sample components at the test and control lines. Readings (whether visual or colorimetric) were taken after 20 mins. The lateral flow strips were scanned with an HP Deskjet 2720 scanner (Hewlett Packard, Palo Alto, CA, USA) at high resolution (1200 dpi). Scanned images were processed by applying fluorescence and brilliance filters and the intensity of the color in the test line for each OTA concentration was converted to green mode (RGB mode) and measured using Inkscape. For calibration purposes, color intensities were graphically plotted against OTA concentrations. The visual detection limit of the assay is defined as the minimum OTA concentration that can be visually differentiated from the blank solution. The quantitative LOD was determined from the calibration plot as the OTA concentration that produced a signal equal to $-3 \times SD$ (where SD is the standard deviation of the blank signal).

4. Conclusions

A fast, sensitive and low-cost colorimetric aptamer-based lateral flow assay was developed for OTA detection. The aptasensor achieves a visual LOD of 0.05 ng mL^{-1} and an instrumental LOD of 0.02 ng mL^{-1}. The assay time is 20 min while other mycotoxins do not interfere. The cost of each strip is calculated to be < 0.10 in terms of materials. The aptasensor was applied to the determination of OTA in beer, wine, apple juice and milk samples with recoveries in the range from 91 to 114% and relative standard deviations <14%. These biosensors are fit for purpose for the detection of OTA and compete favorably with existing aptasensing strips in terms of analytical features, simplicity, cost and speed.

Supplementary Materials: The following supporting information can be downloaded at https://www.mdpi.com/article/10.3390/molecules28248135/s1, Preparation of the AuNPs; Figure S1: Absorption spectra of AuNPs with size of (a) 22 nm prepared using the modified Frens method and (b) 14 nm prepared using the modified Turkevich method; Figure S2. Selection of (a) the NC membrane, (b) the glass fiber conjugate pad and (c) the sample pad; Figure S3. Effect of (a) the type of buffer, (b) the concentration of PEG, (c) the Ficoll 400 concentration and (d) the Ficoll 70 concentration in the running buffer; Figure S4. Effect of the (a) the concentration of sucrose, (b) the concentration of BSA, (c) the type of surfactant, (d) the concentration of Tween in the running buffer and (e) the concentration of PVP to form the flow barrier.

Author Contributions: Conceptualization, A.E., C.K. and V.P.; methodology, A.E. and V.P.; validation, V.P. and E.M.; investigation, E.M. and V.P.; data curation, E.M., V.P. and A.E.; writing—original draft preparation, E.M. and C.K.; writing—review and editing, A.E. and C.K.; project administration, A.E.; funding acquisition, A.E. All authors have read and agreed to the published version of the manuscript.

Funding: This work was funded under the European Union's Horizon 2020 Research and Innovation Program through the Marie Skłodowska-Curie Safemilk Project (Grant Agreement No 101007299).

Institutional Review Board Statement: Not applicable.

Informed Consent Statement: Not applicable.

Data Availability Statement: Data provided in the Supplementary Materials.

Conflicts of Interest: The authors declare no conflict of interest.

Abbreviations

OTA	ochratoxin A
AuNPs	gold nanoparticles
NC	nitrocellulose
IARC	International Agency for Research on Cancer
MLs	maximum limits
TLC	thin-layer chromatography
HPLC	high-performance liquid chromatography
GC	gas chromatography
SELEX	Systematic Evolution of Ligands by EXponential enrichment
AFB1	aflatoxin B1
FB1	fumonisin B1
DON	deoxynivalenol
PEG	polyethylene glycol

PVP	polyvinylpyrrolidone
PVA	poly vinyl alcohol
BSA	bovine serum albumin
RSD	relative standard deviation
LOD	limit of detection
QDs	quantum dots
AgNPs	silver nanoparticles

References

1. Van der Merwe, K.J.; Steyn, P.S.; Fourie, L.; Scott, D.B.; Theron, J.J. A toxic metabolite produced by *Aspergillus ochraceus* Wilh. *Nature* **1965**, *205*, 1112–1113. [CrossRef] [PubMed]
2. Atumo, S. A Review of Ochratoxin A Occurrence, Condition for the Formation and Analytical Methods. *Int. J. Agric. Sc. Food Technol.* **2020**, *6*, 180–185.
3. Raiola, A.; Tenore, G.C.; Manyes, L.; Meca, G.; Ritieni, A. Risk analysis of main mycotoxins occurring in food for children: An overview. *Food Chem. Toxicol.* **2015**, *84*, 169–180. [CrossRef] [PubMed]
4. Chen, X.; Gao, D.; Sun, F. Nanomaterial-based aptamer biosensors for ochratoxin A detection: A review. *Anal. Bioanal. Chem.* **2022**, *414*, 2953–2969. [CrossRef] [PubMed]
5. Roland, A.; Bros, P.; Bouisseau, A.; Cavelier, F.; Schneider, R. Analysis of ochratoxin A in grapes, musts and wines by LC–MS/MS: First comparison of stable isotope dilution assay and diastereomeric dilution assay methods. *Anal. Chim. Acta* **2014**, *818*, 39–45. [CrossRef] [PubMed]
6. Yamada, K.; Shibata, H.; Suzuki, K.; Citterio, D. Toward practical application of paper-based microfluidics for medical diagnostics: State-of-the-art and challenges. *Lab Chip* **2017**, *17*, 1206–1249. [CrossRef] [PubMed]
7. Bahadir, E.; Sezgintürk, M. Lateral flow assays: Principles, designs and labels. *Trends Anal. Chem.* **2016**, *82*, 286–306. [CrossRef]
8. Meulenberg, E.P. Immunochemical methods for ochratoxin A detection: A review. *Toxins* **2012**, *4*, 244–266. [CrossRef]
9. Goud, K.Y.; Reddy, K.K.; Satyanarayana, M.; Kummari, S.; Gobi, K.V. A review on recent developments in optical and electrochemical aptamer-based assays for mycotoxins using advanced nanomaterials. *Microchim. Acta* **2019**, *187*, 29. [CrossRef]
10. Sun, H.; Zu, Y. A highlight of recent advances in aptamer technology and its application. *Molecules* **2015**, *20*, 11959–11980. [CrossRef]
11. Bazin, I.; Nabais, E.; Lopez-Ferber, M. Rapid visual tests: Fast and reliable detection of ochratoxin A. *Toxins* **2010**, *2*, 2230–2241. [CrossRef] [PubMed]
12. Ha, T.H. Recent Advances for the Detection of Ochratoxin A. *Toxins* **2015**, *7*, 5276–5300. [CrossRef] [PubMed]
13. Tuerk, C.; Gold, L. Systematic evolution of ligands by exponential enrichment: RNA ligands to bacteriophage T4 DNA polymerase. *Science* **1990**, *249*, 505–510. [CrossRef] [PubMed]
14. Ellington, A.D.; Szostak, J.W. In vitro selection of RNA molecules that bind specific ligands. *Nature* **1990**, *346*, 818–822. [CrossRef] [PubMed]
15. Tombelli, S.; Minunni, M.; Mascini, M. Analytical applications of aptamers. *Biosens. Bioelectron.* **2005**, *20*, 2424–2434. [CrossRef] [PubMed]
16. Ilgu, M.; Nilsen-Hamilton, M. Aptamers in analytics. *Analyst* **2016**, *141*, 1551–1568. [CrossRef] [PubMed]
17. Rozenbluma, G.T.; Lopeza, V.G.; Vitullo, A.D.; Radrizzani, M. Aptamers: Current challenges and future prospects. *Expert Opin. Drug Dis.* **2016**, *11*, 127–135. [CrossRef]
18. Lee, J.; So, H.; Jeon, E.; Chang, H.; Won, K.; Kim, Y. Aptamers as molecular recognition elements for electrical nanobiosensors. *Anal. Bioanal. Chem.* **2008**, *390*, 1023–1032. [CrossRef]
19. Majdinasab, M.; Badea, M.; Marty, J.L. Aptamer-Based Lateral Flow Assays: Current Trends in Clinical Diagnostic Rapid Tests. *Pharmaceuticals* **2022**, *15*, 90. [CrossRef]
20. Song, S.; Wang, L.; Li, J.; Zhao, J.; Fan, C. Aptamer-based biosensors. *Trends Anal. Chem.* **2008**, *27*, 108–117. [CrossRef]
21. Alam, N.; Tong, L.; He, Z.; Tang, R.; Ahsan, L.; Ni, Y. Improving the sensitivity of cellulose fiber-based lateral flow assay by incorporating a water-dissolvable polyvinyl alcohol dam. *Cellulose* **2021**, *28*, 8641–8651. [CrossRef] [PubMed]
22. Harpaz, D.; Axelrod, T.; Yitian, A.; Eltzov, E.; Marks, R.S.; Tok, A. Dissolvable Polyvinyl-Alcohol Film, a Time-Barrier to Modulate Sample Flow in a 3D-Printed Holder for Capillary Flow Paper Diagnostics. *Materials* **2019**, *12*, 343. [CrossRef] [PubMed]
23. Christopoulou, N.; Kalogianni, D.; Christopoulos, T. Macromolecular crowding agents enhance the sensitivity of lateral flow immunoassays. *Biosens. Bioelectron.* **2022**, *218*, 114737. [CrossRef] [PubMed]
24. Amini, M.; Pourmand, M.R.; Faridi-Majidi, R.; Heiat, M.; Mohammad Nezhady, M.A.; Safari, M.; Noorbakhsh, F.; Baharifar, H. Optimising effective parameters to improve performance quality in lateral flow immunoassay for detection of PBP2a in methicillin-resistant *Staphylococcus aureus* (MRSA). *J. Exp. Nanosci.* **2020**, *15*, 266–279. [CrossRef]
25. Lu, S.Y.; Lin, C.; Li, Y.S.; Zhou, Y.; Meng, X.M.; Yu, S.Y.; Li, Z.H.; Li, L.; Ren, H.L.; Liu, Z.S. A screening lateral flow immunochromatographic assay for on-site detection of okadaic acid in shellfish products. *Anal. Biochem.* **2012**, *422*, 59–65. [CrossRef] [PubMed]

26. Wang, S.; Zong, Z.; Xu, J.; Yao, B.; Xu, Z.; Yao, L.; Chen, W. Recognition-Activated Primer-Mediated Exponential Rolling Circle Amplification for Signal Probe Production and Ultrasensitive Visual Detection of Ochratoxin A with Nucleic Acid Lateral Flow Strips. *Anal. Chem.* **2023**, *95*, 16398–16406. [CrossRef] [PubMed]
27. Wang, L.; Ma, W.; Chen, W.; Liu, L.; Ma, W.; Zhu, Y.; Xu, L.; Kuang, H.; Xu, C. An aptamer-based chromatographic strip assay for sensitive toxin semi-quantitative detection. *Biosens. Bioelectron.* **2011**, *26*, 3059–3062. [CrossRef]
28. Zhou, W.; Kong, W.; Dou, X.; Zhao, M.; Ouyang, Z.; Yang, M. An aptamer based lateral flow strip for on-site rapid detection of ochratoxin A in *Astragalus membranaceus*. *J. Chromatogr. B* **2016**, *1022*, 102–108. [CrossRef]
29. Liu, Y.; Liu, D.; Cui, S.; Li, C.; Yun, Z.; Zhang, J.; Sun, F. Design of a Signal-Amplified Aptamer-Based Lateral Flow Test Strip for the Rapid Detection of Ochratoxin A in Red Wine. *Foods* **2022**, *11*, 1598. [CrossRef]
30. Velu, R.; DeRosa, M. Lateral flow assays for Ochratoxin A using metal nanoparticles: Comparison of "adsorption–desorption" approach to linkage inversion assembled nano-aptasensors (LIANA). *Analyst* **2018**, *143*, 4566–4574. [CrossRef]
31. Zhang, G.; Zhu, C.; Huang, Y.; Yan, J.; Chen, A. A Lateral Flow Strip Based Aptasensor for Detection of Ochratoxin A in Corn Samples. *Molecules* **2018**, *95*, 291. [CrossRef] [PubMed]
32. Wang, L.; Chen, W.; Ma, W.; Liu, L.; Ma, W.; Zhao, Y.; Zhu, Y.; Xu, L.; Kuang, H.; Xu, C. Fluorescent strip sensor for rapid determination of toxins. *Chem. Commun.* **2011**, *47*, 1574–1576. [CrossRef] [PubMed]
33. Jin, B.; Yang, Y.; He, R.; Park, Y.; Lee, A.; Bai, D.; Li, F.; Lu, T.; Xu, F.; Lin, M. Lateral flow aptamer assay integrated smartphone-based portable device for simultaneous detection of multiple targets using upconversion nanoparticles. *Sens. Actuat. B Chem.* **2018**, *276*, 48–56. [CrossRef]
34. Wu, S.; Liu, L.; Duan, N.; Wang, W.; Yu, Q.; Wang, Z. A test strip for ochratoxin A based on the use of aptamer-modified fluorescence upconversion nanoparticles. *Microchim. Acta* **2018**, *185*, 497. [CrossRef]
35. Fu, Q.; Yuan, L.; Cao, F.; Zang, L.; Ji, D. Lateral flow strip biosensor based on streptavidin-coated gold nanoparticles with recombinase polymerase amplification for the quantitative point-of-care testing of Salmonella. *Microchem. J.* **2021**, *171*, 106859. [CrossRef]
36. He, Y.; Liu, S.; Kong, L.; Liu, Z. A study on the sizes and concentrations of gold nanoparticles by spectra of absorption, resonance Rayleigh scattering and resonance non-linear scattering. *Spectrochim. Acta A Mol. Biomol. Spectrosc.* **2005**, *61*, 2861–2866. [CrossRef]

Disclaimer/Publisher's Note: The statements, opinions and data contained in all publications are solely those of the individual author(s) and contributor(s) and not of MDPI and/or the editor(s). MDPI and/or the editor(s) disclaim responsibility for any injury to people or property resulting from any ideas, methods, instructions or products referred to in the content.

Article

Longitudinal Plant Health Monitoring via High-Resolution Mass Spectrometry Screening Workflows: Application to a Fertilizer Mediated Tomato Growth Experiment

Anthi Panara, Evagelos Gikas, Anastasia Koupa and Nikolaos S. Thomaidis *

Laboratory of Analytical Chemistry, Department of Chemistry, National and Kapodistrian University of Athens, Panepistimiopolis Zografou, 15771 Athens, Greece; panaranthi@chem.uoa.gr (A.P.); vgikas@chem.uoa.gr (E.G.); ankoupa@chem.uoa.gr (A.K.)
* Correspondence: ntho@chem.uoa.gr

Abstract: Significant efforts have been spent in the modern era towards implementing environmentally friendly procedures like composting to mitigate the negative effects of intensive agricultural practices. In this context, a novel fertilizer was produced via the hydrolysis of an onion-derived compost, and has been previously comprehensively chemically characterized. In order to characterize its efficacy, the product was applied to tomato plants at five time points to monitor plant health and growth. Control samples were also used at each time point to eliminate confounding parameters due to the plant's normal growth process. After harvesting, the plant leaves were extracted using aq. MeOH (70:30, v/v) and analyzed via UPLC-QToF-MS, using a C18 column in both ionization modes (±ESI). The data-independent (DIA/bbCID) acquisition mode was employed, and the data were analyzed by MS-DIAL. Statistical analysis, including multivariate and trend analysis for longitudinal monitoring, were employed to highlight the differentiated features among the controls and treated plants as well as the time-point sequence. Metabolites related to plant growth belonging to several chemical classes were identified, proving the efficacy of the fertilizer product. Furthermore, the efficiency of the analytical and statistical workflows utilized was demonstrated.

Keywords: onion-based fertilizer; tomato leaves; plant growth; HRMS; time series; chemometrics

Citation: Panara, A.; Gikas, E.; Koupa, A.; Thomaidis, N.S. Longitudinal Plant Health Monitoring via High-Resolution Mass Spectrometry Screening Workflows: Application to a Fertilizer Mediated Tomato Growth Experiment. *Molecules* **2023**, *28*, 6771. https://doi.org/10.3390/molecules28196771

Academic Editors: Paraskevas D. Tzanavaras and Victoria Samanidou

Received: 8 August 2023
Revised: 15 September 2023
Accepted: 15 September 2023
Published: 22 September 2023

Copyright: © 2023 by the authors. Licensee MDPI, Basel, Switzerland. This article is an open access article distributed under the terms and conditions of the Creative Commons Attribution (CC BY) license (https://creativecommons.org/licenses/by/4.0/).

1. Introduction

The agri-food needs of contemporary society have led to an international explosion of intensive agricultural practices. However, the necessity for environmentally friendly methods to ensure agricultural biosafety and protect human health has become eminent. The implementation of fertilizers produced via composting offers an alternative method for increasing the agricultural production yield in a less invasive way, taking advantage of their beneficial effect on soil and plant nutrition [1] and contributing to the cyclic economy by recycling organic waste [2].

In the terms of the current study, an onion-based fertilizer manufactured via the hydrolysis of an onion compost product is evaluated. The fertilizer was applied to tomato plants at five time points, covering all anticipated developmental stages of the tomato plant's life cycle, in order to investigate plant health and growth. This fertilizer was comprehensively chemically characterized in a previous study by our research team [3]. The selection of tomatoes, as a model organism, was driven by the plant's financial importance, as well as its widespread adoption as a reference organism for omics analysis [4–9]. Tomato (*Solanum lycopersicum*) is a *Solanaceae* family member and an exceedingly well-known and financially significant crop globally [10]. Interestingly, the tomato market is expected to grow at a CAGR of 4.76% from USD 197.76 billion in 2023 to USD 249.53 billion by 2028 [11].

A wide range of analytical techniques have been used in the literature to identify a diverse range of compounds in tomato plants. High resolution (HR) techniques, such as nuclear magnetic resonance (NMR) [7,12,13] and mass spectrometry-based (MS-based) methods such as gas chromatography–HRMS (GC-HRMS) [5,14] or liquid chromatography–HRMS (LC-HRMS) [14–17] have been employed for the determination of metabolites in tomato plants.

In order to evaluate the efficacy of a fertilizer, the plant subject's phenotypic characteristics should first be evaluated. Equally important is the assessment of the biochemical alterations that occur in the plants due to the application of fertilizer, which should be monitored via chemical analysis. The determination of specific molecular species is not as effective as holistic fingerprinting, with the latter providing access to the widest possible array of biochemical changes induced by the implementation of the novel soil compost improvement product. Therefore, metabolomics is employed in such cases [18,19], as this field has the potential to uncover the underlying biochemical processes in a hypothesis-free manner. Nevertheless, the majority of metabolomic workflows analyze a control group versus a treated one, or, at most, a limited number of treated groups. A small number of studies can be found in the literature concerning time-series metabolomics. The main drawback in such cases is the very limited number of observations, which cannot deal with autocorrelation effects (seasonal variation), unevenly spaced sampling intervals, or the mathematical complexity of evaluating a vast number of metabolites according to their longitudinal variation.

Currently, two approaches have arisen in the literature for treating time-dependent metabolomic data. For the first approach, several algorithms have been proposed to deal with the issue based on the data treatment via the incorporation of smoothing functions (i.e., cubic splines, polynomials, LOESS). Two software packages, namely, Time-omics [20] and MetaboClust [21], are available for the abovementioned approach. The second approach involves ANOVA-based algorithms such as the ANOVA–Simultaneous Component Analysis (ASCA) methodology [22].

The aim of this research was to examine the plausible differentiations between plants that had been irrigated with the novel onion-based fertilizer versus untreated plants (control samples). The examination of the fertilizer's biochemical effect on the plant will provide an overview of its efficacy in a hypothesis-free experiment. Thus, the connection between the fertilizer application and its effect on plant growth demonstrated that alternative soil treatment products could benefit plants, paving the way for a new field of fertilizer preparation with a focus on a circular economy. According to our best knowledge, this is the first study to monitor the effect of a compost fertilizer on plant growth. Metabolomics, in conjunction with chemometrics, is the appropriate tool for unveiling the chemical spaces produced by plants. Another equally important objective of the current study was the efficient combination of multivariate chemometrics with time trend analysis in order to contribute to this field, which is largely unexplored.

2. Results

2.1. Chemometric Results

A series of five time points was utilized for the investigation of the induced alterations on the plants' metabolome. The time point t0, which was the initial point of the sampling, was set on the 38th day after the planting of the seeds. At this time point, the plants have four to five leaves. The time points t15, t30, t45, and t60 correspond to 15 days, 30 days, 45 days, and 60 days, respectively, after the starting time point (t0). The leaves of tomato plants that were irrigated with the onion-based fertilizer will be referred to as "onion samples" in this manuscript. On the other hand, the leaves of the tomato plants that were not been treated with the fertilizer will be coded as "control samples". Six replicates were collected for each matrix (control and onion samples) at each time point.

2.1.1. Multivariate Analysis of Data Obtained by +ESI

Multivariate statistical analysis was deemed necessary for the data analysis due to the complexity of the experimental design. Therefore, Principal Component Analysis

(PCA) was employed. The results showed no outliers, but various time trends could be easily noticed, whereas the validity of the analysis was verified as the QC's formed a tight cluster close to the center of the plot. The locations of the QC samples are illustrated in Figures S1 and S2 in the Supplementary Materials for the control and onion samples in positive and negative ionization mode, respectively. O2PLS-DA was employed to better understand the clustering observed ($R^2Y = 0.532$, $Q^2 = 0.391$). The model's validity was assessed using permutation testing and was found to be acceptable, with permuted values of R^2 and Q^2 of 0.349 and (−0.207), respectively. The model's efficiency for proper classification was verified by CV-ANOVA and was evaluated by the corresponding p-values (p-value = 1.97×10^{-5}) and the Fisher probability test (Fisher probability practically tends to zero) resulting from the misclassification table, showing no false classifications.

The number of clusters should be assigned correctly to verify that our model is valid. Therefore, an orthogonal unbiased methodology for assigning the proper clustering was employed via the hierarchical analysis of the O2PLS-DA results. As illustrated in Figure 1b, 10 clusters corresponding to the time points from each matrix (control and onion samples) were accurately assigned using this methodology. The pruning towards the formation of the clusters was carried out according to the length of the branching, which revealed the main trends related to plant growth. Hence, the time points t30, t45, and t60 for the onion-irrigated samples are markedly different (Figure 1b). This proves that the effect of the fertilizer starts to be evident after t15. The different time points for the control and onion samples are classified properly, based on the score plots (Figure 1a). Additionally, no statistically significant difference between the aforementioned samples at time point t0 was observed, which was expected, as t0 was sampled before the application of the product.

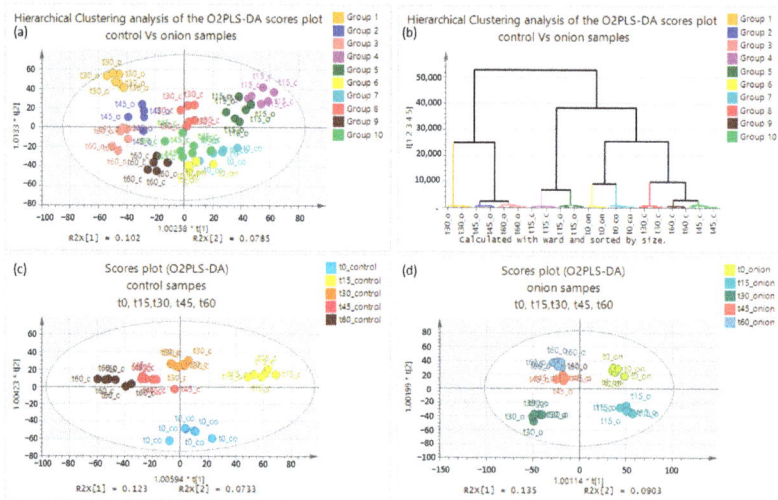

Figure 1. Plots of O2PLS-DA and hierarchical clustering analysis: (**a**) score plot of control and onion samples; (**b**) hierarchical clustering analysis O2PLS-DA of control and onion samples; (**c**) score plot O2PLS-DA discrimination using UV scaling between control samples; (**d**) score plot O2PLS-DA discrimination using UV scaling for onion samples at five time points (t0, t15, t30, t45, t60) in positive ionization mode.

In order to compare the metabolic differences imposed on the plant, excluding the mutual cross-contributions from the opposite group, O2PLS-DA was performed separately for the onion and control samples, including all of the investigated time points, as shown in Figure 1c,d. At time points t30, t45, and t60, the plants irrigated with the onion-based fertilizer showed no actual differentiation at the first component, whereas t45 and t60 can only be distinguished from t30 at the second principal component. Coalescence at t45 and t60 was observed, indicating that the metabolome does not change significantly at these time points, implying that their growth process is constrained and indicating that plant growth was completed up to the former time point (i.e., t45). A clear discrimination is observed between three time points (t0, t15, and t30), as each group is in a different quartile, a fact that is depicted in Figure 1d. Furthermore, the maximum difference between the time points can be observed between the t0 onion and the t30 onion as the maximum distance occurs, and they can be distinguished at the first and second principal components (Figure 1d).

Regarding the control samples, differentiation can be observed for t15, t30, t45, and t60 at the first principal component, whereas t0 is differentiated from the others at the second principal component. After the 45th day, slow differentiation is still visible, indicating that the metabolism continues to evolve and is consistent with the growth procedure. Furthermore, the time points in the control samples are more tightly clustered than their counterparts in the onion samples. This difference could be attributed to the contribution of fertilizer-induced growth enhancement. The difference between the various time points of the control samples is related to the plants' normal growth, whereas the difference between the various time points of the onion samples is caused by the additive effect of both their natural growth and the use of the fertilizer (Figure 1c), which may also include an interaction term.

Pairwise comparison was employed for obtaining interpretable chemometric results, revealing important variables between time points (i.e., t0 vs. t15). The pairwise comparisons (t0 vs. t15, t0 vs. t30, and t15 vs. t30), alongside their corresponding S-plots and their permutation testing, are shown in Figure 2. The Figures of Merit of the classification model for control and onion samples for all time points, as well as pairwise comparisons, are provided in Table S1 in the Supplementary Materials. Additionally, the pairwise comparisons for the time points t30 vs. t45, the S-plots, and their corresponding permutation testing in both ionization modes are presented in Figure S3 in the Supplementary Materials.

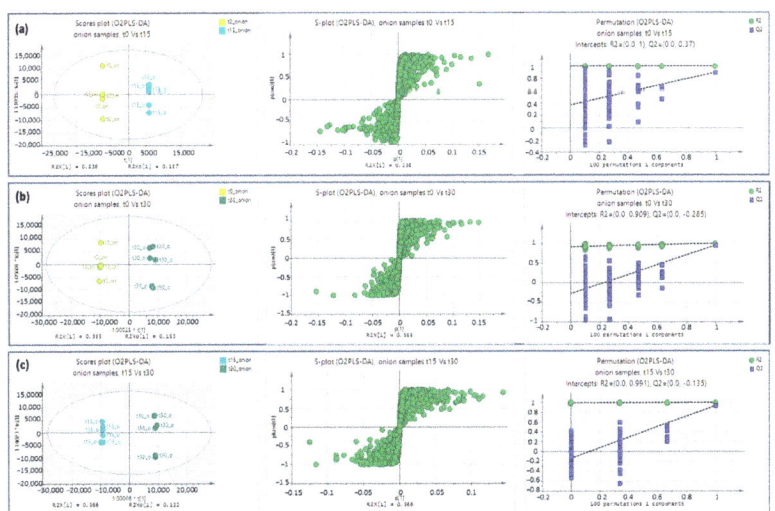

Figure 2. Score plot O2PLS-DA discrimination using pareto scaling, S-Plot, and permutation testing for onion samples at (**a**) t0, t15, (**b**) t0, t30, (**c**) t15, and t30 in positive ionization mode.

2.1.2. Multivariate Analysis of Data Obtained by ESI−

The above-described procedure was followed for the negative ionization mode. Score plots of the control and onion samples (Figure S4a), hierarchical clustering analysis for the two groups (Figure S4b), O2PLS-DA score plots for the control samples (Figure S4c), and O2PLS-DA score plots for the onion samples (Figure S4d) using UV scaling at five timepoints (t0, t15, t30, t45, t60) are presented in the Supplementary Materials.

The statistical treatment exhibits that the influence of the fertilizer differentiates the development process of the plant (Figure S4a,b). Additionally, as illustrated in Figure S4c for the control samples, t0 is discriminated from the other time points at both the first and second principal components, while t15 is only distinguished from the others at the first component. No obvious difference is seen among the time points t30, t45, and t60. For the onion-irrigated samples, as shown in Figure S4d, the time points t0, t15, and t30 are differentiated at both principal components, although t45 and t60 coalesce. Similar trends were observed from the data derived from both ionization modes. However, the number of features mined from the alignment list in ESI+ is notably higher, which presumably leads to models exhibiting higher confidence.

Pairwise O2PLS-DA comparisons were performed for the time points (a) t0 vs. t15 (b) t0 vs. t30 (c) t15 vs. t30; their score plots, as well as their corresponding S-plots, are presented in Figure S5a–c, respectively, in the Supplementary Materials. The validity of the classification models is proven by the Figures of Merit tabulated in Table S2.

2.2. Metabolites Annotation in Both Ionization Modes

The most influential variables derived from the S-plots in both ionization modes were annotated with the aid of the freely available annotation tool Sirius. The metabolites that were identified can be classified to several categories and may have an impact on plant growth. Specifically, three steroidal alkaloids and their glucosides (tomatidine, tomatidine galactoside, and solasodine), one fatty acid (stearic acid), two organic acids (chlorogenic acid, quinic acid), two flavonoids and their metabolites (quercetin, rutin (quercetin rutinoside)), one lipidglycerol (diacyl glycerol 32:2), and one chlorophyl degradation product (epoxypheophorbide a2) were identified.

The trend of these metabolites and their clustering were investigated via MetaboClust (version 1.2.2.0). The procedure described permits compounds with similar time profiles to be grouped together. The compound's name, chemical formula, experimental retention time, experimental and theoretical m/z values, and ionization mode are presented in Table 1. The CSI finger ID score, similarity score derived from SIRIUS, identification level of confidence based on Schymanski et al. [23], and clustering as provided by MetaboClust are also tabulated in this table.

Table 1. Most influential variables identified via non-target screening.

Compound Name	Chemical Formula	Exp. t_R (min)	Exp. m/z Values	Theor. m/z	ESI Mode	CSI Finger ID Score	Similarity	Level [a]	Cluster k
tomatidine	$C_{27}H_{45}NO_2$	8.0	416.3548	416.3523	+ESI	−110.14	59.8	2a	3
tomatidine galactoside	$C_{33}H_{55}NO_7$	7.3	578.4055	578.4051	+ESI	−94.64	67.8	2a	6
solasodine	$C_{27}H_{43}NO_2$	7.7	414.3384	414.3367	+ESI	−150.3	54.1	2a	2
quercetin	$C_{15}H_{10}O_7$	6.8	303.0506	303.0499	+ESI	−31.2	84.3	1	2
diacyl glycerol 32:2	$C_{35}H_{66}O_4$	15.6	551.5031	551.5034	+ESI	Lipid map		3	3
epoxypheo phorbide a(2−)	$C_{35}H_{34}N_4O_6$	14.0	607.2529	607.2551	+ESI	TomatoCyc		3	7
quinic acid	$C_7H_{12}O_6$	1.3	191.0567	191.0561	−ESI	−67.6	47.7	1	5
stearic acid	$C_{18}H_{36}O_2$	14.5	283.2654	283.2643	−ESI	−11.3	100	2a	3
chlorogenic acid	$C_{16}H_{18}O_9$	2.9	353.0873	353.0878	−ESI	−8.1	100	1	1
rutin	$C_{27}H_{30}O_{16}$	5.7	609.1459	609.1461	−ESI	−57.98	89.2	1	9

[a] Level of identification confidence based on Schymanski et al. [23].

The overlaid time-series diagrams of the annotated metabolites for the control (in red) and onion samples (in green) at five time points (t0, t15, t30, t45, t60) in both ionization modes are depicted in Figure 3. The cluster inclusion of the annotated compounds in ESI+ are illustrated in Figure S6a–d, while the corresponding ones for the annotated compounds in ESI- are depicted in Figure S7a–d.

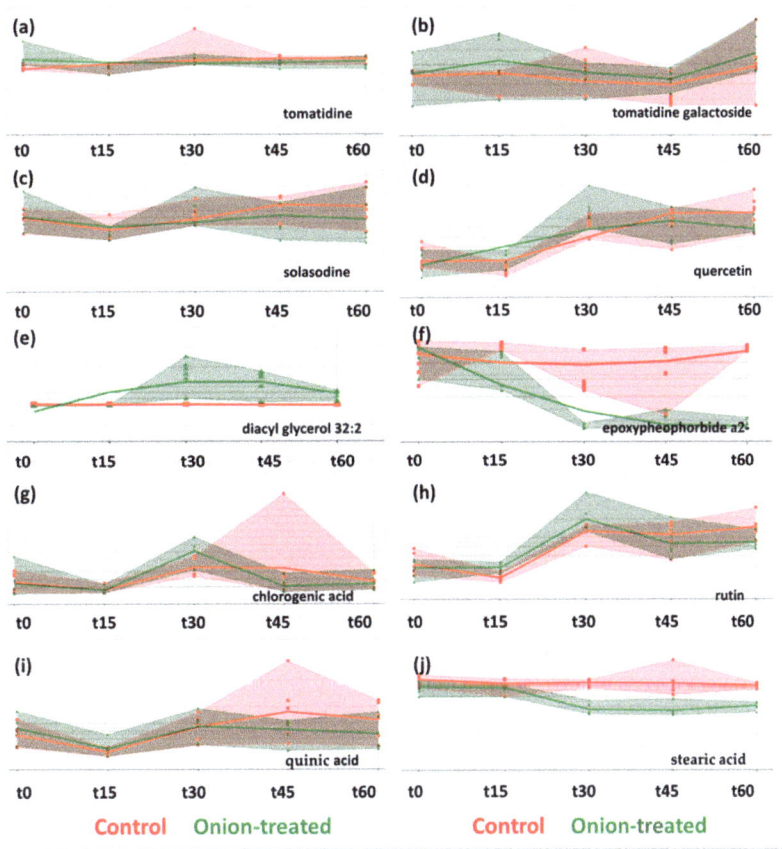

Figure 3. Overlaid time-series diagrams of control (in red) and onion samples (in green) using MetaboClust for (**a**) tomatidine, (**b**) tomatidine galactoside, (**c**) solasodine, (**d**) quercetin, (**e**) diacyl glycerol 32:2, (**f**) epoxypheophorbide a2-, (**g**) chlorogenic acid, (**h**) rutin, (**i**) quinic acid, and (**j**) stearic acid at five time points (t0, t15, t30, t45, t60). The bold red and green lines are the result of using the LOESS smoothing algorithm (span = 5). The red and green dots correspond to the replicates ($n = 6$) at each time point. The red and green shaded areas correspond to the convex hulls of the respective control and onion samples.

3. Discussion

3.1. Optimization of Sample Preparation

Metabolomic experiments should be focused on detecting as many metabolites in a sample as possible. In this direction, the maximum number of features acquired was used as a criterion to select the most efficient extraction. To decipher the influence of the extraction solvent, the same sample preparation was applied differing only for this parameter. The solvents examined were MeOH, H_2O:MeOH, 70:30 (v/v), and Chloroform: MeOH:H_2O, 20:60:20 ($v/v/v$). The samples were homogenized and lyophilized. The procedure employed was as follows: Fifty milligrams of the lyophilized samples were weighed, and 1 mL of the abovementioned solvents was added to the samples in each case. The samples were vortexed vigorously for 1 min and then placed in an ultrasonication bath for 15 min at room temperature. Following a 5 min centrifugation at 4000 rpm, the supernatants were collected and filtered through RC syringe filters. The extracts were further diluted (10-fold) with ultra-pure water and transferred to 2 mL autosampler glass vials. The extracts (5 µL) were injected into the UPLC-QToF-MS in both ionization modes. The raw data were calibrated via the Data Analysis software (version 6.0) (Bruker Daltonics, Bremen, Germany), converted by the ABF converter to the abf format [24], and then uploaded to the open-source MS-DIAL software (version 4.9.221218) [25,26], which was used for the estimation of the number of features in each case. The number of features in the positive ionization mode was 804, 1036, 1061 for MeOH, H_2O:MeOH in a 70:30 (v/v) ratio, and Chloroform: MeOH:H_2O 20:60:20, $(v/v/v)$, respectively, whereas in the negative ionization mode, the number of features was 71, 145, and 130 for MeOH, H_2O:MeOH in a 70:30 (v/v) ratio, and Chloroform: MeOH:H_2O 20:60:20, $(v/v/v)$, respectively. Taking into consideration the abovementioned process, H_2O:MeOH in a proportion of 70:30 (v/v) was selected as the most efficient extraction solvent.

A second experiment was conducted to investigate the effects of lyophilization, shaking, and sonication. The optimum extraction solvent H_2O:MeOH, 70:30 (v/v) was used for the extraction. The features obtained for each sample preparation were 898, 984, and 1173 in ESI+ for the lyophilized samples that had undergone sonication (30 min), for the non-lyophilized samples that were sonicated (30 min), and for the non-lyophilized samples that were only shaken (45 min), respectively. Furthermore, the number of features was 528, 566, and 555 for the sonicated lyophilized, sonicated non-lyophilized, and shaken non-lyophilized samples, respectively, in ESI-. The moisture percentage of the leaves (77 ± 1.9%) was taken into consideration to adjust the initial weight of the fresh samples. Taking into consideration the aforementioned results, it can be assumed that the removal of the solvent during lyophilization may have diminished the number of features that could be detected, since substances may interact irreversibly in the solid phase. This either causes the formation of insoluble or difficult-to-detect complexes. Additionally, based on the obtained results, it is evident that the sonication procedure affects the process negatively, as it leads to a lower number of features. This effect could presumably occur due to the dissipation of excess energy to the molecules, which leads to their thermal decomposition. It could be recommended, at least for such samples, that sonication or use under controlled conditions, i.e., using an ice bath, should be avoided.

3.2. Chemometric Strategy

The chemometric approach of the time-series studies was complex. Performing all of the pairwise comparisons (i.e., between the time points of the control and onion samples) may be ineffective and could lead to an enormous comparison data matrix that is difficult to interpret. Therefore, a reduced approach was selected with the aim of providing meaningful results. Thus, the maximum growth data point was located from the O2PLS-DA score plot diagrams (i.e., the time point with the largest distance from the preceding and following one derived from pairwise comparisons). These comparisons were performed for the fertilizer-treated samples. The S-plots from these comparisons showed the metabolites with the largest deviations. We attributed this to the fact that these metabolites were responsible for the developmental evolution of the tomato plants. In order to incorporate the participation of the control samples, trend analysis was performed for all data points (t0, t15, t30, t45, and t60). Finally, the trend analysis results were interrogated for their statistically significant importance for the variable highlighted in the previous procedure.

3.3. Role of the Identified Compounds in Plant Health and Their Trends

The primary goal of the current study was to demonstrate how fertilizer irrigation resulted in differentiating plant growth. A metabolomic approach was employed in an attempt to comprehend the mechanism underlying the growth, which may have demonstrated the metabolites that might be expressed differently in the control and onion groups. This is a complementary approach to the assessment of plants' phenotype characterization. Generally, the trend is focused on the time point t30, as highlighted by the O2PLS-DA. However, some differentiation was noticed for the other time points using MetaboClust. Finally, the time points that were statistically differentiated were examined.

A series of substances were highlighted as markers of growth. Tomatidine acts as a defensive agent, protecting the plant from insects, bacteria, parasites, viruses, and fungi [27]. In accordance with scientific research, tomatidine levels decrease during plant growth as they are converted to their glucoside analogs [28,29]. Based on the diagram in Figure 3a, the systematic decrease in substance levels demonstrates that the plant is developing faster as a result of the fertilizer application. At time points t45 and t60, they are differentiated statistically at 90 percent confidence level, with obtained p-values of 0.066 and 0.105, respectively.

The galactoside of tomatidine follows the opposite trend of the corresponding aglucone, as it is found to be at higher levels in the irrigated fertilizer plant (Figure 3b). Tomatidine galactoside is produced by the activity of the GlycoAlkaloid MEtabolism 1 (GAME1) glycosyltransferase [30]. This result verifies the observation that the novel fertilizer enhances the growth rate of the tomato plant compared to the controls, as it demonstrates faster anabolism of the substance compared to the control.

Solasodine decreased in an analogous matter similar to tomatidine. Based on the biosynthesis pathway "biosynthesis of alkaloids derived from terpenoid and polyketides of KEGG" (accessed on 13 July 2023) (map010666), the *Solanum* alkaloids pathway map shows that tomatidine belongs to the same route as solasodine, only being differentiated in one unsaturation to the b-ring of the steroidal skeleton. Therefore, the same trend in the time series is followed, which might partly justify the conclusion that the onion-irrigated plants had faster growth compared to the untreated ones (Figure 3c).

According to the obtained trend diagram, it appears that at time point t30, which was highlighted from O2PLS-DA as the point of maximum plant growth acceleration, the quercetin levels showed a statistically significant difference (increase) compared to the control ones (Figure 3d). Quercetin promotes plant growth [31]: it was found that spraying quercetin-3-O-rutinoside (rutin) boosts plant development and fruit production, while the antioxidant effects of the substance improve the plants' health and agility [32–34]. It is also a general remark that the antioxidant capacity is tightly linked to the enhancement of plant growth [35]. Finally, it was found that flavonoids, such as quercetin, may act as attractants, contributing to the germination of the plants [36].

The 16:3 moiety of lipids is quite widespread in the *Solanaceae* family [37]. Lipids are used in mitochondrial evolution pathways for energy storage and mobilization. An interesting trend with a quite significant increase in onion samples was observed for the compound diacyl glycerol 32:2 at time point t30. This indicates the mobilization of the lipids, presumably in order to cover the increased energy demands of the plant, which can be correlated with the fast growth of the investigated organism. It is noteworthy that this trend was maintained until the time point t60, which was defined as the end of the experiment. The above-described results are illustrated in Figure 3e.

Epoxy pheophorbide, a colorless product of chlorophyll degradation connected to the loss of green color, followed a declining trend from the initial time point (t0) of the plants' irrigation with the fertilizer and throughout the experiment (t60), as illustrated in Figure 3f. Based on the chlorophyll pathway, we come to the conclusion that the abovementioned degradation product is diminished by the application of the novel fertilizer. This may indicate the slower catabolism of chlorophyll and the contribution of fertilizer in improving plant health [38].

Chlorogenic acid, a well-known antioxidant, is vital for the chemical defense against insect herbivores [39]. Furthermore, a plant's antioxidant capacity is crucial during the early stages of development. Moreover, there are some indications that chlorogenic acid's conversion to caffeic acid may produce lignans [40], which are essential for the formation of the cell wall. Chlorogenic acid levels only increased at t30 and declined to levels below that of the control at the other time points, according to the trends shown for this compound. The only data that show significant differences between the two groups is t30 (i.e., at t30, the concentration of chlorogenic acid was higher in the onion samples than in the control ones), as illustrated in Figure 3g. The trend analysis performed by MetaboClust revealed that this trend may be significant, since it may be linked to the enhancement of plant growth.

The use of rutin has been proposed as a means of enhancing tomato plant growth, as it promotes the photosynthetic procedure as well as the levels of vital primary plant metabolites such as chlorophyll, carbohydrates, and protein content [32]. Rutin is also a powerful antioxidant, which, as mentioned before, enhances the plant growth [41]. According to the diagram in Figure 3h, a transient, statistically significant increase was observed for t15 and t30 for rutin, whereas a consistent decrease was noticed till the end of the experiment. Statistically significant important differences were observed for the time points t15 (p-value = 0.0015, confidence level 95%) and t30 (p-value = 0.10, confidence level 90%) between the control and the onion samples. Furthermore, at t60, statistically significantly lower levels (p-value = 0.021, confidence level 95%) were noticed for the onion samples.

Quinic acid exhibits antioxidant and synergistic activity in conjunction with numerous compounds such as quercetin [42] and undecanoic acid [43]. Furthermore, quinic acid is shown to chelate metals, which is in accordance with previous observations [44]. There are literature references correlating the growth and the levels of quinic acid in kiwi fruit, supporting that fruit size is increased with increasing substance levels, eventually reaching a plateau [45]. Therefore, quinic acid is considered to play a significant role in plant growth. According to Figure 3i, a statistically significant difference was noticed between t15 and t30 (p-value = 0.011) for the onion-irrigated plants, whereas for the rest of the timepoints, there appeared to be a plateau, indicating the completion of the growth. On the other hand, for the control samples, the peak was reached at the t45 time point, demonstrating a slower rate of growth.

Fatty acids are essential components of cellular metabolism, as they are used for the production of energy via beta-oxidation. Furthermore, they are used as building blocks of the cellular membrane [46]. According to the diagram in Figure 3j, stearic acid was found to decrease in the onion samples while remaining constant in the control samples. Statistically, significant differences were observed for the time points t30, t45, and t60 (p-value = 6.92×10^{-6}, 0.0018, and 2.64×10^{-6}, respectively) between the investigated groups (i.e., the control samples vs. the onion samples at the respective time points). This indicates that the onion-irrigated plants, which showed lower levels of the acid, used stearic acid intensively. This is in accordance with the use of the molecule as fuel for the increased energy demands of the process, as well as its consumption for constructing the cellular membranes of newly formed cells.

4. Materials and Methods

4.1. Leaf Harvesting and Time Point Definition

The tomato seeds were planted using supplied soil with perlite in proportion (4:1) in a greenhouse. The temperature was set at 25 ± 2 °C and the plants were watered on a daily basis using an automatic watering mechanism. Based on the life cycle of the tomato plant, five time points (t0, t15, t30, t45, and t60) were selected for plant health/growth monitoring. The application of the fertilizer was performed without any further processing of the product. The procedure for obtaining the onion-based fertilizer was described in detail in the authors' previous research [3]. Control samples were used at each time point to eliminate cofounding parameters due to the plant's normal growth. The initial time point for leaf ripening (t0) was defined 38 days after plant cultivation, when a sufficient quantity of leaves was available for the experiment to proceed. The time points t15, t30, t45, and t60 correspond to 15 days, 30 days, 45 days, and 60 days, respectively, after the starting time point (t0).

Special care was given to the termination of metabolomic functions in plants. The leaves were gently harvested, immediately placed in liquid nitrogen, and then stored at −80 °C until analysis, according to literature recommendations [47]. At each time point, leaves from three different heights of the plant (low, medium, and high) were harvested, pooled, and place in alumina foils. The harvesting of the leaves took place before the daily watering.

4.2. Reagents and Materials

All standards and reagents used were of analytical grade purity (<95%), unless differently stated. Ammonium acetate and ammonium formate were purchased from Fisher Scientific (Geel, Belgium). Methanol (MeOH) (LC–MS grade) was obtained from Merck (Darmstadt, Germany), while formic acid 99% and acetic acid were provided by Fluka (Buchs, Switzerland). The ultrapure water (H_2O) was provided by a Milli-Q device (Millipore Direct-Q UV, Bedford, MA, USA). Regenerated cellulose syringe filters (RC filters, pore size 0.2 μm, diameter 15 mm) were acquired from Macherey-Nagel (Düren, Germany). Yohimbine, reserpine, and 4-aminosalicylic acid were purchased from Sigma-Aldrich (Stenheim, Germany). Stock solutions of the reference standards (1g L^{-1}) were prepared in MeOH (LC-MS grade) and stored at −20 °C in ambient glass containers. Working solutions at a concentration of 40 mg L^{-1} were prepared by appropriate dilutions of the stock solutions with MeOH.

4.3. Sample Preparation for HRMS Analysis

The leaves were homogenized, and 0.25 g were weighted into 15 mL centrifuge tubes. A volume of 3.5 mL MeOH, including the internal standards (resperidine, yohimbine, and 4-amino salicylic acid at a concentration of 12.0 mg L^{-1}) was added to the samples. The samples were vortexed for 30 s and the addition of 6.5 mL ultrapure water followed. Consequently, they were vortexed for 30 s and shaken for 30 min on a horizontal shaker. The samples were centrifuged at 4000 rpm for 5 min, and the supernatants were filtered through RC syringe filters. The extracts were diluted two-fold with the addition of a H_2O: MeOH, 70:30 (v/v) mixture and 5 μL of the extract was injected into UPLC-QToF-MS in both ionization modes.

4.4. Instrumental Analysis

The analysis for the elucidation of the differentiated metabolites in the plant leaves was carried out using an ultra-high-performance liquid chromatograph (UHPLC) equipped with an HPG-3400 pump (Dionex Ultimate 3000 RSLC, Thermo Fisher Scientific, Dreieich, Germany) coupled to a time-of-flight mass analyzer (Hybrid Quadrupole Time of Flight Matic Bruker Daltonics, Bremen, Germany). An Acclaim RSLC 120 C18 column (2.2 μm, 2.1 × 100 mm, Thermo Fisher Scientific, Dreieich, Germany) was used, equipped with a pre-column (Van guard Acquity UPLC BEH C18 (1.7 μm, 2.1 × 5 mm, Waters, Ireland). The column temperature was set at 30 °C throughout the chromatographic analysis, while the injection volume was 5 μL. The analysis was performed in both ESI modes (positive and negative). The mobile phases in the positive ionization mode consisted of: (A) aq. 5 mM ammonium formate: MeOH (90:10 v/v) acidified with 0.01% formic acid, and (B) 5 mM ammonium formate in MeOH acidified with 0.01% formic acid, and, in negative ionization mode, were (A) aq. 10 mM ammonium acetate: MeOH (90:10 v/v) and (B) 10 mM ammonium acetate in MeOH. The LC and MS settings are described in detail in previous work by our group [3]. Both data-independent (broad-band Collision Induced Dissociation -bbCID: DIA) and data-dependent acquisition (AutoMS, DDA) modes were utilized. The MS and MS/MS spectra were obtained using two different collision energies (4 eV and 25 eV) in the bbCID mode. In AutoMS mode, the five most abundant ions per MS scan were selected and fragmented using ramp collision energy.

Quality control (QC) samples were prepared and analyzed at the beginning as well as during the sequence in both acquisition modes. The QC samples were analyzed at the beginning of the sequence as an indicator of instrument stability concerning the retention time and the signal intensity. Additionally, QC samples were used during the sequence for batch correction due to the drift of the instrument. Two QCs were prepared related to the matrix investigated, receiving equal quantities of each matrix at each time point. Therefore, one QC sample for the control samples and one QC sample for the leaves irrigated with the onion-based fertilizer were prepared. Additionally, one procedure blank at each time point was prepared.

4.5. Mass Spectrometry Data Analysis

The workflow developed for the current study is depicted in Figure 4. Each step is analyzed below.

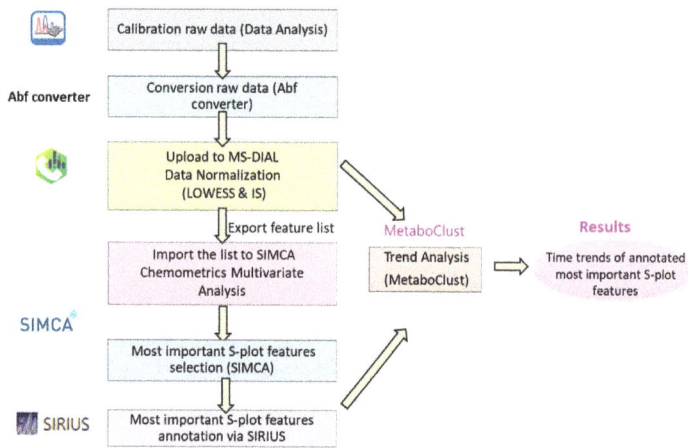

Figure 4. Workflow of the developed methodology for the identification of the most important metabolites for plant growth.

4.5.1. Screening Methodology

The raw data were calibrated via Data Analysis software (Bruker Daltonics, Bremen, Germany), converted using the ABF converter [24], and then uploaded to the open-source MS-DIAL software (version 4.9.221218) [25,26]. The data derived from data-dependent (autoMS) and data-independent (bbCID) acquisition modes were processed separately. The sequence of the samples was noted, as normalization was conducted with QC samples and internal standards (LOWESS and internal standard normalization). The exported list (peak area, retention time, and m/z) of features was imported to SIMCA 14.1 for chemometric analysis.

4.5.2. Chemometrics Methodology

The annotation list was derived from MS-DIAL, and the information (m/z, retention time, peak area, class) was used to build the SIMCA dataset. PCA hierarchical analysis and O2PLS-DA were carried out, with various scaling parameters (UV, pareto, no scaling) employed. The predictive value (Q^2) was calculated using seven-fold cross validation, while R^2 was estimated using all of the observations. The model's credibility was assessed via Figures of Merit (R^2, Q^2, CV-ANOVA, p-value, permutation testing for 100 random permutations, and confusion matrix). O2PLS-DA was performed using UV scaling on the control and onion samples (t0, t15, t30, t45, t60), and pairwise O2PLS-DA was performed for the onion samples (t0 vs. t15, t15 vs. t30, and t0 vs. t30) using pareto scaling. Their corresponding S-plots were constructed to elucidate the model's most highly contributing features. To investigate the trend of these features, MetaboClust was used.

4.5.3. Time Series

MetaboClust is an open-source software program used to analyze time-course metabolomics [21]. The peak intensities, observations, and the peak information table, that is, the data inputs for the software, were organized in csv files. The peak-picking process, which was carried out using the MS-DIAL, generated an msp file including information such as the peak areas per sample for the associated feature (m/z and retention time). These pieces of information were included in the csv file called "IntensityData" based on the requirement of the software. Furthermore, the "Observation Info" csv file comprised experimental information such as sample names, the experimental group (e.g., control, onion), the time point (e.g., t0, t15, t30, t45, t60), the number of replicates per group and time point, as well as the batch and acquisition order. The "Peak Info" csv file, which contained information regarding m/z, retention time, and the ionization mode, was also required. Finally, the annotation file contained the name of the metabolite alongside its related m/z and retention time.

Batch correction was accomplished using UV scaling and centering, while trend fitting was realized using LOESS, where the span was set to 5. Metabolite clustering was achieved using the k-means Hartigan Wong algorithm based on Euclidean distance, where the number of clusters was defined to 10 (k = 10).

4.5.4. Metabolite Annotation

Sirius was utilized for metabolite annotation [48]. The list of the features was exported from MS-DIAL (msp format) and imported to Sirius 5.8.1. Briefly, the identification was based on the accurate mass and the MS/MS spectra. Thus, MS-DIAL software (version 4.9.221218) was used to construct the extracted ion chromatograms, but mainly to deconvolute the MS/MS spectra, which were then subjected to the Sirius 5.8.1-based analysis. Various databases were implemented, such as PlantCyc, BioCyc, Coconut, GNPS, PubChem, Natural Products, KNApSacK, and KEGG, for molecular formula and structure identification. CANOPUS was used for predicting the compound class using the MS/MS spectra [49]. CSI:FingerID [50] and the similarity score implemented were used for the evaluation of the annotated metabolites. The identification level of confidence was ranked based on Schymanski et al.'s scheme.

5. Conclusions

A comprehensively characterized onion-based fertilizer was utilized for the irrigation of tomato plants to investigate its effect on plant growth. Differentiation between the control and the onion-irrigated plants was highlighted via the investigation of a five-time-point experiment based on the tomato cycle life. The identified compounds, belonging to various categories such as steroidal alkaloids and their glucosides, organic acids, fatty acids, flavonoids, and their metabolites, act beneficially for plant growth. These metabolites were derived by applying a newly developed workflow combining multivariate chemometrics (O2PLS-DA) and time trend analysis of the most important variables. In order to facilitate of the adoption of the proposed workflow from the scientific community, open-source software was employed for peak picking (MS-DIAL), annotation (SIRIUS), and time-series monitoring (MetaboClust). Employing advanced analytical methods (i.e., HRMS-based metabolomics) provides the opportunity to gain holistic information about plant development, as these approaches reveal how the underlying biochemistry is affected by the corresponding intervention. This kind of information acts in a complementary and synergistic way alongside the established methods of agricultural experimentation, such as phenotype monitoring.

The results indicate that the presence of the identified compounds could ameliorate plant health, paving the way for the holistic monitoring of plant growth. Finally, this research highlights the significance of developing by-product-based fertilizers, exploiting otherwise neglected materials and ultimately contributing to the cyclic economy.

Supplementary Materials: The following supporting information can be downloaded at: https://www.mdpi.com/article/10.3390/molecules28196771/s1: Figure S1: Score plot O2PLS-DA of (a) control samples and (b) onion samples at five time points (t0, t15, t30, t45, t60) with their corresponding QC samples using UV scaling in positive ionization mode; Figure S2: Score plot O2PLS-DA of (a) control samples and (b) onion samples at five timepoints (t0, t15, t30, t45, t60) with their corresponding QC samples using UV scaling in negative ionization mode; Figure S3: Score plot O2PLS-DA discrimination using pareto scaling, S-Plot, and permutation testing for onion samples at t30 vs. t45 in (a) positive ionization mode and (b) negative ionization mode; Figure S4: Plots of O2PLS-DA and hierarchical clustering analysis: (a) score plots of control and onion samples; (b) hierarchical clustering analysis O2PLS-DA of control and onion samples: score plot O2PLS-DA discrimination between (c) control samples and (d) onion samples at five timepoints (t0, t15, t30, t45, t60) using UV scaling in negative ionization mode; Figure S5: Score plot O2PLS-DA discrimination using pareto scaling, S-Plot, and permutation testing for onion samples at (a) t0 vs. t15, (b) t0 vs. t30, and (c) t15 vs. t30 in negative ionization mode; Figure S6: Examples of compound clustering between control (in red) and onion samples (in green) in positive ionization mode using MetaboClust: (a) Cluster 2, (b) Cluster 3, (c) Cluster 6, and (d) Cluster 7; Figure S7: Examples of compound clustering between control (in red) and onion samples (in green) using MetaboClust: (a) Cluster 1, (b) Cluster 3, (c) Cluster 5, and (d) Cluster 9 in negative ionization mode. Table S1: Performance characteristics of each classification model in positive ionization mode. Table S2: Performance characteristics of each classification model in negative ionization mode.

Author Contributions: Conceptualization, N.S.T.; methodology, A.P.; validation; A.P. and E.G.; formal analysis, A.P. and E.G.; investigation, A.P. and A.K.; resources, N.S.T.; data curation, A.P. and E.G.; writing—original draft preparation, A.P.; writing—review and editing, A.P., N.S.T. and E.G.; supervision, N.S.T. and E.G.; project administration, N.S.T.; funding acquisition, N.S.T. All authors have read and agreed to the published version of the manuscript.

Funding: This research was co-financed by the European Regional Development Fund of the European Union and Greek national funds through the Operational Program of Competitiveness, Entrepreneurship and Innovation, under the call RESEARCH—CREATE—INNOVATE (project code: T2EDK-00965).

Institutional Review Board Statement: Not applicable.

Informed Consent Statement: Not applicable.

Data Availability Statement: Data are available upon request.

Conflicts of Interest: The authors declare no conflict of interest.

Sample Availability: Samples are not available for further analysis.

References

1. Adugna, G. A review on impact of compost on soil properties, water use and crop productivity. *Acad. Res. J. Agric. Sci. Res.* **2016**, *4*, 93–104. [CrossRef]
2. Mohammad, H.; Golabi, M.J.D.; Iyekar, C. Use of Composted Organic Wastes As Alternative to Synthetic Fertilizers for Enhancing Crop Productivity and Agricultural Sustainability on the Tropical Island of Guam. In Proceedings of the 13th International Soil Conservation Organisation Conference, Brisbane, Australia, 4–8 July 2004; Conserving Soil and Water for Society: Sharing Solutions. Volume 234.
3. Panara, A.; Gikas, E.; Thomaidis, N.S. From By-Products to Fertilizer: Chemical Characterization Using UPLC-QToF-MS via Suspect and Non-Target Screening Strategies. *Molecules* **2022**, *27*, 3498. [CrossRef]
4. Chaudhary, J.; Khatri, P.; Singla, P.; Kumawat, S.; Kumari, A.; Vikram, A.; Jindal, S.K.; Kardile, H.; Kumar, R.; Sonah, H.; et al. Advances in Omics Approaches for Abiotic Stress Tolerance in Tomato. *Biology* **2019**, *8*, 90. [CrossRef]
5. Kazmi, R.H.; Willems, L.A.J.; Joosen, R.V.L.; Khan, N.; Ligterink, W.; Hilhorst, H.W.M. Metabolomic analysis of tomato seed germination. *Metabolomics* **2017**, *13*, 145. [CrossRef]
6. Pentimone, I.; Colagiero, M.; Rosso, L.C.; Ciancio, A. Omics applications: Towards a sustainable protection of tomato. *Appl. Microbiol. Biotechnol.* **2020**, *104*, 4185–4195. [CrossRef] [PubMed]

7. Afifah, E.N.; Murti, R.H.; Nuringtyas, T.R. Metabolomics Approach for the Analysis of Resistance of Four Tomato Genotypes (*Solanum lycopersicum* L.) to Root-Knot Nematodes (Meloidogyne Incognita). *Open Life Sci.* **2019**, *14*, 141–149. [CrossRef]
8. Zhu, G.; Wang, S.; Huang, Z.; Zhang, S.; Liao, Q.; Zhang, C.; Lin, T.; Qin, M.; Peng, M.; Yang, C.; et al. Rewiring of the Fruit Metabolome in Tomato Breeding. *Cell* **2018**, *172*, 249–261.e12. [CrossRef]
9. De Vos, R.C.; Hall, R.D.; Moing, A. *Metabolomics of a Model Fruit: Tomato*; Blackwell: Oxford, UK, 2011; Volume 43, pp. 109–156.
10. Knapp, S.; Peralta, I. *The Tomato (Solanum lycopersicum L., Solanaceae) and Its Botanical Relatives*; Springer: Berlin/Heidelberg, Germany, 2016. [CrossRef]
11. Intelligence, M. Tomato Market Size & Share Analysis—Growth Trends & Forecasts (2023–2028). Available online: https://www.mordorintelligence.com/industry-reports/tomato-market (accessed on 22 June 2023).
12. Abreu, A.C.; Fernandez, I. NMR Metabolomics Applied on the Discrimination of Variables Influencing Tomato (*Solanum lycopersicum*). *Molecules* **2020**, *25*, 3738. [CrossRef]
13. Mazzei, P.; Vinale, F.; Woo, S.L.; Pascale, A.; Lorito, M.; Piccolo, A. Metabolomics by Proton High-Resolution Magic-Angle-Spinning Nuclear Magnetic Resonance of Tomato Plants Treated with Two Secondary Metabolites Isolated from Trichoderma. *J. Agric. Food Chem.* **2016**, *64*, 3538–3545. [CrossRef] [PubMed]
14. Mun, H.I.; Kwon, M.C.; Lee, N.R.; Son, S.Y.; Song, D.H.; Lee, C.H. Comparing Metabolites and Functional Properties of Various Tomatoes Using Mass Spectrometry-Based Metabolomics Approach. *Front. Nutr.* **2021**, *8*, 659646. [CrossRef]
15. Moco, S.; Bino, R.J.; Vorst, O.; Verhoeven, H.A.; de Groot, J.; van Beek, T.A.; Vervoort, J.; de Vos, C.H. A liquid chromatography-mass spectrometry-based metabolome database for tomato. *Plant Physiol.* **2006**, *141*, 1205–1218. [CrossRef]
16. de Oliveira, A.N.; Bolognini, S.R.F.; Navarro, L.C.; Delafiori, J.; Sales, G.M.; de Oliveira, D.N.; Catharino, R.R. Tomato classification using mass spectrometry-machine learning technique: A food safety-enhancing platform. *Food Chem.* **2023**, *398*, 133870. [CrossRef]
17. Messaili, S.; Qu, Y.; Fougere, L.; Colas, C.; Desneux, N.; Lavoir, A.V.; Destandau, E.; Michel, T. Untargeted metabolomic and molecular network approaches to reveal tomato root secondary metabolites. *Phytochem. Anal. PCA* **2021**, *32*, 672–684. [CrossRef] [PubMed]
18. Cheng, H.; Yuan, M.; Tang, L.; Shen, Y.; Yu, Q.; Li, S. Integrated microbiology and metabolomics analysis reveal responses of soil microorganisms and metabolic functions to phosphorus fertilizer on semiarid farm. *Sci. Total Environ.* **2022**, *817*, 152878. [CrossRef] [PubMed]
19. Sun, J.; Li, W.; Zhang, Y.; Guo, Y.; Duan, Z.; Tang, Z.; Abozeid, A. Metabolomics Analysis Reveals Potential Mechanisms in *Bupleurum* L. (Apiaceae) Induced by Three Levels of Nitrogen Fertilization. *Agronomy* **2021**, *11*, 2291. [CrossRef]
20. Bodein, A.; Scott-Boyer, M.P.; Perin, O.; Le Cao, K.A.; Droit, A. timeOmics: An R package for longitudinal multi-omics data integration. *Bioinformatics* **2022**, *38*, 577–579. [CrossRef]
21. Rusilowicz, M.J.; Dickinson, M.; Charlton, A.J.; O'Keefe, S.; Wilson, J. MetaboClust: Using interactive time-series cluster analysis to relate metabolomic data with perturbed pathways. *PLoS ONE* **2018**, *13*, e0205968. [CrossRef]
22. Bertinetto, C.; Engel, J.; Jansen, J. ANOVA simultaneous component analysis: A tutorial review. *Anal. Chim. Acta X* **2020**, *6*, 100061. [CrossRef] [PubMed]
23. Schymanski, E.L.; Jeon, J.; Gulde, R.; Fenner, K.; Ruff, M.; Singer, H.P.; Hollender, J. Identifying small molecules via high resolution mass spectrometry: Communicating confidence. *Environ. Sci. Technol.* **2014**, *48*, 2097–2098. [CrossRef]
24. Reifycs Abf Converter. Available online: https://www.reifycs.com/AbfConverter (accessed on 12 July 2023).
25. Tsugawa, H.; Cajka, T.; Kind, T.; Ma, Y.; Higgins, B.; Ikeda, K.; Kanazawa, M.; VanderGheynst, J.; Fiehn, O.; Arita, M. MS-DIAL: Data-independent MS/MS deconvolution for comprehensive metabolome analysis. *Nat. Methods* **2015**, *12*, 523–526. [CrossRef]
26. MS Dial. Available online: http://prime.psc.riken.jp/compms/msdial/main.html (accessed on 12 July 2023).
27. Bailly, C. The steroidal alkaloids alpha-tomatine and tomatidine: Panorama of their mode of action and pharmacological properties. *Steroids* **2021**, *176*, 108933. [CrossRef]
28. Nakayasu, M.; Ohno, K.; Takamatsu, K.; Aoki, Y.; Yamazaki, S.; Takase, H.; Shoji, T.; Yazaki, K.; Sugiyama, A. Tomato roots secrete tomatine to modulate the bacterial assemblage of the rhizosphere. *Plant Physiol.* **2021**, *186*, 270–284. [CrossRef]
29. Nakabayashi, R.; Sawada, Y.; Yamada, Y.; Suzuki, M.; Hirai, M.Y.; Sakurai, T.; Saito, K. Combination of liquid chromatography-Fourier transform ion cyclotron resonance-mass spectrometry with 13C-labeling for chemical assignment of sulfur-containing metabolites in onion bulbs. *Anal. Chem.* **2013**, *85*, 1310–1315. [CrossRef]
30. Itkin, M.; Heinig, U.; Tzfadia, O.; Bhide, A.J.; Shinde, B.; Cardenas, P.D.; Bocobza, S.E.; Unger, T.; Malitsky, S.; Finkers, R.; et al. Biosynthesis of antinutritional alkaloids in solanaceous crops is mediated by clustered genes. *Science* **2013**, *341*, 175–179. [CrossRef]
31. Singh, P.; Arif, Y.; Bajguz, A.; Hayat, S. The role of quercetin in plants. *Plant Physiol. Biochem.* **2021**, *166*, 10–19. [CrossRef] [PubMed]
32. Gorni, P.H.; de Lima, G.R.; Pereira, L.M.d.O.; Spera, K.D.; Lapaz, A.d.M.; Pacheco, A.C. Increasing plant performance, fruit production and nutritional value of tomato through foliar applied rutin. *Sci. Hortic.* **2022**, *294*, 110755. [CrossRef]
33. Parvin, K.; Hasanuzzaman, M.; Bhuyan, M.; Mohsin, S.M.; Fujita, A.M. Quercetin Mediated Salt Tolerance in Tomato through the Enhancement of Plant Antioxidant Defense and Glyoxalase Systems. *Plants* **2019**, *8*, 247. [CrossRef] [PubMed]
34. Aharoni, A.; O'Connell, A.P. Gene expression analysis of strawberry achene and receptacle maturation using DNA microarrays. *J. Exp. Bot.* **2002**, *53*, 2073–2087. [CrossRef] [PubMed]

35. Rodrigues de Queiroz, A.; Hines, C.; Brown, J.; Sahay, S.; Vijayan, J.; Stone, J.M.; Bickford, N.; Wuellner, M.; Glowacka, K.; Buan, N.R.; et al. The effects of exogenously applied antioxidants on plant growth and resilience. *Phytochem. Rev.* **2023**, *22*, 407–447. [CrossRef]
36. Mierziak, J.; Kostyn, K.; Kulma, A. Flavonoids as important molecules of plant interactions with the environment. *Molecules* **2014**, *19*, 16240–16265. [CrossRef]
37. Christie, W.W. The Lipid Web, Glycosyldiacylglycerols. Available online: https://lipidmaps.org/resources/lipidweb/lipidweb_html/lipids/complex/mg-dgdg/index.htm (accessed on 20 July 2023).
38. Hortensteiner, S.; Wuthrich, K.L.; Matile, P.; Ongania, K.H.; Krautler, B. The key step in chlorophyll breakdown in higher plants. Cleavage of pheophorbide a macrocycle by a monooxygenase. *J. Biol. Chem.* **1998**, *273*, 15335–15339. [CrossRef]
39. Kundu, A.; Vadassery, J. Chlorogenic acid-mediated chemical defence of plants against insect herbivores. *Plant Biol.* **2019**, *21*, 185–189. [CrossRef]
40. Volpi, E.S.N.; Mazzafera, P.; Cesarino, I. Should I stay or should I go: Are chlorogenic acids mobilized towards lignin biosynthesis? *Phytochemistry* **2019**, *166*, 112063. [CrossRef]
41. Yang, J.; Guo, J.; Yuan, J. In vitro antioxidant properties of rutin. *LWT—Food Sci. Technol.* **2008**, *41*, 1060–1066. [CrossRef]
42. Arya, A.; Al-Obaidi, M.M.; Shahid, N.; Bin Noordin, M.I.; Looi, C.Y.; Wong, W.F.; Khaing, S.L.; Mustafa, M.R. Synergistic effect of quercetin and quinic acid by alleviating structural degeneration in the liver, kidney and pancreas tissues of STZ-induced diabetic rats: A mechanistic study. *Food Chem. Toxicol. Int. J. Publ. Br. Ind. Biol. Res. Assoc.* **2014**, *71*, 183–196. [CrossRef]
43. Muthamil, S.; Balasubramaniam, B.; Balamurugan, K.; Pandian, S.K. Synergistic Effect of Quinic Acid Derived from Syzygium cumini and Undecanoic Acid Against *Candida* spp. Biofilm and Virulence. *Front. Microbiol.* **2018**, *9*, 2835. [CrossRef] [PubMed]
44. Menelaou, M.; Mateescu, C.; Salifoglou, A. D-(-)-Quinic acid: An efficient physiological metal ion ligand. *J. Agroaliment. Process. Technol.* **2011**, *17*, 344–347.
45. Marsh, K.B.; Boldingh, H.L.; Shilton, R.S.; Laing, W.A. Changes in quinic acid metabolism during fruit development in three kiwifruit species. *Funct. Plant Biol.* **2009**, *36*, 463–470. [CrossRef]
46. Lim, G.H.; Singhal, R.; Kachroo, A.; Kachroo, P. Fatty Acid- and Lipid-Mediated Signaling in Plant Defense. *Annu. Rev. Phytopathol.* **2017**, *55*, 505–536. [CrossRef] [PubMed]
47. Kim, H.K.; Verpoorte, R. Sample preparation for plant metabolomics. *Phytochem. Anal. PCA* **2010**, *21*, 4–13. [CrossRef]
48. Duhrkop, K.; Fleischauer, M.; Ludwig, M.; Aksenov, A.A.; Melnik, A.V.; Meusel, M.; Dorrestein, P.C.; Rousu, J.; Bocker, S. SIRIUS 4: A rapid tool for turning tandem mass spectra into metabolite structure information. *Nat. Methods* **2019**, *16*, 299–302. [CrossRef] [PubMed]
49. Duhrkop, K.; Nothias, L.F.; Fleischauer, M.; Reher, R.; Ludwig, M.; Hoffmann, M.A.; Petras, D.; Gerwick, W.H.; Rousu, J.; Dorrestein, P.C.; et al. Systematic classification of unknown metabolites using high-resolution fragmentation mass spectra. *Nat. Biotechnol.* **2021**, *39*, 462–471. [CrossRef] [PubMed]
50. Duhrkop, K.; Shen, H.; Meusel, M.; Rousu, J.; Bocker, S. Searching molecular structure databases with tandem mass spectra using CSI:FingerID. *Proc. Natl. Acad. Sci. USA* **2015**, *112*, 12580–12585. [CrossRef] [PubMed]

Disclaimer/Publisher's Note: The statements, opinions and data contained in all publications are solely those of the individual author(s) and contributor(s) and not of MDPI and/or the editor(s). MDPI and/or the editor(s) disclaim responsibility for any injury to people or property resulting from any ideas, methods, instructions or products referred to in the content.

Article

Electrochemical Determination of the Drug Colchicine in Pharmaceutical and Biological Samples Using a 3D-Printed Device

Maria Filopoulou, Giorgios Michail [†], Vasiliki Katseli [†], Anastasios Economou [†] and Christos Kokkinos *

Laboratory of Analytical Chemistry, Department of Chemistry, National and Kapodistrian University of Athens, 157 71 Athens, Greece; mari.filop3@gmail.com (M.F.); geomixe1@gmail.com (G.M.); lilikats0@gmail.com (V.K.); aeconomo@chem.uoa.gr (A.E.)
* Correspondence: christok@chem.uoa.gr; Tel.: +30-2107274312
[†] These authors contributed equally to this work.

Abstract: In this work, a simple, fast, and sensitive voltammetric method for the trace determination of the alkaloid drug colchicine (Colc) using a 3D-printed device is described. The electrochemical method was based on the adsorptive accumulation of the drug at a carbon-black polylactic acid (CB/PLA) working electrode, followed by voltammetric determination of the accumulated species. The plastic sensor was printed in a single step by a low-cost dual extruder 3D-printer and featured three CB/PLA electrodes (serving as working, reference, and counter electrodes) and a holder, printed from a non-conductive PLA filament. The electrochemical parameters that affected the response of the device towards Colc determination, such as accumulation time and potential, solution pH, and other variables, were optimized. Under the selected conditions, the oxidation current of Colc was proportional to the concentration of Colc, and its quantification was conducted in the concentration range of 0.6–2.2 μmol L^{-1} with a limit of detection of 0.11 μmol L^{-1} in phosphate buffer (pH 7.0). Both within-device and between-device reproducibility were lower than 9%, revealing satisfactory operational and fabrication reproducibility. Furthermore, the 3D-printed device was employed for the voltammetric determination of Colc in pharmaceutical tablets and in human urine with satisfactory results, justifying its suitability for low-cost routine analysis of Colc.

Keywords: 3D-printing; voltammetry; colchicine; urine; sensor

Citation: Filopoulou, M.; Michail, G.; Katseli, V.; Economou, A.; Kokkinos, C. Electrochemical Determination of the Drug Colchicine in Pharmaceutical and Biological Samples Using a 3D-Printed Device. *Molecules* 2023, 28, 5539. https://doi.org/10.3390/molecules28145539

Academic Editors: Victoria Samanidou and Paraskevas D. Tzanavaras

Received: 5 July 2023
Revised: 18 July 2023
Accepted: 19 July 2023
Published: 20 July 2023

Copyright: © 2023 by the authors. Licensee MDPI, Basel, Switzerland. This article is an open access article distributed under the terms and conditions of the Creative Commons Attribution (CC BY) license (https://creativecommons.org/licenses/by/4.0/).

1. Introduction

Colchicine (Colc), N-[(7S)-1,2,3,10-tetramethoxy-9-oxo-5,6,7,9-tetrahydrobenzo(a) heptalen-7-yl] acetamide, is a member of the proto-alkaloids group, prepared from the dried corns and seeds of the liliaceous plant meadow saffron (*Colchicum autumnale* L.). Its chemical structure is shown in Figure 1A. Colc demonstrates a multimodal mechanism of action and continues to be at the center of very recent biomedical, clinical, and toxicological research. Colc is a very old drug used for the relief of joint pain and serves as a specific anti-inflammatory agent in acute attacks of gout by inhibiting the migration of leucocytes to inflammatory areas, thus interrupting the inflammatory response that sustains the acute attack. In general, Colc impairs the liberation of enzymes from lysosomes and so retards the development of inflammation [1,2]. It is also applied to polyarthritis associated with sarcoidosis, and it has been approved for the treatment of Behcet's disease and familial Mediterranean fever. Colc can be used in gastroenterology, especially in conditions of cirrhosis, as it reduces the formation of fibrous tissue in the liver [3,4]. The ability of Colc to bind tubulins to inhibit mitosis has made it a promising candidate for the treatment of cancer. Although Colc is not used clinically to treat cancer due to its toxicity, research has shown that it produces anti-vascular effects, leading to a bigger reduction in blood flow in tumors than in normal tissues and its ability to overcome P-glycoprotein efflux pump-mediated multidrug resistance [5]. These features render Colc a model compound for the development of novel anticancer drugs with a better toxicological profile, many of which are currently undergoing clinical studies. Additionally, there are numerous studies dealing with the potential role of Colc as an HIV inhibitor to treat AIDS [6]. Last but not least, Colc has been used in patients with COVID-19 as an immunomodulatory drug. The use of Colc in COVID-19 relies on the NLRP3 inflammasome activation caused by viroporin E, which is a component of COVID-19 and has an inflammatory response. As Colc reduces the inflammasome of NLRP3, it has been suggested for application in infections caused by COVID-19. Nevertheless, Colc did not decrease mortality or the duration of hospitalization in contrast to standard care for patients who were affected by COVID-19. In addition, the published reports are insufficient to suggest Colc as a therapy in patients affected by COVID-19 [7].

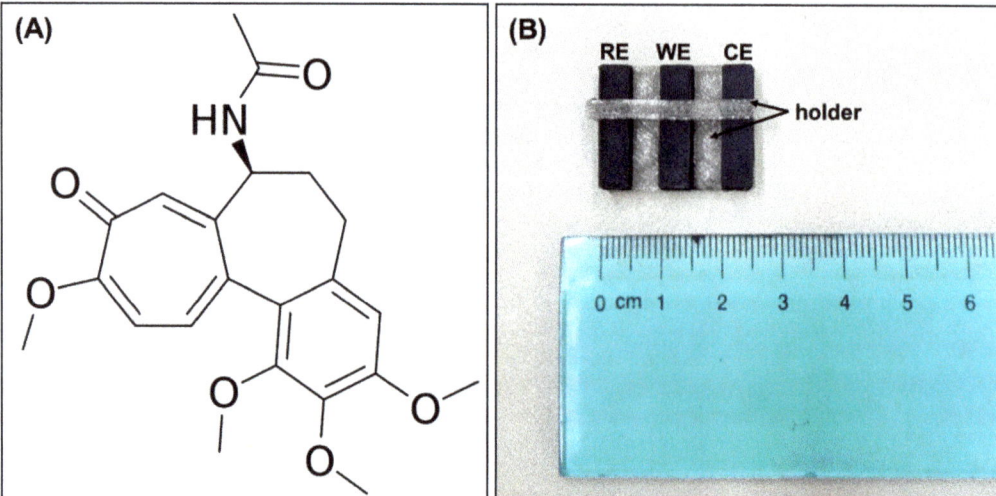

Figure 1. (**A**) Chemical structure of colchicine; (**B**) photograph of the 3D-printed device.

Despite its pharmacological utility, Colc can cause toxicity and significant side effects. Doses of Colc between 0.5 and 0.8 mg kg^{-1} can cause poisoning, while doses higher than 0.8 mg kg^{-1} may be fatal. The metabolism of Colc takes place partly in the liver, and then it is excreted in the urine and feces. It has been demonstrated that a low concentration of Colc (1.2 mg per day) causes a reduction in pain and gout symptoms, whereas a high dose of Colc (4.8 mg over 6 h) can cause common side effects, such as gastrointestinal upset including nausea, diarrhea, vomiting, low blood cell count, and rhabdomyolysis, as well as bone marrow damage, anemia, and hair loss. Colc causes oxidative stress in animals, leading to cognitive impairment, and its therapeutic use has been linked to sporadic Alzheimer's disease in humans [8–10].

It is obvious that the quantification of Colc content in pharmaceutical and biological samples is very important in order to monitor the treatment of patients and perform pharmacological studies. Several analytical methods have been reported for Colc quantification, such as high-performance liquid chromatography (HPLC), mass spectroscopy (MS), spectrophotometry, HPLC-MS, and HPLC-UV [11–15]. Despite the advantages of these techniques, they require time-consuming sample preparation procedures and expensive instrumentation. In the quest for rapid, simple, and cheap methods for Colc determination, electrochemical methods exhibit their predominance, offering miniaturized portable instrumentation and point-of-need devices with extremely low cost and practical use. Different working electrodes have already been reported for electrochemical determination of Colc, mainly via voltammetry, including the hanging mercury drop electrodes [16–18], graphite-based screen-printed electrodes [19], boron-doped diamond electrodes [20,21], gold electrodes [22–24], bare glassy carbon electrodes [25], and modified glassy carbon electrodes with poly(o-phenylenediamine)/single-wall carbon nanotubes [26], or with acetylene black–dihexadecyl hydrogen phosphate composite film [27], or with magnetic ionic liquid/CuO nanoparticles/carbon nanofibers [28], and also carbon paste electrodes modified with multiwall carbon nanotubes [29]. All these electrochemical methods of Colc quantification make use of separate conventional "large-size" electrodes and do not present any degree of miniaturization, integration, or ease of manufacturing. On the contrary, three-dimensional (3D) printing gives the opportunity for the in-house production of completed electrochemical small systems. Fused deposition modeling (FDM) is an advanced 3D printing procedure in which an electrochemical device is CAD-designed with open-source software and printed from thermoplastic filaments. The filaments are heated to a semi-molten state and extruded on a platform, where they solidify, forming the device. This digital fabrication procedure makes use of low-cost and portable printers and does not require laboratory facilities, provides great flexibility in the size and geometry of the printed devices, involves fast fabrication speed, uses low-cost filaments with different properties, and is environmentally friendly as it does not use chemicals and does not produce waste. Moreover, FDM involves e-transferability of the device, as the design file format can be sent through e-mail and printed on every 3D-printer [30–36].

In this work, we exploited the advantages of FDM in the fabrication of a miniaturized and fully integrated 3D-printed device, which was applied for the first time to the voltammetric determination of Colc. The 3D-printed device was printed in a single step using a dual extruder 3D printer, and it was composed of three electrodes (printed by a carbon black–polylactic acid (CB/PLA)) filament and a holder (printed by a PLA filament) (Figure 1B). The 3D-printed device can be considered a ready-to-use sensor as it integrates the working, counter, and reference electrodes (WE, CE, and RE, respectively) and does not require any modification step with materials for the voltammetric determination of Colc. The developed voltammetric method was successfully applied to the analysis of pharmaceutical tablets and human urine.

2. Results and Discussion

2.1. Electrochemical Behavior of Colchicine at the 3D-Printed Device

The electrochemical behavior of Colc on the surface of a 3D-printed working electrode (WE) was studied using cyclic voltammetry (CV) and differential pulse voltammetry (DPV) in 0.1 mol L^{-1} phosphate buffer (PB) pH (7.0). In addition, the electrochemical properties of the 3D-printed electrodes were examined through electrochemical impedance spectroscopy (EIS) and cyclic voltammetry in a 0.1 mol L^{-1} KCl solution containing K$_3$[Fe(CN)$_6$]/K$_4$[Fe(CN)$_6$] as a redox probe. As shown in the cyclic voltammograms (Figure 2A) and differential pulse voltammograms (Figure 2B), when the 3D-printed electrode was used as-printed (red traces in Figure 2), the response towards Colc was poor, while when the WE was electrochemically activated (black traces in Figure 2), the sensitivity of WE was significantly enhanced, and Colc gave rise to a well-defined oxidation peak at about +1.3 V. Also, the cyclic voltammograms (Figure 2A) revealed that the Colc oxidation was irreversible at the 3D-printed device, showing similar behavior to that of other carbon-based electrodes [21,26–28]. The electrochemical activation of the 3D-printed WE included anodic polarization of the WE at +1.8 V for 200 s, followed by cathodic polarization of the WE at −1.8 V for 200 s. It has been shown before that after anodic polarization of carbonaceous electrodes, graphene oxide and functional groups are electrogenerated in situ on the surface of the electrodes, while the subsequent cathodic polarization converts the electrochemically produced graphene oxide to electrochemically active reduced graphene oxide, improving further the electrochemical properties of the electrodes [37–39]. Moreover, it has already been demonstrated that the electrochemical treatment of 3D-printed PLA carbonaceous electrodes removes parts of the polymeric material from the electrode surface, resulting in the exposure of the active carbonaceous material on the electrode surface [38,39]. As depicted in Figure 2B, the electrochemical activation causes an enhancement of the WE response towards the oxidation of Colc of about 73% compared to that of the as-printed 3D-printed WE. The Nyquist plots of the as-printed and activated WEs are presented in Figure 2C. The Nyquist plot of the as-printed 3D-printed WE (red trace) suggested a high charge transfer resistance (R_{ct}), which was due to its relatively low conductivity. On the other hand, at the electrochemically activated 3D-printed WE (black trace), the R_{ct} decreased significantly, suggesting that the charge transfer resistance of the electrode was lowered and the conductivity enhanced. Hence, the significantly reduced R_{ct} can be attributed to the electro-etching effect induced by the electrochemical activation, which provides a faster electron transfer rate compared to the as-printed WE [37]. This fact was also confirmed by cyclic voltammetric studies in a 0.1 mol L^{-1} KCl solution containing 10 mmol L^{-1} K$_3$[Fe(CN)$_6$]/K$_4$[Fe(CN)$_6$] (Figure 2D), which revealed a significant increase in cathodic and anodic peak currents of the ferro/ferri probe at the electrochemically activated 3D-printed WE [37,39].

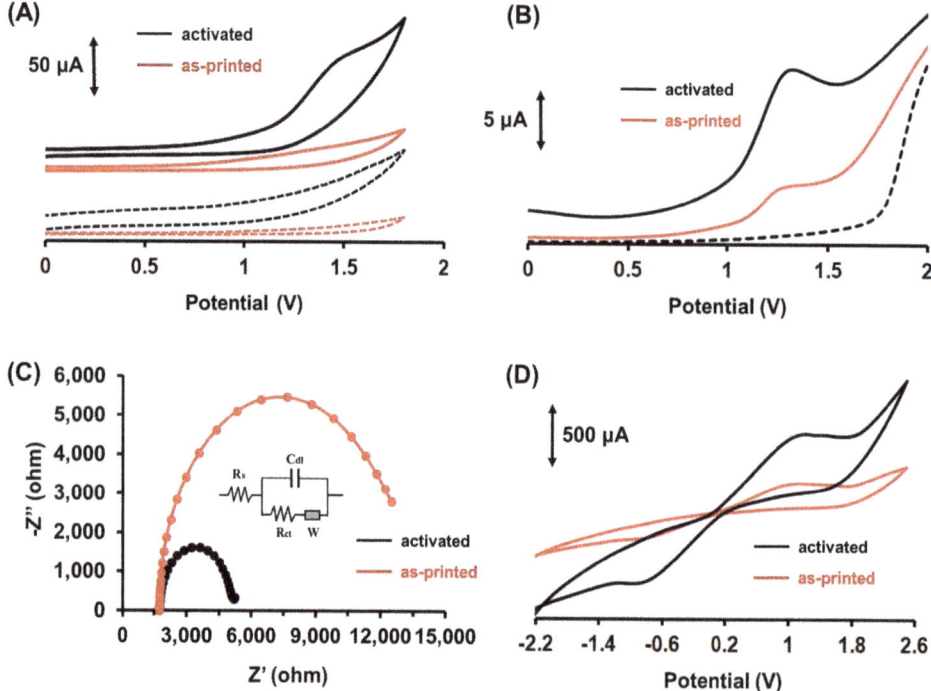

Figure 2. (A) Cyclic voltammograms in the absence (dot trace) and presence (solid trace) of 2 µmol L^{-1} Colc in 0.1 mol L^{-1} PB pH (7.0) at the as-printed 3D-printed CB/PLA electrode (red trace) and at the electrochemically activated 3D-printed CB/PLA electrode (black trace). (B) Differential pulse voltammograms in the absence (dot trace) and presence (solid trace) of 1 µmol L^{-1} Colc in 0.1 mol L^{-1} PB (pH 7.0) at the as-printed 3D-printed CB/PLA electrode (red trace) and at the electrochemically activated 3D-printed CB/PLA electrode (black trace). Accumulation for 120 s at −1.2 V. (C) Nyquist plots of the as-printed 3D-printed CB/PLA electrode (red trace) and of the electrochemically activated 3D-printed CB/PLA electrode (black trace) in 0.1 mol L^{-1} KCl solution containing 25 mmol L^{-1} K$_3$[Fe(CN)$_6$]/K$_4$[Fe(CN)$_6$]. (D) Cyclic voltammograms of the as-printed 3D-printed CB/PLA electrode (red trace) and of the electrochemically activated 3D-printed CB/PLA electrode (black trace) in the presence of 10 mmol L^{-1} K$_3$[Fe(CN)$_6$]/K$_4$[Fe(CN)$_6$] in 0.1 mol L^{-1} KCl.

2.2. Influence of the pH, the Accumulation Conditions, the Voltammetric Waveforms on the Determination of Colc

The electrochemical method for the determination of Colc was based on the adsorptive accumulation of the drug on the surface of the electrochemically activated 3D-printed electrode, followed by the anodic voltammetric determination of the accumulated species. The adsorptive voltammetric response was evaluated with respect to the pH of the working solution, the voltammetric waveform, and the accumulation time and potential.

The electrochemical responses of Colc at the 3D-printed CB/PLA activated electrode were examined in solutions covering the pH range from 1.0 to 9.0, using HCl solutions with pH 1.0 and 3.0 and 0.1 mol L^{-1} PB with pH 5.0, 7.0, and 9.0. As depicted in Figure 3A, the solution pH had a strong effect on the voltammetric response, and the experiment results showed that the highest oxidation peak current was obtained at 0.1 mol L^{-1} PB with pH 7. 0, which was finally selected. The oxidation peak current shifted to more negative potentials by increasing the pH of the supporting electrolyte, and the inset of Figure 3A shows the linear relationship between the anodic peak potential (E_{pa}) and the pH with a slope of 0.053 (V/pH), which is close to the anticipated theoretical value of 0.059 for the electrochemical reaction of Colc, involving two protons and two electrons [28]. Moreover, the differential pulse (DP) and square wave (SW) waveforms were tested (Figure 3B). The DP mode presented a higher oxidation peak current than the SW mode, and thus DP was selected. The adsorption of Colc on the electrode surface was affected by the duration of the accumulation process. As depicted in Figure 3C, the longer the accumulation period, the more Colc was adsorbed on the electrode surface and the larger the oxidation peak current. Considering the sensitivity and the analysis time, an accumulation period of 120 s was selected as a satisfactory compromise. The dependence of the oxidation peak current of Colc on the accumulation potential was examined over the range from −1.4 to +0.6 V. Equal adsorption efficiency was observed over the entire tested potential range, showing that the accumulation potential had no influence on the oxidation peak current of Colc. Therefore, the electrochemical method for the voltammetric determination of Colc was selected to be carried out in 0.1 mol L^{-1} PB (pH 7.0), applying an accumulation potential of −1.2 V for 120 s, and using the DP waveform.

2.3. Analytical Performance and Interferences

Next, the analytical features of the method were evaluated. The calibration curve was constructed by the oxidation peak current of Colc, obtained under optimized experimental parameters (i.e., in 0.1 mol L^{-1} PB (pH 7.0) applying an accumulation potential of −1.2 V for 120 s), versus different concentrations of Colc. The DP voltammetric response at different Colc concentrations in the range 0.6–2.2 µmol L^{-1} and the respective calibration plot are shown in Figure 4. The oxidation peak height showed a linear concentration dependence in the examined concentration range, with R^2 = 0.997. The limit of detection (LOD) was 0.11 µmol L^{-1} Colc and was calculated by the equation LOD = 3 sd/a (where sd was the standard deviation of the intercept of the calibration plot and a was the slope of the calibration plot). The LOD of the developed method utilizing the 3D-printed device was comparable to that obtained with other voltammetric methods utilizing unmodified carbon-based electrodes, such as graphite screen-printed electrodes, boron-doped diamond electrodes, and glassy carbon electrodes (ranging from 0.1 to 2.0 µmol L^{-1}) [17,19,21]. The within-device reproducibility in terms of % relative standard deviation (%RSD) was 5.4%, and it was estimated by measuring the DP voltammetric responses of 1.6 µmol L^{-1} Colc via eight repeated measurements. The between-device reproducibility (in terms of % RSD at five different devices) was 8.2% at 1.6 µmol L^{-1} of Colc. These electroanalytical results revealed the significant repeatability of the proposed 3D-printed device for the determination of Colc.

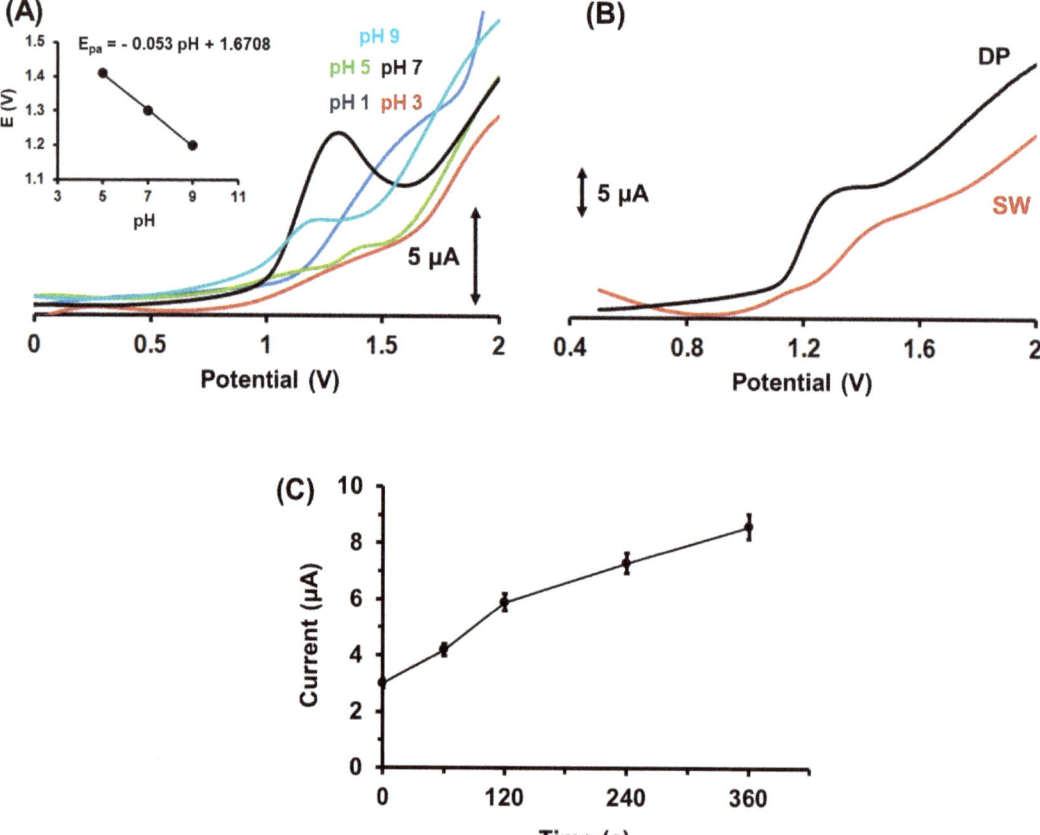

Figure 3. (**A**) Differential pulse voltammograms of 1 μmol L^{-1} Colc at an electrochemically activated 3D-printed CB/PLA electrode in solutions with different pH (HCl with pH 1.0 and 3.0, 0.1 mol L^{-1} PB with pH 5.0, 7.0 and 9.0). Inset: The linear relationship between the anodic peak potential of Colc oxidation and the pH. Accumulation for 120 s at −1.2 V. (**B**) The effect of the scanning waveform (DP and SW) on the voltammetric response of 1 μmol L^{-1} Colc at electrochemically activated 3D-printed CB/PLA electrode. Accumulation for 120 s at −1.2 V. (**C**) The effect of accumulation time on the DPV response of 1 μmol L^{-1} Colc at electrochemically activated 3D-printed CB/PLA electrode. Error bars are the mean value ± sd (n = 3). Accumulation potential at −1.2 V.

The effect of various possible interfering species, which can be included in pharmaceutical and biological samples, was examined by the addition of the substances to a solution containing 1 μmol L^{-1} Colc (Figure 4C). Glucose, sucrose, urea, uric acid, lactic acid, and K$^+$, Na$^+$, Ca^{2+}, NH$_4$$^+$, Cl$^-$, PO$_4$$^{3-}$, and SO$_4$$^{2-}$ did not interfere at a concentration ratio of 1:100 (Colc solution: interference compound), revealing adequate selectivity of the presented electrochemical method for the determination of Colc in real samples.

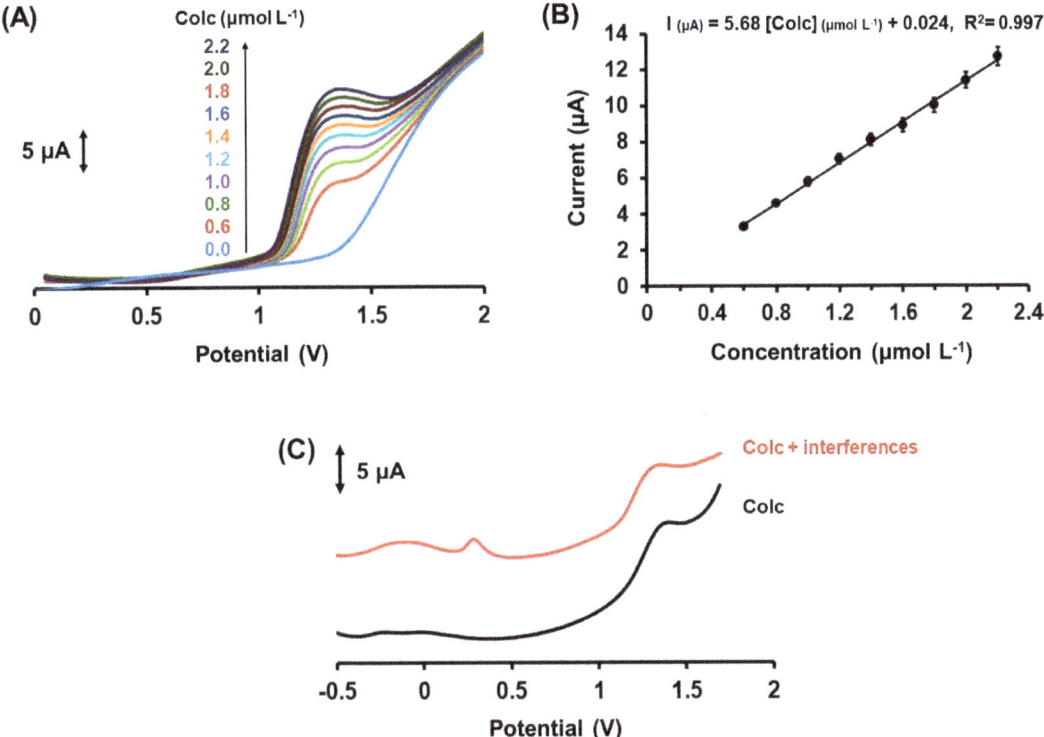

Figure 4. (**A**) Differential pulse voltammetric responses of Colc at 3D-printed CB/PLA sensor in 0.1 mol L^{-1} PB (pH 7) at different concentrations. From bottom to top: 0.0, 0.6, 0.8, 1.0, 1.2, 1.4, 1.6, 1.8, 2.0, and 2.2 µmol L^{-1} Colc. Accumulation for 120 s at −1.2 V. (**B**) The respective calibration plot. The points in the calibration plot are the mean value ± sd (n = 3). (**C**) The black trace is the DP voltammogram of 1 µmol L^{-1} Colc at the 3D-printed CB/PLA sensor in 0.1 mol L^{-1} PB (pH 7) and the red trace is the respective voltammogram of 1 µmol L^{-1} Colc in 0.1 mol L^{-1} PB (pH 7) containing 100 µmol L^{-1} of glucose, sucrose, urea, uric acid, lactic acid, and K$^+$, Na$^+$, Ca^{2+}, NH$_4$$^+$, Cl$^-$, PO$_4$$^{3-}$, and SO$_4$$^{2-}$. Accumulation for 120 s at −1.2 V.

2.4. Applications

In order to verify the accuracy of the presented electrochemical method, the electrochemically activated 3D-printed device was applied to the determination of Colc in pharmaceutical tablets and in spiked human urine. Colc is mainly excreted via urine and feces, and 10–20% of Colc remains unchanged in urine; thus, measuring the quantity of discarded Colc is crucial to the calculation of its bioavailability and other pharmaceutical characteristics. The preparation procedures for both samples are described in Section 3.3. In human urine and pharmaceutical tablets, the determination of Colc quantity was carried out by the method of standard additions. Particularly in the pharmaceutical tablets, the Colc content was determined as 1.01 ± 0.05 mg per tablet, and the respective recovery was 101 ± 5% (n = 3), while in the case of a urine sample spiked with Colc, the obtained recovery was 97 ± 6% (n = 3) (Figure 5). These recovery values indicated that the accuracy of the presented voltammetric method utilizing the 3D-printed device was satisfactory and confirmed its usefulness for Colc determination in practical samples.

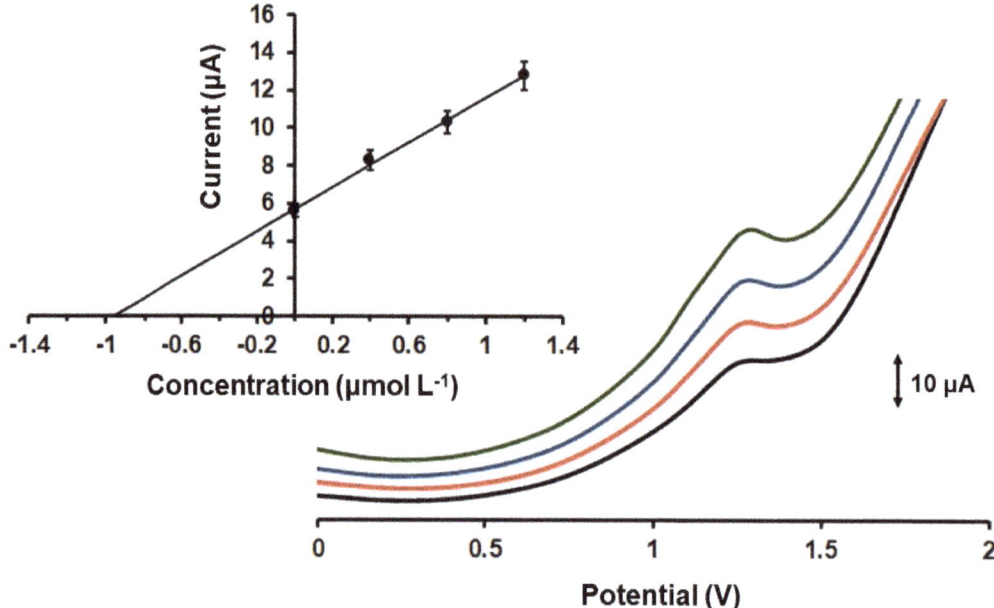

Figure 5. Differential pulse voltammograms obtained at the 3D-printed device for the determination of Colc in a human urine sample using the method of standard additions. Traces from below: urine sample spiked with Colc (black trace) and 3 standard additions of 0.4 μmol L^{-1} Colc (red, blue and green traces). Inset: the respective standard additions plot. The points in the plot are the mean value ± sd (n = 3). Accumulation potential −1.2 V for 120 s.

3. Materials and Methods

3.1. Reagents and Apparatus

The transparent non-conductive filament was polylactic acid (PLA) from 3DEdge, while the conductive filament was a carbon-black-loaded PLA filament from Proto-Pasta. The diameter of each filament was 1.75 mm. All the other reagents were purchased from Sigma-Aldrich. The pharmaceutical tablets (commercial name: COLCHICINA/ACARPIA from Acarpia) contained an average of 1 mg of Colc and were obtained from a local drug store. Besides, the pharmaceutical tablets contained as excipients the following materials: povidone, sucrose, microcrystalline cellulose, sodium starch glycollate, magnesium stearate, and purified talc. For the preparation of the phosphate buffer (PB), the appropriate amounts of Na_2HPO_4 and NaH_2PO_4 were mixed, and the pH value was adjusted by the addition of a 0.1 mol L^{-1} solution of HCl or NaOH. The electrochemical experiments were conducted by the PalmSense4 potentiostat (Palm Sens, Houten, The Netherlands) using the PS Trace 5.5 software (Palm Sens).

3.2. Fabrication of the 3D-Printed Device

The Tinkercad software was used for the design of the device, and the file was saved in .STL format. Then, the Flashprint software was used to open the file and to set the conditions of the 3D printing process, which were: 60 °C for the platform, 200 °C for the head dispensers for the printing of both PLA filaments (conductive and non-conductive) and a printing speed of 40 mm s^{-1}. The respective file was saved in .3x format and transferred to an SD card. Next, the SD card was inserted in the appropriate slot of the Creator Pro dual extruder 3D-printer (Flashforge), and the printing process was started. In Figure 1A, a photograph of the 3D-printed device is displayed.

3.3. Electrochemical Measurements and Samples Analysis

The connection of the 3D-printed device to the portable potentiostat was accomplished with crocodile clips. For the differential pulse voltammetric measurements, the solution was stirred at 1000 rpm and a potential of -1.2 V for 120 s was applied to the working electrode, followed by a differential pulse scan in a static solution, and the voltammogram was recorded. The differential pulse parameters were: modulation amplitude, 50 mV; increment, 10 mV; pulse width, 75 ms; and pulse repeat time, 50 ms. The cyclic voltammograms were obtained in 0.1 mol L^{-1} PB (pH 7.0) at a scan rate of 50 mV s^{-1}. The EIS studies were performed using a 0.1 mol L^{-1} KCl solution containing 25 mmol L^{-1} K$_3$[Fe(CN)$_6$]/K$_4$[Fe(CN)$_6$] as the electrochemical probe. The frequency ranged from 10^5 to 1 Hz with the application of 0.1 V of potential (DC) and 0.01 V of amplitude (AC). For the differential pulse voltammetric analysis of pharmaceutical tablets, 4 tablets of COLCHICINA/ACARPIA (1 mg Colc per tablet) were pulverized and dissolved in 100 mL of doubly distilled water. Next, 100 µL of the resulting solution was transferred into the electrochemical cell, which had previously been filled with 9.9 mL of 0.1 mol L^{-1} PB (pH 7.0). The urine sample was obtained from a healthy 22-year-old male volunteer with his informed written consent, and the measurements were conducted in accordance with GCP regulations. The urine sample was spiked with Colc to a final concentration of 10 µmol L^{-1} Colc, and then it was diluted 1:10 with 0.1 mol L^{-1} PB (pH 7.0). In human urine and pharmaceutical tablets, the determination of Colc was carried out by the method of standard additions to minimize matrix effects. For the interference study, possible substances that can be present in pharmaceutical tablets and human urine, such as glucose, sucrose, urea, uric acid, lactic acid, and K$^+$, Na$^+$, Ca^{2+}, NH$_4^+$, Cl$^-$, PO$_4^{3-}$, and SO$_4^{2-}$ were added at a concentration of 100 µmol L^{-1} in 0.1 mol L^{-1} PB (pH 7.0) solution that contained 1 µmol L^{-1} Colc. All potentials at the 3D-printed device are referred to with respect to the carbon-black-loaded PLA reference electrode.

4. Conclusions

In summary, a 3D-printed device consisting of three carbon-black-loaded PLA electrodes was successfully applied to the DPV quantification of the drug Colc. The electrochemical device exploited the advantages of fused deposition modeling in terms of fabrication speed and operation simplicity, while electrochemical activation via anodic and cathodic polarization significantly improved the electrochemical response of the device towards Colc. The voltammetric method offered excellent sensitivity (LOD of 0.11 µmol L^{-1}) and reproducibility (%RSD < 9%) and allowed the determination of Colc in real samples with satisfactory selectivity and without the requirement of any separation or complex sample pretreatment. These features made the proposed method of utilizing the 3D-printed device practical for the simple and low-cost routine determination of Colc.

Author Contributions: Conceptualization, C.K.; investigation, M.F., G.M. and V.K.; resources, C.K. and A.E.; writing—original draft preparation, C.K.; writing—review and editing, C.K. and A.E.; visualization, C.K.; supervision, C.K. All authors have read and agreed to the published version of the manuscript.

Funding: This research received no external funding.

Institutional Review Board Statement: Not applicable.

Informed Consent Statement: Not applicable.

Data Availability Statement: All the used data have been provided in the text.

Conflicts of Interest: The authors declare no conflict of interest.

Sample Availability: Not applicable.

References

1. Chabert, J.F.D.; Vinader, V.; Santos, A.R.; Horcajo, M.R.; Dreneau, A.; Basak, R.; Cosentino, L.; Marston, G.; Rahman, H.A.; Loadman, P.M.; et al. Synthesis and biological evaluation of colchicine C-ring analogues tethered with aliphatic linkers suitable for prodrug derivatization. *Bioorg. Med. Chem. Lett.* **2012**, *22*, 7693–7696. [CrossRef] [PubMed]
2. Dalbeth, N.; Lauterio, T.J.; Wolfe, H.R. Mechanism of action of colchicine in the treatment of gout. *Clin. Ther.* **2014**, *36*, 1465–1479. [CrossRef]
3. Sonmez, A.O.; Sonmez, H.E.; Çakan, M.; Yavuz, M.; Keskindemirci, G.; Ayaz, N.A. The evaluation of anxiety, depression and quality of life scores of children and adolescents with familial Mediterranean fever. *Rheumatol. Int.* **2020**, *40*, 757–763. [CrossRef] [PubMed]
4. Ozen, S.; Kone-Paut, I.; Gul, A. Colchicine resistance and intolerance in familial mediterranean fever: Definition, causes, and alternative treatments. *Semin. Arthritis Rheum.* **2017**, *47*, 115–120. [CrossRef] [PubMed]
5. Lu, Y.; Chen, J.; Xiao, M.; Li, W.; Miller, D.D. An overview of tubulin inhibitors that interact with the colchicine binding site. *Pharmaceut. Res.* **2012**, *29*, 2943–2971. [CrossRef] [PubMed]
6. Worachartcheewan, A.; Songtawee, N.; Siriwong, S.; Prachayasittikul, S.; Nantasenamat, C.; Prachayasittikul, V. Rational Design of Colchicine Derivatives as anti-HIV Agents via QSAR and Molecular Docking. *Med. Chem.* **2019**, *15*, 328–340. [CrossRef]
7. Toro-Huamanchumo, C.J.; Benites-Meza, J.K.; Mamani-Garcia, C.S.; Bustamante-Paytan, D.; Gracia-Ramos, A.E.; Diaz-Velez, C.; Barboza, J.J. Efficacy of Colchicine in the Treatment of COVID-19 Patients: A Systematic Review and Meta-Analysis. *J. Clin. Med.* **2022**, *11*, 2615. [CrossRef]
8. Slobodnick, A.; Shah, B.; Krasnokutsky, S.; Pillinger, M.H. Update on colchicine, 2017. *Rheumatology* **2018**, *57*, i4–i11. [CrossRef]
9. Zhang, D.; Li, L.; Li, J.; Wei, Y.L.; Tang, J.; Man, X.; Liu, F. Colchicine improves severe acute pancreatitis-induced acute lung injury by suppressing inflammation, apoptosis and oxidative stress in rats. *Biomed. Pharmacother.* **2022**, *153*, 113461. [CrossRef]
10. Hammadi, S.H.; Hassan, M.A.; Allam, E.A.; Elsharkawy, A.M.; Shams, S.S. Effect of sacubitril/valsartan on cognitive impairment in colchicine-induced Alzheimer's model in rats. *Fundam. Clin. Pharmacol.* **2023**, *37*, 275–286. [CrossRef] [PubMed]
11. Maciel, T.R.; de Oliveira Pacheco, C.; Ramos, P.F.; Ribeiro, A.C.F.; dos Santos, R.B.; Haas, S.E. Simultaneous determination of chloroquine and colchicine co-nanoencapsulated by HPLC-DAD. *J. Appl. Pharm. Sci.* **2023**, *13*, 106–112. [CrossRef]
12. Dimitrijevic, D.; Fabian, E.; Nicol, B.; Funk-Weyer, D.; Landsiedel, R. Toward Realistic Dosimetry In Vitro: Determining Effective Concentrations of Test Substances in Cell Culture and Their Prediction by an In Silico Mass Balance Model. *Chem. Res. Toxicol.* **2022**, *35*, 1962–1973. [CrossRef] [PubMed]
13. Sajitha, T.P.; Siva, R.; Manjunatha, B.L.; Rajani, P.; Navdeep, G.; Kavita, D.; Ravikanth, G.; Shaanker, R.U. Sequestration of the plant secondary metabolite, colchicine, by the noctuid moth Polytela gloriosae (Fab.). *Chemoecology* **2019**, *29*, 135–142. [CrossRef]
14. Izidoro, M.A.; Cecconi, A.; Panadero, M.I.; Mateo, J.; Godzien, J.; Vilchez, J.P.; Lopez-Gonzalvez, A.; Ruiz-Cabello, J.; Ibanez, B.; Barbas, C.; et al. Plasma Metabolic Signature of Atherosclerosis Progression and Colchicine Treatment in Rabbits. *Sci. Rep.* **2020**, *10*, 7072. [CrossRef]
15. Cankaya, N.; Bulduk, I.; Colak, A.M. Extraction, development and validation of HPLC-UV method for rapid and sensitive determination of colchicine from *Colchicum autumnale* L. Bulbs. *Saudi J. Biol. Sci.* **2019**, *26*, 345–351. [CrossRef]
16. Wang, J.; Ozsoz, M. Trace measurements of colchicine by adsorptive stripping voltammetry. *Talanta* **1990**, *37*, 783–787. [CrossRef]
17. Bodoki, E.; Sandulescu, R.; Roman, L. Method validation in quantitative electrochemical analysis of colchicine using glassy carbon electrode. *Cent. Eur. J. Chem.* **2007**, *5*, 766–778. [CrossRef]
18. Kasim, E.A. Voltammetric behavior of anti-inflammatory alkaloid colchicine at a glassy carbon electrode and a hanging mercury electrode and its determination at ppb levels. *Anal. Lett.* **2002**, *35*, 1987–2004. [CrossRef]
19. Bodoki, E.; Laschi, S.; Palchetti, I.; Sandulescu, R.; Mascini, M. Electrochemical behavior of colchicine using graphite-based screen-printed electrodes. *Talanta* **2008**, *76*, 288–294. [CrossRef]
20. Moreira, D.A.R.; de Oliveira, F.M.; Pimentel, D.M.; Guedes, T.J.; Luz, R.C.S.; Damos, F.S.; Pereira, A.C.; da Silva, R.A.B.; dos Santos, W.T.P. Determination of Colchicine in Pharmaceutical Formulations and Urine by Multiple-Pulse Amperometric Detection in an FIA System Using Boron-Doped Diamond Electrode. *J. Braz. Chem. Soc.* **2018**, *29*, 1796–1802. [CrossRef]
21. Stankovic, D.M.; Svorc, L.; Mariano, J.F.M.L.; Ortner, A.; Kalcher, K. Electrochemical Determination of Natural Drug Colchicine in Pharmaceuticals and Human Serum Sample and its Interaction with DNA. *Electroanalysis* **2017**, *29*, 2276–2281. [CrossRef]
22. Bishop, E.; Hussein, W. Anodic voltammetry of colchicine. *Analyst* **1984**, *109*, 623–625. [CrossRef]
23. Pang, T.T.; Zhang, X.Y.; Xue, Y.B. Supramolecular Detoxification of Colchicine through Electrochemical Recognition. *Anal. Lett.* **2017**, *50*, 1743–1753. [CrossRef]
24. Pang, T.T.; Zhang, X.Y. Detection of Colchicine using p-Sulfonated Calix[4]arene Modified Electrode. *Asian J. Chem.* **2019**, *31*, 2663–2668.
25. Bodoki, E.; Chira, R.; Zaharia, V.; Sandulescu, R. Mechanistic study of colchicine's electrochemical oxidation. *Electrochim. Acta* **2015**, *178*, 624–630. [CrossRef]
26. Zhang, X.H.; Wang, S.M.; Jia, L.; Xu, Z.X.; Zeng, Y. Electrochemical properties of colchicine on the PoPD/SWNTs composite-modified glassy carbon electrode. *Sens. Actuators B Chem.* **2008**, *134*, 477–482. [CrossRef]
27. Zhang, H. Electrochemistry and voltammetric determination of colchicine using an acetylene black-dihexadecyl hydrogen phosphate composite film modified glassy carbon electrode. *Bioelectrochemistry* **2006**, *68*, 197–201. [CrossRef]

28. Afzali, M.; Mostafavi, A.; Shamspur, T. Sensitive detection of colchicine at a glassy carbon electrode modified with magnetic ionic liquid/CuO nanoparticles/carbon nanofibers in pharmaceutical and plasma samples. *J. Iran. Chem. Soc.* **2020**, *17*, 1753–1764. [CrossRef]
29. Zhang, K.; Zhou, J.; Liu, J.; Li, K.; Li, Y.; Yang, L.; Ye, B. Sensitive determination of colchicine at carbon paste electrode doped with multiwall carbon nanotubes. *Anal. Methods* **2013**, *5*, 1830–1836. [CrossRef]
30. Carrasco-Correa, E.J.; Simo-Alfonso, E.F.; Herrero-Martínez, J.M.; Miro, M. The Emerging Role of 3D Printing in the Fabrication of Detection Systems. *Trends Anal. Chem.* **2021**, *136*, 116177. [CrossRef]
31. Sharafeldin, M.; Jones, A.; Rusling, J.F. 3D-printed biosensor arrays for medical diagnostics. *Micromachines* **2018**, *9*, 394. [CrossRef] [PubMed]
32. Abdalla, A.; Patel, B.A. 3D-Printed Electrochemical Sensors: A New Horizon for Measurement of Biomolecules. *Curr. Opin. Electrochem.* **2020**, *20*, 78–81. [CrossRef]
33. Cardoso, R.M.; Kalinke, C.; Rocha, R.G.L.; dos Santos, P.; Rocha, D.P.; Oliveira, P.R.; Janegitz, B.C.; Bonacin, J.A.; Richter, E.M.; Munoz, R.A.A. Additive-Manufactured (3D-printed) Electrochemical Sensors: A Critical Review. *Anal. Chim. Acta* **2020**, *1118*, 73–91. [CrossRef] [PubMed]
34. Chang, Y.; Cao, Q.; Venton, B.J. 3D printing for customized carbon electrodes. *Curr. Opin. Electrochem.* **2023**, *38*, 101228. [CrossRef]
35. Omar, M.H.; Razak, K.A.; Wahab, M.N.A.; Hamzah, H.H. Recent progress of conductive 3D-printed electrodes based upon polymers/carbon nanomaterials using a fused deposition modelling (FDM) method as emerging electrochemical sensing devices. *RSC Adv.* **2021**, *11*, 16557. [CrossRef] [PubMed]
36. Ambrosi, A.; Bonanni, A. How 3D printing can boost advances in analytical and bioanalytical chemistry. *Microchim. Acta* **2021**, *188*, 265. [CrossRef] [PubMed]
37. Li, Y.; Zhou, J.; Song, J.; Liang, X.; Zhang, Z.; Men, D.; Wang, D.; Zhang, X.E. Chemical nature of electrochemical activation of carbon electrodes. *Biosens. Bioelectron.* **2019**, *144*, 111534. [CrossRef]
38. dos Santos, P.L.; Katic, V.; Loureiro, H.C.; dos Santos, M.F.; dos Santos, D.P.; Formiga, A.L.B.; Bonacin, J.A. Enhanced performance of 3D printed graphene electrodes after electrochemical pre-treatment: Role of exposed graphene sheets. *Sens. Actuators B Chem.* **2019**, *281*, 837–848. [CrossRef]
39. Rocha, D.P.; Squissato, A.L.; da Silva, S.M.; Richter, E.M.; Munoz, R.A.A. Improved electrochemical detection of metals in biological samples using 3D-printed electrode: Chemical/electrochemical treatment exposes carbon-black conductive sites. *Electrochim. Acta* **2020**, *335*, 135688. [CrossRef]

Disclaimer/Publisher's Note: The statements, opinions and data contained in all publications are solely those of the individual author(s) and contributor(s) and not of MDPI and/or the editor(s). MDPI and/or the editor(s) disclaim responsibility for any injury to people or property resulting from any ideas, methods, instructions or products referred to in the content.

Article

In Vitro Assessment of the Physiologically Relevant Oral Bioaccessibility of Metallic Elements in Edible Herbs Using the Unified Bioaccessibility Protocol

Tatiana G. Choleva, Charikleia Tziasiou [†], Vasiliki Gouma [†], Athanasios G. Vlessidis and Dimosthenis L. Giokas *

Department of Chemistry, University of Ioannina, 45110 Ioannina, Greece
* Correspondence: dgiokas@uoi.gr
[†] These authors contributed equally to this work.

Abstract: In this work, the total content of seven metallic elements (Fe, Cu, Zn, Mg, Pb, Ni, and Co) in common edible herbs was determined and related to their bioaccessibility by an in vitro human digestion model. Specifically, the unified bioaccessibility protocol developed by the BioAccessibility Research Group of Europe (BARGE) was used to determine the release of each element during gastric and gastrointestinal digestion. The results show that Fe, Zn, and Mg are released during gastric digestion (34–57% Fe, 28–80% Zn, 79–95% Mg), but their overall bioaccessibility is reduced in the gastrointestinal tract (<30%). On the contrary, Cu is more bioaccessible during gastrointestinal digestion (38–60%). Pb, Ni, and Co exhibited similar bioaccessibility in both gastric and gastrointestinal fluids. Principle component analysis of the data shows that the classification of the nutritional value of herbs differs between the total and the gastrointestinal concentration, suggesting that the total concentration alone is not an adequate indicator for drawing secure conclusions concerning the nutritional benefits of edible plant species.

Keywords: herbs; bioaccessibility; micronutrients; herbs; heavy metals; essential elements

1. Introduction

Plants and plant-based products (such as vegetables, herbs, seeds, etc.) are among the most important food commodities in the diet of most countries and play a significant role in human nutrition. This is because they are available in large quantities and they are rich in a variety of important constituents, such as vitamins, polyphenols, antioxidants, metallic elements, proteins, etc., which are necessary for human nutrition and act protectively against serious clinical disorders and diseases [1]. In addition, many plant species contain pharmaceutically active compounds that have been used for centuries to treat diseases and improve human health [2]. The bioactive compounds in many plants have also been the basis for the design of functional foods and nutraceuticals that offer additional benefits to the human organism.

The composition of edible plants has been thoroughly investigated, and a vast amount of information is currently available for almost all known species. However, the presence of bioactive compounds in plants is not directly related to their uptake in the human organism. To obtain the benefits of these compounds, they must be released from the plant matrix and modified in the gastrointestinal tract to become bioaccessible and bioavailable [3]. Bioaccessibility refers to the amount of a compound that is released from the food matrix in the gastrointestinal tract and potentially becomes available for absorption, while bioavailability refers to the fraction of the compound that reaches systemic circulation and is utilized by the human organism [4]. Therefore, to evaluate the actual benefits of plant consumption, it is necessary to evaluate both the bioaccessibility and bioavailability of its active components.

Since the determination of bioavailability is an intricate and time-consuming procedure that requires in vivo tests and extensive clinical evaluation, bioaccessibility is most frequently tested, usually utilizing in vitro experimental procedures [5,6]. These procedures do not aim to accurately replicate the conditions found in the various compartments of the human digestive system, but they aim to simulate the physiological conditions and the main sequence of physicochemical events that occur during digestion. To accomplish this task, various components found in the mouth, the stomach, and the intestine, such as bile salts and enzymes (mucin, pepsin, pancreatin, etc.), organic molecules (such as glucose, urea, amino acids, etc.), and inorganic electrolytes, are used to mimic the composition of gastric and gastrointestinal fluids. Based on this rationale, several experimental protocols have been developed, such as the physiologically based extraction test (PBET), the simulation of the human intestinal microbial ecosystem (SHIME) procedure, the dynamic gastrointestinal model (TIM), and the in vitro gastrointestinal method, among others [3,4,7]. These methods, although designed to simulate the human digestive tract, exhibit many dissimilarities that do not enable a direct data comparison [8]. To standardize operational procedures and harmonize biomimetic extraction tests, the BioAccessibility Research Group of Europe (BARGE), developed the standardized unified BARGE method (UBM), using surrogate digestive fluids, (saliva, gastric juice, duodenal juice, and bile), for which their chemical composition is adjusted to be similar to that of human physiology [9]. The experimental procedure of the UBM method involves two sequential extraction steps: gastric digestion, in which the saliva and the gastric juice are added to the solid substrate (mouth and stomach compartments), followed by gastrointestinal digestion, in which the duodenal fluid and the bile salts are also included along with the gastric fluid. This harmonized method, although initially developed to evaluate the risks associated with the unintentional ingestion of soil, has been also applied to several food commodities, such as seeds [10,11], nutritional supplements [12], seaweeds [13], and recently roots [14], fruits, and vegetables [15].

Among foods of plant origin, fresh herbs are commonly used as food seasonings in small quantities. However, they are physically, biochemically, and nutritionally similar to vegetables and contain a variety of bioactive phytochemicals, such as vitamins, terpenoids, flavonoids, phenolic acids, and essential mineral elements [16,17]. However, due to the low amount that is consumed compared with other green leafy vegetables, studies on the bioaccessibility of their bioactive components in the gastrointestinal environment are scant. Among the most important bioactive components found in plant and herb species are metallic elements, which play an important role in various biological functions. For example, elements such as Fe, Cu, Zn, and Mg are involved in enzymatic reactions, the transport of oxygen, the functioning of nerve and muscle cells, the formation of DNA, etc., and their deficiency is associated with major clinical disorders and diseases. However, apart from essential elements, plant species may also contain heavy metal ions as a result of contamination during cultivation, transportation, and storage. These metals may accumulate in the human organism and cause adverse health effects. Although some of these heavy metals (e.g., Ni, Co, Mn) are involved in biochemical functions at low concentration levels, they exhibit negative effects at higher concentration levels. On the other hand, heavy metals such as Pb, As, Cr, etc., pose significant risks to human health [18]. In this regard, this work aims to investigate the nutritional value of fresh herbs through the determination of the gastric and intestinal bioaccessible fractions of four essential elements (Fe, Mg, Zn, and Cu) using the validated UBM method. Moreover, the presence and bioaccessibility of heavy metal ions (Pb, Ni, and Co) are also assessed to evaluate the role of fresh herbs in the uptake of heavy metal ions. Importantly, since edible herbs belong to the same species worldwide, evaluating the bioaccessibility of metallic elements can serve as a foundation for assessing their contribution to the uptake of these elements.

2. Results and Discussion

2.1. Total Content of Essential Elements and Heavy Metals

The total content of the examined metal ions was first determined to get insight into the quality of herbs. In addition, the total concentration of metal ions was used for mass balance calculations that were necessary for the quality control of the experimental data. Table 1 summarizes the results (average from three independent sampling campaigns) from the analysis of seven metal ions in the selected herbs. The concentration of metal ions changed according to the order: Mg (454–922.5 mg/Kg) > Fe (19.4–156.2 mg/Kg) > Zn (16.8–138.9 mg/Kg) > Cu (0.3–6.1 mg/Kg) > Ni (0–6.3 mg/Kg) > Co (0–0.3 mg/Kg) > Pb (0–1 mg/Kg). The total content of metal ions was within the range of values reported in other studies [19–21]. Interestingly, the concentration of Fe in the samples was reversibly related to that of Zn in most herbs, which may be attributed to the cultivation conditions. Fe is related to respiration activities, N fixation, and electron transfer [22], and increases with photosynthetic activity while Zn competes with the ligands that promote Fe accumulation, reducing Fe transport to the aerial parts of the plants when photosynthetic activity is lower [23]. Since Fe in parsley and oregano is lower than Zn, we can conclude that Fe is less accumulated, indicating a possible lower photosynthetic activity. For parsley, this can also be related to the fact it has been grown in mixed cultivations, where agronomic conditions are adjusted to the needs of all species and not to specific species.

Table 1. The total content of metallic elements in mg/Kg (average results from triplicate measurements in three independent sampling campaigns).

	Fe	Mg	Zn	Cu	Co	Pb	Ni
Parsley	35.9 ± 5.2	719.7 ± 88	111.4 ± 17	2.4 ± 0.8	0.2 ± 0.1	0.9 ± 0.1	2.0 ± 0.08
Dill	39.8 ± 4.7	768.2 ± 101	17.6 ± 2.6	5.4 ± 0.8	0.3 ± 0.1	<LOD	0.8 ± 0.08
Spearmint	156.2 ± 25	454.0 ± 76	46.7 ± 5.5	6.1 ± 1.1	<LOD	<LOD	6.3 ± 0.9
Oregano	19.4 ± 3.8	830.1 ± 91	138.9 ± 21	3.9 ± 1.0	0.2 ± 0.1	1.0 ± 0.2	0.8 ± 0.1
Thyme	58.1 ± 5.4	922.5 ± 98	16.8 ± 3.3	3.9 ± 0.8	<LOD	<LOD	<LOD
Rosemary	236.0 ± 27	682.1 ± 84	28.4 ± 4.2	0.3 ± 0.1	0.2 ± 0.1	0.1 ± 0.1	1.6 ± 0.5

<LOD: Lower than the detection limit.

To better visualize the results, a cluster analysis plot was built up using the raw data (i.e., total metal content in the plant) as shown in Figure 1a. Metal ions such as Fe, Mg, Zn, and Cu (which are all essential elements) form separate clusters from Ni, Co, and Pb, indicating their different sources, while Cu is grouped between these species, indicating that part of Cu might also stem from the agricultural activity. A similar observation can also be made for Mg, which is grouped separately from all metallic elements. Utilizing PC1 × PC2, ca. 70% of the data variance was explained (Figure 1b), allowing for a partial discriminatory classification of the herbs in four categories: (1) parsley and oregano, depending on the concentrations of Pb and Zn; (2) spearmint, which shows a large dependence on the concentrations of Ni; (3) dill and thyme based mainly on the concentrations of Mg and Co; and (4) rosemary, which seems to have a different profile in Fe and Cu.

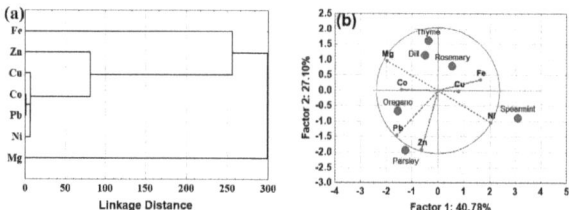

Figure 1. (a) Hierarchal cluster analysis (single linkage Euclidean distances) between metal ions. (b) Graphic of scores (blue) and loadings (red) of PCA in the evaluation of the contents of Cu, Fe, Mg, Zn, Cu, Co, Pb, and Ni in edible herbs. The raw data (total concentration of metal ions in the herbs) were used for the classifications.

2.2. Bioaccessibility of Metallic Elements

The above results confirm the importance of herbs as a source of metallic elements in human nutrition. However, as previously discussed, the total content of metal ions in each herb does not reflect the actual benefits or risks associated with its consumption. Therefore, the in vitro UBM digestion protocol was used to estimate the human bioaccessibility of nutrients and toxic metal ions from herb consumption. The bioaccessible fraction was defined as the ratio between the content of a metal ion in the bioaccessible fraction and the total metal content in the plant [12] as follows:

$$BF(\%) = (\text{bioaccessible content})/(\text{total content}) \times 100.$$

The results for bioaccessible and residual fractions in herbs following gastric and gastrointestinal digestion are depicted in Figure 2. The recovery of metal ions, determined from mass balance calculations (bioaccessible and residual fraction compared with the total content), yielded satisfactory results that ranged from 70–122% for Fe, Cu, Zn, and Mg and 62–113% for Pb, Ni, and Co, indicating the acceptable accuracy of the overall experimental procedure. The bioaccessible fractions for each metal ion are depicted in the bar plots of Supplementary Materials: Figure S1. From these plots, it can be seen that the bioaccessibility of Fe, Mg, Zn, and Ni was higher in the gastric phase, while Cu exhibited higher bioaccessibility in the gastrointestinal phase. On the other hand, the bioaccessibility of Pb and Co was comparable in the gastric and gastrointestinal phases. It should be noted that the concentration of Mg in the gastrointestinal phase was not determined due to the high concentration of Mg salts that were added to simulate the composition of the gastrointestinal tract.

The bioaccessibility of metal ions in the gastric phase can be mainly attributed to the low pH, which increases the solubility of metals, as well as to the presence of pepsin, which is more effective in acid conditions to break down proteins [24,25]. The lower bioaccessibility of Fe and Zn in the gastrointestinal phase may be due to the higher pH of the duodenal fluid (approximately 6.3), which may cause the formation of insoluble Fe and Zn oxyhydroxide complexes and hamper their intestinal adsorption [11,26]. Moreover, the presence of fibers, oxalates, tannins, polyphenols, and mainly phytates (which is a main source of phosphorous) plays a significant role in the reduced bioaccessibility of metal ions in the gastrointestinal phase by complexing metal ions or forming metal precipitates that reduce their bioaccessibility [15,26–29]. In contrast to other elements, Cu bioaccessibility was higher in the gastrointestinal phase. This is attributed to the fact that Cu, due to its unpaired electrons, forms stable complexes with proteins that also include pepsin [30,31], which is the main enzyme during gastric digestion. On the other hand, the increase in the bioaccessibility of Cu in the gastrointestinal fluid is in agreement with previous reports and can be attributed to the almost neutral pH and the presence of low-molecular-weight organic acids (e.g., ascorbic, malic, etc.), which have been associated with the improved bioaccessibility of Cu in the gastrointestinal tract [11,30,32,33].

Regarding toxic metal ions, Pb exhibited similar bioaccessibility in the gastric and gastrointestinal phases. This can be attributed to the fact that pepsin in the gastric fluid may not be able to release most of Pb, but pancreatin in the duodenal fluid can decompose cell walls and increase the release of Pb into intestinal juice [34]. However, the action of pancreatin in the gastrointestinal tract competes with the coprecipitation of Pb by iron oxides, which restricts the bioaccessibility of Pb [14].

The presence of Co in the samples is attributed to exogenous sources (i.e., human activities) since these herbs do not naturally contain cobalamin. The low pH and the presence of pepsin in the gastric fluid releases Co, which exhibits a high affinity for proteins and amino acids [35]. The high bioaccessibility of Co in the gastrointestinal phase may be ascribed to the formation of low-molecular-mass cobalt complexes [36] or the formation of chloride complexes (due to the presence of high concentrations of chloride salts), which exhibit high solubility in both gastric and gastrointestinal fluids [37]. Finally, the bioaccessibility of Ni in the gastric and gastrointestinal phases was similar to that reported in previous studies [19]. The bioaccessible concentration of Ni in the gastric phase is slightly higher than in the gastrointestinal phase, possibly due to Ni coprecipitation and adsorption on Fe oxides at neutral pH (6.3–7.0) of the duodenal fluid [38]. However, the chelation of Ni by some enzymes, such as pepsin, bile, and mucin, may inhibit Ni adsorption by Fe oxides in the intestine and maintain part of Ni as bioaccessible [38].

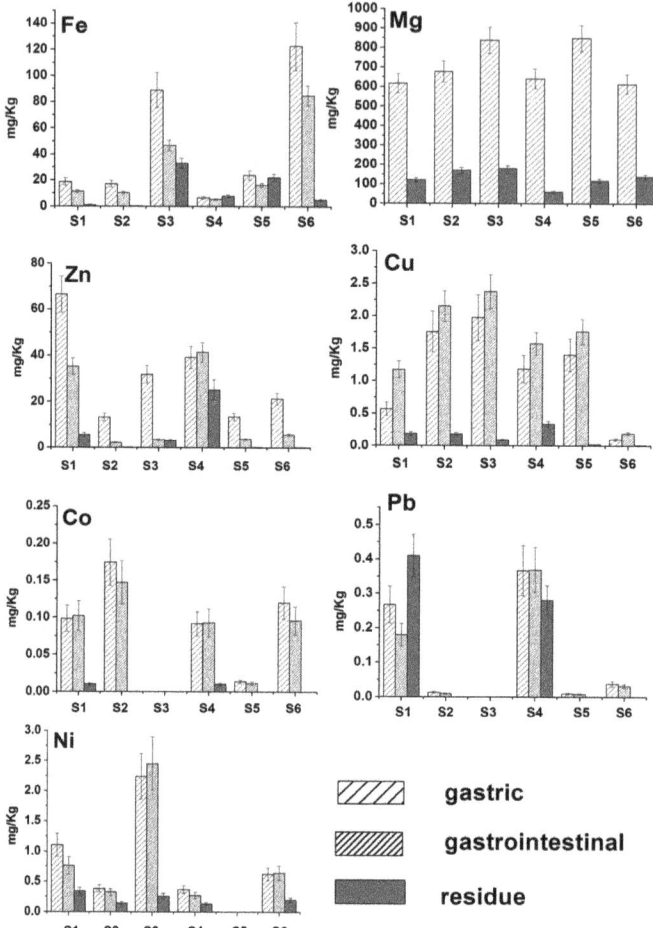

Figure 2. Gastric, gastrointestinal, and residual concentrations of essential elements and toxic metal ions in herbs. S1: Parsley; S2: Dill; S3: Spearmint; S4: Oregano; S5: Thyme; S6: Rosemary.

Using the total concentration of elements and their concentration in the gastrointestinal fluid, PCA was used to identify potential similarities between different herbs in terms of element bioaccessibility. This could be used to design functional products with improved bioaccessibility of essential elements. The score plots in Figure 3 show that the total concentrations and the concentrations in the bioaccessible fraction do not lead to similar classifications, which means that the total concentration does not provide a representative evaluation of the bioaccessibility of elements. Based on the results of Figure 3b, oregano and parsley exhibit the highest bioaccessibility of Zn while dill and thyme offer improved bioaccessibility of Cu. Finally, spearmint seems to offer improved bioaccessibility of Fe. However, spearmint was found to have the highest total concentration of Cu. Therefore, although the bioaccessibility of Cu in the gastrointestinal fluid (38.9%) was lower than thyme (≈45.2%), it still offered a high content of bioaccessible Cu in terms of absolute concentrations (ca. 2.38 mg/Kg in spearmint and 1.75 mg/Kg in thyme). Similarly, although only 30% of Zn in oregano was bioaccessible in the gastrointestinal tract, the total concentration was high, contributing to its high bioaccessibility. Hence, both the total concentration of elements and their bioaccessibility in the gastrointestinal tract should be evaluated to identify potential herb synergies and design functional foods that maximize the bioaccessibility of metallic elements.

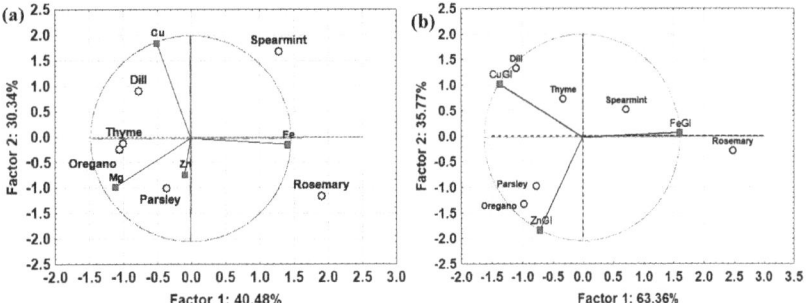

Figure 3. Classification of herbs by comparing element bioaccessibility and total concentration: (**a**) the total concentration of four essential elements and (**b**) the bioaccessible (gastrointestinal) concentration of four essential elements. The open black circles represent the graphic scores and the blue squares represent the factor loadings.

3. Materials and Methods

3.1. Reagents and Solutions

All reagents (salts and organic chemicals) were at least of analytical grade and purchased from major suppliers. Ultrapure grade HNO_3 and H_2O_2 for inorganic trace analysis were a product of Supelco. Alpha amylase (*Bacillus* sp. ≥ 1500 units/mg protein, Merck, Darmstadt, Germany), mucin from porcine stomach (type II, Sigma Aldrich, Steinheim, Germany), pepsin from porcine gastric mucosa (≥500 U/mg, Merck, Darmstand, Germany), bovine serum albumin (Sigma Aldrich, Steinheim, Germany), lipase from porcine pancreas (>150 units/mg protein, Sigma-Aldrich, Steinheim, Germany), pancreatin from porcine pancreas (350 FIP-U/g protease, 6000 FIP-U/g lipase, 7500 FIP-U/g amylase, Merck, Darmstand, Germany), and bile from porcine pancreas (Sigma-Aldrich, Steinheim, Germany) were used to simulate the enzyme composition of gastric and gastrointestinal fluids. The multielement standard solution 6 for ICP (Supelco TraceCERT®, 100 mg/ L, Sigma-Aldrich, Steinheim, Germany) was used to prepare standard metal ion solutions for calibration. Before use, all glassware was rinsed with acetone and water, soaked in 2 M ultrapure HNO_3 overnight, thoroughly washed with distilled water (<2 µS/cm), and dried in a ventilated oven.

The preparation of the digestive fluids used in the UBM test (saliva, gastric, duodenal, and bile) was performed the day before use. Each fluid consisted of a mixture of an inorganic electrolyte solution, a solution of organic compounds, and a solution containing the appropriate enzymes [9]. The composition of each fluid is compiled in Table S1. Each solution was prepared separately, mixed, and stirred for 4 h. The pH of each solution was then recorded and, if necessary, adjusted with the dropwise addition of NaOH or HCl, as follows: salivary fluid 6.5 ± 0.5, gastric fluid 1.1 ± 0.1, duodenal fluid 7.4 ± 0.2, and bile 8 ± 0.2. Before the application of the UBM protocol, all fluids were incubated at 37 °C for 1 h.

3.2. Instrumentation

An ICP OES (Shimadzu, Kyoto, Japan, ICPE-9800), equipped with a semiconductor CCD detector, was used for the measurements. The readouts were recorded with axial view mode, at an exposure time of 30 s. The plasma torch was operated with an RF power and frequency of 1.2 kW and 27 MHz, respectively, a coolant argon flow rate of 10 L/min, an auxiliary argon flow rate of 0.6 L/min^{-1}, and a carrier flow rate of 1.0 mL/min, with simultaneous recording of analytical signals at wavelengths of 239.349, 238.204, and 259.940 nm for Fe, 213.598, 224.700 and 324.754 nm for Cu, 202.548, 206.200, 213.856 nm for Zn, 279.553, 280.270 and 285.213 nm for Mg, 228.616, 237.862 and 238.892 nm for Co, 221.647, 231.604 and 341.476 nm for Ni, and 216.999, 220.353, and 405.783 nm for Pb as a quality control measures.

3.3. Samples

Samples of fresh herbs (except for oregano that was obtained air-dried) were purchased from the same suppliers in three independent sampling campaigns. All herbs were cultivated in organized cultivations in different parts of Greece and were nonorganic. Some species (parsley, spearmint, and dill) were reported to be grown in mixed cultivations with other species, but no specific information was given.

3.4. Sample Preparation and Determination of Metal Ions

Fresh herbs were manually cut into small pieces using a ceramic knife, macerated in liquid nitrogen, and lyophilized for 72 h in a benchtop freeze dryer (Alpha 1-2 LD Plus) (Christ, Osterode am Harz, Germany). The dry solids were ground to powder with a ball mill and stored in a desiccator until analysis and for no longer than 3 days. For determination of the total concentration of metal ions, 0.15 ± 0.02 g of the dry (lyophilized) material was weighted and transferred to PFA microwave digestion vessels (Savillex, Eden Prairie, MN, USA). Then 1 mL of ultrapure concentrated HNO_3 and 0.6 mL of ultrapure H_2O_2 (30%, w/w) were added. The solution was predigested at 50 °C for 30 min to decompose organic matter and release gases. After cooling, the vessels were tightly sealed and exposed to microwave-assisted extraction. Specifically, in the first step, power was held at 430 W for 5 min to aid the mineralization of the organic components. Then, energy was increased to 720 W for 4 min, and decomposition was completed by exposure at 900 W for 5 min, twice. The vessels were cooled in a water bath, and the extracts were transferred to volumetric flasks by filtering through Whatman No. 40 (ashless, metal-free) filters using dilute ultrapure HNO_3 (2 M).

For the determination of metal ions calibration plots were prepared by analyzing standard solutions in the range of 20–500 µg/L. For each element, the part of the calibration plot that enabled the best linear range was used. Then, the sample solutions were analyzed either directly or after appropriate dilution with 2 M ultrapure HNO_3.

3.5. In Vitro Simulation of Gastrointestinal Digestion

For each solid sample, the UBM test was performed in triplicate, including one blank sample (without solid) as control. Therefore, for every herb sample, 8 liquid samples were generated (3 replicates of the gastric phase and 1 gastric fluid blank, and 3 replicates of the gastro-intestinal phase plus 1 gastro-intestinal fluid blank). First, 0.8–1.0 g of samples was weighed in polypropylene flasks (50 mL), and 4.5 mL of salivary solution was added followed by manual agitation for 10 s. Then, 6.75 mL of gastric fluid was added, and the pH was adjusted to 1.2 ± 0.05, using small volumes of concentrated HCl. The mixture was then incubated for 60 min at 37 °C in a RotoFlex Plus end-over-end rotator (Agros, Chicago, IL, USA) at 40 rpm. The pH of the gastric phase was measured to verify that it is lower than 1.5. The gastric samples were centrifuged at 4500 rpm for 15 min, and the supernatant solution was carefully retrieved with a glass pastier pipette and acidified with HNO_3 to stop the enzymatic reactions and preserve the samples until analysis.

For the gastrointestinal digestion phase, 13.5 mL of duodenal and 4.5 mL of biliary fluids were added to the gastric samples, and the pH was adjusted to 6.3 ± 0.4 with 5 mol/L NaOH. The flasks were mixed in the end-over-end rotator for 4 h at 37 °C at 40 rpm, centrifuged at 4500 g for 15 min, and the supernatant was acidified with HNO_3. All samples were sequentially filtered through Whatman No 40 and Nylon filters of 0.45 μm pore size. All filters were also rinsed with 2 M HNO_3 to avoid metal sorption.

The solid material (residual fraction after digestion) was dried in an oven and decomposed by microwave irradiation using the same procedure that was used for the extraction of total metal content from the raw (lyophilized) herb samples.

3.6. Principle Components and Cluster Analysis

Principle component analysis (PCA) was used to unravel relationships between the profile of metallic elements and herb species, as well as their bioaccessibility. We used the herb species as a grouping variable and the concentration of metal ions (total, gastric, and gastrointestinal) as variables for the analysis and classification of the herbs. Cluster analysis (CA) was also used to identify relationships between the metal ions as expressed by their concentration levels. Two sets of data were used: the first consisted of the raw data containing the total concentration of metallic elements in the herbs, while the second set was the concentration of metallic elements determined in the gastric and gastrointestinal extracts.

4. Conclusions

The assessment of the elemental composition of common edible herbs shows that they constitute a supplementary source of metallic elements in the human body and that they may also contribute to the uptake of heavy metals. The bioaccessible (gastric and gastrointestinal) concentrations of essential (Fe, Zn, Mg, and Cu) investigated by the UBM method show that the most bioaccessible element is Cu, followed by Fe and Zn. Moreover, metal ions, such as Pb, Ni, and Co, have been found to be readily bioaccessible and potentially bioavailable in the gastrointestinal tract. Therefore, although the concentration of metals may not be high, special attention must be given to the cultivation conditions and processing of herbs to minimize contamination from heavy metals. The results from bioaccessibility testing also show the potential for designing functional foods and nutraceuticals with improved bioaccessibility of essential elements.

Supplementary Materials: The following supporting information can be downloaded at: https://www.mdpi.com/article/10.3390/molecules28145396/s1, Table S1: Composition of the digestive fluids used in the in vitro UBM bioaccessibility method. Concentrations are in g/L unless otherwise stated. Figure S1: Recovery (%) of essential elements and toxic metal ions in herbs. Patterns inside the bar blots show the bioaccessible fraction (BF, %) in the gastric and gastrointestinal phases and the residual amount remaining after digestion.

Author Contributions: Conceptualization, D.L.G.; investigation, T.G.C., C.T. and V.G.; methodology, A.G.V. and D.L.G.; writing—original draft, T.G.C., C.T. and V.G.; Writing—review and editing, A.G.V. and D.L.G.; All authors have read and agreed to the published version of the manuscript.

Funding: We acknowledge the support of this work by the project "Development of research infrastructure for the design, production, development of quality characteristics and safety of agro foods and functional foods (RI-Agrofoods)" (MIS 5047235), which is implemented under the action "Reinforcement of the Research and Innovation Infrastructure", funded by the Operational Programme "Competitiveness, Entrepreneurship, and Innovation" (NSRF 2014-2020) and co-financed by Greece and the European Union (European Regional Development Fund).

Institutional Review Board Statement: Not applicable.

Informed Consent Statement: Not applicable.

Data Availability Statement: The authors confirm that the data supporting the findings of this study are available within the article and its supplementary materials.

Conflicts of Interest: The authors declare no conflict of interest. The funders had no role in the design of the study; in the collection, analyses, or interpretation of data; in the writing of the manuscript; or in the decision to publish the results.

References

1. Van Wyk, B.-E. *Food Plants of the World Identification, Culinary Uses, and Nutritional Value*; CABI: Wallingford, UK, 2019; ISBN 978-1-78924-130-3.
2. Hoffmann, D. *Medical Herbalism: The Science Principles and Practices of Herbal Medicine*; Healing Arts Press: Rochester, VT, USA, 2003; ISBN 978-0892817498.
3. Fernández-García, E.; Carvajal-Lérida, I.; Pérez-Gálvez, A. In Vitro Bioaccessibility Assessment as a Prediction Tool of Nutritional Efficiency. *Nutr. Res.* **2009**, *29*, 751–760. [CrossRef] [PubMed]
4. Carbonell-Capella, J.M.; Buniowska, M.; Barba, F.J.; Esteve, M.J.; Frígola, A. Analytical Methods for Determining Bioavailability and Bioaccessibility of Bioactive Compounds from Fruits and Vegetables: A Review. *Compr. Rev. Food Sci. Food Saf.* **2014**, *13*, 155–171. [CrossRef] [PubMed]
5. Dima, C.; Assadpour, E.; Dima, S.; Jafari, S.M. Bioavailability and Bioaccessibility of Food Bioactive Compounds; Overview and Assessment by in vitro Methods. *Compr. Rev. Food Sci. Food Saf.* **2020**, *19*, 2862–2884. [CrossRef] [PubMed]
6. Rein, M.J.; Renouf, M.; Cruz-Hernandez, C.; Actis-Goretta, L.; Thakkar, S.K.; da Silva Pinto, M. Bioavailability of Bioactive Food Compounds: A Challenging Journey to Bioefficacy. *Br. J. Clin. Pharmacol.* **2013**, *75*, 588–602. [CrossRef] [PubMed]
7. Hur, S.J.; Lim, B.O.; Decker, E.A.; McClements, D.J. In Vitro Human Digestion Models for Food Applications. *Food Chem.* **2011**, *125*, 1–12. [CrossRef]
8. Ng, J.C.; Juhasz, A.; Smith, E.; Naidu, R. Assessing the Bioavailability and Bioaccessibility of Metals and Metalloids. *Environ. Sci. Pollut. Res.* **2015**, *22*, 8802–8825. [CrossRef]
9. BARGE–The Bioaccessibility Research Group of Europe. BARGE Unified Bioaccessibility Method. 2009. Available online: https://www2.bgs.ac.uk/barge/ubm.html (accessed on 12 June 2023).
10. Souza, L.A.; Souza, T.L.; Santana, F.B.; Araujo, R.G.O.; Teixeira, L.S.G.; Santos, D.C.M.B.; Korn, M.G.A. Determination and in vitro Bioaccessibility Evaluation of Ca, Cu, Fe, K, Mg, Mn, Mo, Na, P and Zn in Linseed and Sesame. *Microchem. J.* **2018**, *137*, 8–14. [CrossRef]
11. Herrera-Agudelo, M.A.; Miró, M.; Arruda, M.A.Z. In Vitro Oral Bioaccessibility and Total Content of Cu, Fe, Mn and Zn from Transgenic (through Cp4 EPSPS Gene) and Nontransgenic Precursor/Successor Soybean Seeds. *Food Chem.* **2017**, *225*, 125–131. [CrossRef]
12. Tokalıoğlu, Ş.; Clough, R.; Foulkes, M.; Worsfold, P. Bioaccessibility of Cr, Cu, Fe, Mg, Mn, Mo, Se and Zn from Nutritional Supplements by the Unified BARGE Method. *Food Chem.* **2014**, *150*, 321–327. [CrossRef]
13. Intawongse, M.; Kongchouy, N.; Dean, J.R. Bioaccessibility of Heavy Metals in the Seaweed *Caulerpa racemosa* Var. Corynephora: Human Health Risk from Consumption. *Instrum. Sci. Technol.* **2018**, *46*, 628–644. [CrossRef]
14. Xu, X.; Wang, J.; Wu, H.; Lu, R.; Cui, J. Bioaccessibility and Bioavailability Evaluation of Heavy Metal(Loid)s in Ginger in vitro: Relevance to Human Health Risk Assessment. *Sci. Total Environ.* **2023**, *857*, 159582. [CrossRef]
15. Tokalıoğlu, Ş. Bioaccessibility of Cu, Mn, Fe, and Zn in Fruit and Vegetables by the In Vitro UBM and Statistical Evaluation of the Results. *Biol. Trace Elem. Res.* **2023**, *201*, 1538–1546. [CrossRef] [PubMed]
16. Hedges, L.J.; Lister, C.E. *Nutritional Attributes of Herbs*; Crop & Food Research Confidential Report No. 1891; New Zealand Institute for Crop & Food Research Limited: Christchurch, New Zealand, 2007.
17. Tsogas, G.Z.; Giokas, D.L.; Kapakoglou, N.I.; Efstathiou, D.E.; Vlessidis, A.G.; Dimitrellos, G.N.; Georgiadis, T.D.; Charchanti, A.V. Land-Based Classification of Herb's Origin Based on Supervised and Unsupervised Pattern Recognition of Plant and Soil Chemical Profiling. *Anal. Lett.* **2010**, *43*, 2031–2048. [CrossRef]
18. Munir, N.; Jahangeer, M.; Bouyahya, A.; El Omari, N.; Ghchime, R.; Balahbib, A.; Aboulaghras, S.; Mahmood, Z.; Akram, M.; Ali Shah, S.M.; et al. Heavy Metal Contamination of Natural Foods Is a Serious Health Issue: A Review. *Sustainability* **2021**, *14*, 161. [CrossRef]
19. Hu, J.; Wu, F.; Wu, S.; Cao, Z.; Lin, X.; Wong, M.H. Bioaccessibility, Dietary Exposure and Human Risk Assessment of Heavy Metals from Market Vegetables in Hong Kong Revealed with an in vitro Gastrointestinal Model. *Chemosphere* **2013**, *91*, 455–461. [CrossRef]
20. Yang, L.; Yang, Y.; Tian, W.; Xia, X.; Lu, H.; Wu, X.; Huang, B.; Hu, W. Anthropogenic Activities Affecting Metal Transfer and Health Risk in Plastic-Shed Soil-Vegetable-Human System via Changing Soil PH and Metal Contents. *Chemosphere* **2022**, *307*, 136032. [CrossRef]
21. Intawongse, M.; Dean, J.R. Use of the Physiologically-Based Extraction Test to Assess the Oral Bioaccessibility of Metals in Vegetable Plants Grown in Contaminated Soil. *Environ. Pollut.* **2008**, *152*, 60–72. [CrossRef]
22. Mills, M.M.; Ridame, C.; Davey, M.; La Roche, J.; Geider, R.J. Iron and Phosphorus Co-Limit Nitrogen Fixation in the Eastern Tropical North Atlantic. *Nature* **2004**, *429*, 292–294. [CrossRef]
23. Broadley, M.R.; White, P.J.; Hammond, J.P.; Zelko, I.; Lux, A. Zinc in Plants. *New Phytol.* **2007**, *173*, 677–702. [CrossRef] [PubMed]
24. Intawongse, M.; Dean, J.R. In-Vitro Testing for Assessing Oral Bioaccessibility of Trace Metals in Soil and Food Samples. *TrAC Trends Anal. Chem.* **2006**, *25*, 876–886. [CrossRef]

25. Pelfrêne, A.; Waterlot, C.; Guerin, A.; Proix, N.; Richard, A.; Douay, F. Use of an in vitro Digestion Method to Estimate Human Bioaccessibility of Cd in Vegetables Grown in Smelter-Impacted Soils: The Influence of Cooking. *Environ. Geochem. Health* **2015**, *37*, 767–778. [CrossRef]
26. Sandström, B. Micronutrient Interactions: Effects on Absorption and Bioavailability. *Br. J. Nutr.* **2001**, *85*, S181. [CrossRef]
27. Tuntipopipat, S.; Zeder, C.; Siriprapa, P.; Charoenkiatkul, S. Inhibitory Effects of Spices and Herbs on Iron Availability. *Int. J. Food Sci. Nutr.* **2009**, *60*, 43–55. [CrossRef]
28. Filipiak-Szok, A.; Kurzawa, M.; Szłyk, E. Simultaneous Determination of Selected Anti-Nutritional Components in Asiatic Plants Using Ion Chromatography. *Eur. Food Res. Technol.* **2016**, *242*, 1515–1521. [CrossRef]
29. Konietzny, U.; Greiner, R. *Phytic Acid: Nutritional Impact*, 2nd ed.; Academic Press: Cambridge, MA, USA, 2003; ISBN 9780122270550.
30. Li, Y.; Demisie, W.; Zhang, M. The Function of Digestive Enzymes on Cu, Zn, and Pb Release from Soil in vitro Digestion Tests. *Environ. Sci. Pollut. Res.* **2013**, *20*, 4993–5002. [CrossRef]
31. Steinhart, H.; Beyer, M.G.; Kirchgeßner, M. Zur Komplexbildung von Proteinen Mit Cu^{2+}-Ionen Im Sauren Milieu. *Z. Lebensm Unters Forsch.* **1975**, *159*, 73–77. [CrossRef] [PubMed]
32. Poggio, L.; Vrščaj, B.; Schulin, R.; Hepperle, E.; Ajmone Marsan, F. Metals Pollution and Human Bioaccessibility of Topsoils in Grugliasco (Italy). *Environ. Pollut.* **2009**, *157*, 680–689. [CrossRef]
33. Koplík, R.; Borková, M.; Mestek, O.; Komínková, J.; Suchánek, M. Application of Size-Exclusion Chromatography–Inductively Coupled Plasma Mass Spectrometry for Fractionation of Element Species in Seeds of Legumes. *J. Chromatogr. B* **2002**, *775*, 179–187. [CrossRef]
34. Fu, J.; Cui, Y. In Vitro Digestion/Caco-2 Cell Model to Estimate Cadmium and Lead Bioaccessibility/Bioavailability in Two Vegetables: The Influence of Cooking and Additives. *Food Chem. Toxicol.* **2013**, *59*, 215–221. [CrossRef] [PubMed]
35. Paustenbach, D.J.; Tvermoes, B.E.; Unice, K.M.; Finley, B.L.; Kerger, B.D. A Review of the Health Hazards Posed by Cobalt. *Crit. Rev. Toxicol.* **2013**, *43*, 316–362. [CrossRef] [PubMed]
36. Wojcieszek, J.; Ruzik, L. Study of Bioaccessibility of Cobalt Species in Berries and Seeds by Mass Spectrometry Techniques. *J. Anal. Sci. Technol.* **2020**, *11*, 26. [CrossRef]
37. Danzeisen, R.; Williams, D.L.; Viegas, V.; Dourson, M.; Verberckmoes, S.; Burzlaff, A. Bioelution, Bioavailability, and Toxicity of Cobalt Compounds Correlate. *Toxicol. Sci.* **2020**, *174*, 311–325. [CrossRef] [PubMed]
38. Liang, J.-H.; Lin, X.-Y.; Huang, D.-K.; Xue, R.-Y.; Fu, X.-Q.; Ma, L.Q.; Li, H.-B. Nickel Oral Bioavailability in Contaminated Soils Using a Mouse Urinary Excretion Bioassay: Variation with Bioaccessibility. *Sci. Total Environ.* **2022**, *839*, 156366. [CrossRef] [PubMed]

Disclaimer/Publisher's Note: The statements, opinions and data contained in all publications are solely those of the individual author(s) and contributor(s) and not of MDPI and/or the editor(s). MDPI and/or the editor(s) disclaim responsibility for any injury to people or property resulting from any ideas, methods, instructions or products referred to in the content.

Article

Two Fast GC-MS Methods for the Measurement of Nicotine, Propylene Glycol, Vegetable Glycol, Ethylmaltol, Diacetyl, and Acetylpropionyl in Refill Liquids for E-Cigarettes

Ioanna Dagla [1], Evagelos Gikas [2] and Anthony Tsarbopoulos [1,3,*]

1. Bioanalytical Laboratory, GAIA Research Center, The Goulandris Natural History Museum, 14562 Kifissia, Greece
2. Laboratory of Analytical Chemistry, Faculty of Chemistry, National and Kapodistrian University of Athens, Panepistimioupolis Zografou, 15784 Athens, Greece
3. Department of Pharmacology, Medical School, National and Kapodistrian University of Athens, 11527 Athens, Greece
* Correspondence: atsarbop@med.uoa.gr; Tel.: +30-210-746-2702

Abstract: The use of e-cigarettes (ECs) has become increasingly popular worldwide, even though scientific results have not established their safety. Diacetyl (DA) and acetylpropionyl (AP), which can be present in ECs, are linked with lung diseases. Ethyl maltol (EM)—the most commonly used flavoring agent—can be present in toxic concentrations. Until now, there is no methodology for the determination of nicotine, propylene glycol (PG), vegetable glycerin (VG), EM, DA, and acetylpropionyl in e-liquids that can be used as a quality control procedure. Herein, gas chromatography coupled with mass spectrometry (GC-MS) was applied for the development of analytical methodologies for these substances. Two GC-MS methodologies were developed and fully validated, fulfilling the standards for the integration in a routine quality control procedure by manufacturers. As proof of applicability, the methodology was applied for the analysis of several e-liquids. Differences were observed between the labeled and the experimental levels of PG, VG, and nicotine. Three samples contained EM at higher concentrations compared to the other samples, while only one contained DA. These validated methodologies can be used for the quality control analysis of EC liquid samples regarding nicotine, PG, and VG amounts, as well as for the measurement of the EM.

Keywords: GC-MS; nicotine; ethylmaltol; diacetyl; propylene glycol; quality control procedure; e-liquids; e-cigarettes

Citation: Dagla, I.; Gikas, E.; Tsarbopoulos, A. Two Fast GC-MS Methods for the Measurement of Nicotine, Propylene Glycol, Vegetable Glycol, Ethylmaltol, Diacetyl, and Acetylpropionyl in Refill Liquids for E-Cigarettes. *Molecules* **2023**, *28*, 1902. https://doi.org/10.3390/molecules28041902

Academic Editors: Paraskevas D. Tzanavaras and Victoria Samanidou

Received: 2 February 2023
Revised: 14 February 2023
Accepted: 15 February 2023
Published: 16 February 2023

Copyright: © 2023 by the authors. Licensee MDPI, Basel, Switzerland. This article is an open access article distributed under the terms and conditions of the Creative Commons Attribution (CC BY) license (https://creativecommons.org/licenses/by/4.0/).

1. Introduction

Electronic cigarettes (ECs) are nicotine-delivery products that, instead of tobacco, contain a solution of nicotine benzoate salt in a propylene glycol (PG) and glycerol (vegetable glycerin, VG) base and various flavoring agents. The base is made from PG, VG, or a mixture of the two in various ratios, diluted in purified water. Nicotine concentration varies from 0 mg/mL to 18 mg/mL. The wide variety of tastes (e.g., sweet, cool, bitter, harsh) contributes to a higher likability of EC products and higher initiation rates of vaping [1]. According to the "E-cigarette and Vape Market Size Report, 2022–2030" (https://www.grandviewresearch.com/industry-analysis/e-cigarette-vaping-market, accessed on 27 January 2023), *"The global e-cigarette and vape market size was valued at USD 18.13 billion in 2021 and is expected to expand at a compound annual growth rate (CAGR) of 30.0% from 2022 to 2030"*. The Center for Disease Control and Prevention states that *"about 1 in 5 high school students and 1 in 20 middle school students reported using e-cigarettes in 2020"* (https://www.cdc.gov/tobacco/features/back-to-school/index.html, accessed on 27 January 2023).

Although ECs are likely to be far less harmful than conventional cigarettes, they can be correlated with several health hazards [2]. Concerns have been raised about the potential inhalation toxicity of the flavoring chemicals that are added to ECs to create flavors [3]. Cinnamaldehyde, benzaldehyde, ethyl vanillin, ethyl maltol (EM), and vanillin are specific chemicals that have been linked to cytotoxic effects on respiratory cells [4]. The potential health risk is challenging since the concentration of flavoral chemicals is not known. Manufacturers are not obliged to report the chemical substances or concentrations, and the FDA regulations do not propose guidelines for EC ingredients.

For the determination of nicotine and flavoring chemical concentrations in e-liquid refill samples, a limited number of analytical methodologies exists based on gas chromatography–mass spectrometry (GC-MS) [5–10], liquid chromatography–mass spectrometry (LC-MS) [11], and nuclear magnetic resonance (NMR) [12,13].

A systematic review of refillable e-liquid samples demonstrated that the actual concentration of nicotine might vary considerably from labeled concentrations [14]. Among the flavoring agents, EM, which has a cotton candy fragrance, is the most common component in EC liquids [11]. A study demonstrated that EM was contained in 80% of the tested e-liquids at concentrations 100 times its cytotoxic concentration [15]. The concentration of EM ranges from 0.001 to 10 mg/mL in e-liquid refills [10]. Based on the IC_{50} data, EM is the most toxic ingredient among the flavoring agents, and more interestingly, it has been shown that the cytotoxicity of refill fluids is directly correlated with EM concentrations in the fluids [10]. Furthermore, the cytotoxicity of produced aerosols during vaping has been strongly correlated with nicotine and EM concentrations [9]. EM promotes free radicals in aerosols in a concentration-dependent manner that cause damage to proliferation, survival, and inflammation pathways in the cell [7]. This fact emphasizes the indispensable need for regulations regarding the flavoring chemicals in e-liquids. Diacetyl (DA) is also an ingredient used in e-liquids for its characteristic butter flavor note and was found in more than 60% of samples [16]. Moreover, the formation of DA could be observed during aerosol generation from e-liquids [17]. Unfortunately, DA has been linked to the development of obliterative bronchiolitis, which is an irreversible, life-threatening lung disease [17–20]. In addition, DA has been associated with Alzheimer's disease, as it has been demonstrated to aggregate amyloid-β [21]. Flavoring alternatives to DA have been used in the food industry, e.g., 2,3-pentane-dione, 2,3-heptanedione, and acetoin, but these chemicals cause respiratory hazards as well [16]. FDA suggests that the presence of specific constituents, including DA and acetylpropionyl (AP), should be considered in e-liquids and aerosols to characterize that a product is "appropriate for the protection of public health" [17]. A few analytical methodologies have been developed for the analysis of these chemicals using gas chromatography–electron capture detector [16], GC-MS [22], UPLC-MS [17], and HPLC-UV [23].

Herein, two fast analytical methodologies have been developed for the analysis of nicotine, PG, VG, EM, DA, and AP. These methodologies can be used for the quality control analysis of e-liquids assessing the nicotine, PG, and VG amounts, as well as for the measurement of the toxic flavor chemical EM. The latter (EM concentration) has been directly correlated with the cytotoxicity of e-liquids. The presence of DA and AP can be estimated in a quick analytical procedure. Several methodologies have been previously reported for the quantitation of these substances in e-liquids, but herein, we present a two-stage methodology for the simultaneous determination of PG, VG, nicotine, and EM and the estimation of DA and AP levels. Until now, the reported methodologies for the determination of the PG and VG levels have included NMR technology [24], GC-MS methodologies with long total run time [25,26], and an SPME preparation step prior to analysis [5]. This is the first methodology for the analysis of the combination of these substances using GC-MS. The presented methodology is fast and can be used as a quality control method employing only the GC-MS instrumentation, which is common and familiar to many industries.

2. Results and Discussion

2.1. Method Development

Nicotine, PG, VG, and EM were selected for screening during the quality control of e-liquids. Furthermore, the examination for the potential presence of DA and AP in e-liquids was deemed essential, as these substances are very toxic. Prior to the GC-MS analysis, a derivatization step was necessary. Two derivatization reagents were selected, owing to the different structures of the substances. The analysis of the hydroxy groups of PG, VG, and EM was based on the derivatization with the BSTFA + TMCS reagent (first method), while the analysis of carbonyl groups of DA and AP was performed with derivatization using the o-phenylenediamine (second method). Nicotine was not derivatized.

For the first method, the calibration points of EM, PG, and VG were combined, thus shortening the analysis time. A typical chromatogram of PG, VG, EM, and ISTD is presented in Figure 1.

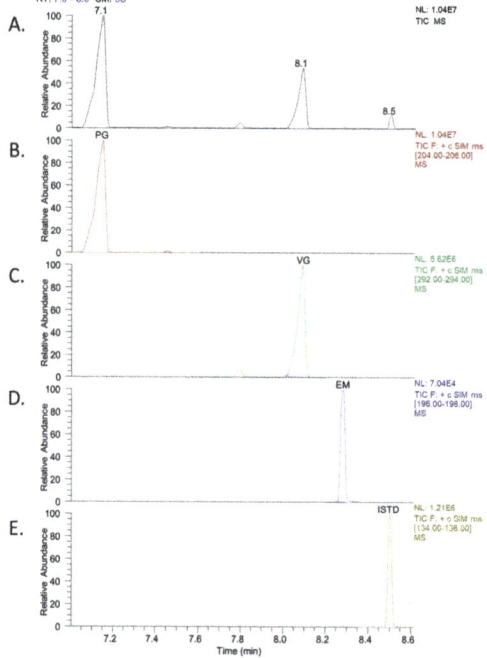

Figure 1. (**A**). Total ion current chromatogram of PG, VG, EM, and ISTD using GC-MS. (**B–E**). Extracted ion chromatograms of PG (m/z 205), VG (m/z 293), EM (m/z 197), and ISTD (m/z 135).

To ensure that the concentration of EM was not affected by the different matrices (different ratios of PG/VG in each calibration point), EM at a concentration of 0.3 mg/mL was spiked in the tested ratios of PG/VG (100/0, 80/20, 70/30, 50/50, 30/70, and 0/100), and the samples were analyzed according to the described methodology. The %RSD of the EM in the examined samples ($n = 6$) was 1.3%, indicating that the ratio of PG/VG did not affect the dilution of EM. Therefore, the combination of calibration points did not result in fault results. Furthermore, the proposed methodology did not require prior knowledge of the PG/VG ratio for the analysis of EM. A blank chromatogram after the addition of derivatization agents and the ISTD is given in Figure S1 (Supplementary Material).

For the second method, the PG/VG ratio of 50/50 was selected as the matrix for the calibration curve of nicotine. The %RSD of nicotine (9 mg/mL, $n = 6$) was 2.1% in different ratios of PG/VG; thus, one ratio of PG/VG was selected to simplify the methodology.

For DA and AP, no calibration curves were constructed. Taking into consideration that the presence of those substances in e-liquids was not desirable, it was deemed that the measurement of their level was beyond the limits of a quality control methodology. The described methodology defines a threshold level of 5 μg/mL to show if an e-liquid contains DA and AP. The dilution of DA and AP in PG/VG was not affected by the ratio of the matrix (%RSD = 1.8, n = 6). A typical chromatogram of nicotine, DA, and AP is presented in Figure 2. A blank chromatogram after the addition of derivatization agents and the ISTD is given in Figure S2.

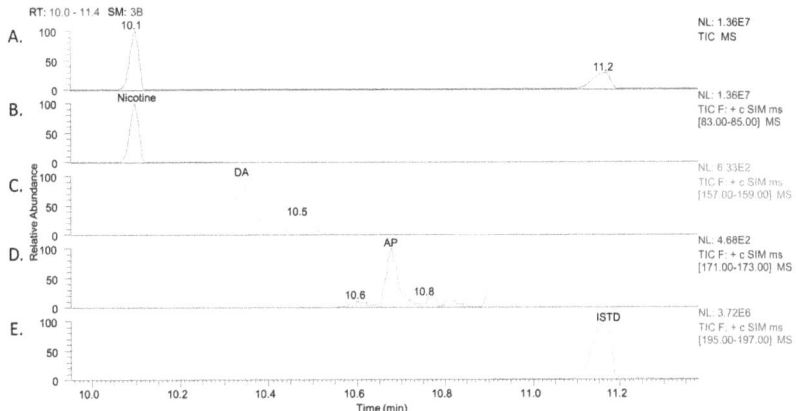

Figure 2. (**A**). Total ion current chromatogram of nicotine, DA, AP, and ISTD using GC-MS. (**B**–**E**). Extracted ion chromatograms of nicotine (m/z 84), 5 μg/mL DA (m/z 158), 5 μg/mL AP (m/z 172), and ISTD (m/z 196).

2.2. Method Validation

2.2.1. Selectivity and Specificity

In both methodologies, it was observed that the matrices (derivatization reagents) did not interfere with the detection of the target analytes. The specificity of the current methodologies was ensured by monitoring the specific ions of nicotine, EM, PG, VG, DA, and AP.

2.2.2. Fitted Models

The calibration curves were constructed for the range of 0–10 mL/mL for PG and VG, 2–20 mg/mL for nicotine, and 0.1–0.5 mg/mL for EM. The areas of targeted compounds were divided by the area of ISTD. The linear and quadratic models were examined. The correlation coefficient (R^2) was better in the quadratic models than in the linear models for all the substances. The percent errors of the back-calculated values were compared for the two models. That error was calculated by the following equation:

$$\%E = \frac{C_{theoretical} - C_{experimental}}{C_{theoretical}} \times 100$$

where $C_{theoretical}$ is the theoretical concentration level of each substance and $C_{experimental}$ is the concentration calculated by the linear or the quadratic regression. The back-calculated values using the quadratic regression presented lower percent errors than those obtained with the linear equation. Generally, the absolute average %E of the back-calculated values with the quadratic and linear regressions were 2.2% and 6.4%, respectively, for nicotine; 2.2% and 10.6% for PG; 2.5% and 4.6% for VG; and 0.1% and 2.2% for EM. Therefore, the calibration curves were established by applying the quadratic regression. The best-fit values of the intercept (B0), the coefficient of the linear term (B1), and the coefficient of the squared term (B2), as well as the standard errors associated with the coefficients, the R^2 and the standard error of estimate (Sy.x), are presented in Table 1. Even though the B2 term has a high standard deviation for some analytes (e.g., PG), the quadratic model was selected over the linear.

Table 1. The best-fit values B0, B1, and B2, the standard errors associated with the coefficients, the correlation coefficient (R^2), and the standard error of the estimate (Sy.x) of the quadratic models for PG, VG, EM, and nicotine.

Compounds	Best-fit Values			Std. Error			R^2	Sy.x
	B0	B1	B2	B0	B1	B2		
PG	0.04949	8.299	−0.06763	0.04882	0.3841	0.04899	0.9994	0.2195
VG	0.2189	0.1991	0.000476	0.3863	0.0183	0.000173	0.9990	0.4298
EM	−0.02828	0.7924	−0.2316	0.01217	0.09273	0.1516	0.9990	0.005673
Nicotine	0.000615	0.2408	−0.00391	0.06438	0.01515	0.000681	0.998	0.05609

2.2.3. Accuracy and Precision

The intra-day and inter-day accuracy expressed as percent standard error from the nominal value (%E) was assessed by analyzing samples at three concentration levels and at three analytical runs. All the models exhibited accuracy lower than 7.5% at the three tested levels. Repeatability (the precision under the same operating condition over a short interval of time) and intermediate precision (the variations between different analytical days, $n = 3$) were expressed as percent relative standard deviation (%RSD). The results are presented in Table 2.

Table 2. Accuracy (intra-day and inter-day), repeatability, intermediate precision, stability, and robustness assessed for PG, VG, EM, nicotine, DA, and AP.

Substance (Levels)	Accuracy (%E, $n = 3$)		Precision (%RSD, $n = 3$)		Stability (%RSD, $n = 3$)	Robustness (%RSD, $n = 3$)
	Intra-day	Inter-day	Repeatability	Intermediate Precision		
PG (mL/10 mL)						
2	1.05	5.35	2.7	1.1	2.4	
5	−5.63	−7.53	7.3	2.8	5.2	<1.2
10	2.55	3.05	3.0	2.5	3.2	
VG (mL/10 mL)						
2	5.47	2.36	3.4	4.3	2.5	
5	2.51	−5.2	4.1	1.2	5.0	<2.0
10	6.24	2.84	0.3	3.5	2.9	
EM (mg/mL)						
0.1	2.1	0.05	1.7	1.9	1.5	
0.3	2.79	2.34	2.3	1.8	1.8	<1.3
0.5	−2.51	−1.58	3.4	3.5	2.1	
Nicotine (mg/mL)						
3	0.02	3.15	0.7	0.0	1.0	
12	3.66	7.53	5.3	1.7	1.6	<1.27
20	−1.95	−5	5.0	1.5	1.2	
DA (μg/mL)						
5	1.23	2.65	2.4	4.5	2.6	<2.4
AP (μg/mL)						
5	−0.59	2.36	1.3	1.1	1.6	<1.9

2.2.4. Stability and Robustness

Stability was examined at the same levels by injecting the same sample ($n = 3$) every three hours (autosampler stability). The results showed that all of the tested compounds were stable at the duration of the data acquisition. The robustness was examined by making deliberate changes (±5%) in the GC parameters (injector temperature and carrier gas flow rate).

2.2.5. Carry-over

Injections of the upper limit of quantitation (ULOQ) were performed. No carry-over effect was observed since a non-detectable amount of the analytes was found in the blank injected samples.

2.2.6. Screening and Quantification of E-Liquid Samples

The validated methodologies were used for the determination of nicotine, EM, PG, VG, DA, and AP in a subset of e-liquid samples. The concentrations of the substances are expressed as % w/v (which is equivalent to mg/100 mL concentration), and the results are presented in Table 3. Apart from sample_1, none of the samples was detected positive for DA (LOD = 5 µg/mL). No other sample was found positive for DA and AP. In most of the tested e-liquids, the results revealed that the levels of PG and VG agreed with those claimed by the manufacturer (<±10.0%E). However, some samples presented >±10.0%E for PG and VG (for example, sample_8 and sample_22). In most cases, the amount of nicotine determined by the developed methodology was about the same as that claimed by the manufacturer (<27.8% E), except for the sample_25 that presented 41.7%E from the labeled value. The amount of EM was not stated in the e-liquid labels; thus, the %E could not be calculated.

Table 3. The levels (expressed as % w/v) of PG, VG, nicotine, and EM in the tested e-liquids calculated by the developed GC-MS methodology. The amounts of PG, VG, and nicotine claimed by the manufacturer are also presented. The error (%E) of the calculated (GC-MS–derived) vs. the claimed values is given in parenthesis. The calculation for the %E was based on the mathematical formula %E = $\frac{C_{claimed} - C_{calcd.}}{C_{claimed}} \times 100$.

Samples	PG % w/v		VG % w/v		Nicotine % w/v		EM % w/v
	Claimed	Calcd. (%E)	Claimed	Calcd. (%E)	Claimed	Calcd. (%E)	Calcd.
Sample_1	70	68.6 (2.0)	30	29.6 (1.2)	1.8	1.7 (5.6)	0.003
Sample_2	70	67.7 (3.2)	30	31.6 (−5.4)	0.6	0.6 (0.0)	0.004
Sample_3	70	75.0 (−7.1)	30	25.0 (16.8)	0.0	0.0 (0.0)	0.001
Sample_4	70	64.4 (8.0)	30	34.4 (−14.8)	1.2	1.1 (8.3)	0.002
Sample_5	60	55.2 (8.0)	40	42.8 (−7.1)	1.8	1.8 (0.0)	0.171
Sample_6	60	56.6 (5.7)	40	41.6 (−4.0)	1.8	1.8 (0.0)	0.006
Sample_7	60	56.5 (5.9)	40	42.4 (−5.9)	1.2	1.2 (0.0)	0.02
Sample_8	60	70.1 (−16.8)	40	29.9 (25.4)	0.0	0.0 (0.0)	0.003
Sample_9	70	69.6 (0.5)	30	29.3 (2.4)	1.2	1.1 (8.3)	0.002
Sample_10	70	70.5 (−0.7)	30	28.9 (3.8)	0.6	0.6 (0.0)	0.006
Sample_11	50	53.0 (−6.0)	50	45.7 (8.6)	1.2	1.3 (−8.3)	0.012
Sample_12	50	46.5 (6.9)	50	53.0 (−5.9)	0.6	0.5 (16.7)	0.000
Sample_13	50	50.5 (−0.9)	50	49.2 (1.5)	0.3	0.3 (0.0)	0.009
Sample_14	50	48.4 (3.2)	50	51.3 (−2.6)	0.3	0.3 (0.0)	0.000
Sample_15	50	47.2 (5.6)	50	52.1 (−4.3)	0.6	0.6 (0.0)	0.11
Sample_16	70	71.7 (−2.5)	30	27.0 (9.9)	1.2	1.2 (0.0)	0.01
Sample_17	50	47.7 (4.7)	50	51.1 (−2.1)	1.2	1.3 (−8.3)	0.000
Sample_18	0	7.8 (0.0)	100	91.7 (8.3)	0.6	0.5 (16.7)	0.013
Sample_19	70	65.5 (6.4)	30	33.0 (−9.9)	1.8	1.5 (16.7)	0.000
Sample_20	70	74.5 (−6.5)	30	23.9 (20.2)	1.8	1.3 (27.8)	0.18
Sample_21	100	100.0 (0.0)	0	0.0 (0.0)	0.0	0.0 (0.0)	0.000
Sample_22	70	81.7 (−16.6)	30	17.0 (43.4)	1.2	1.4 (−16.7)	0.000
Sample_23	70	66.6 (4.8)	30	31.3 (−4.4)	1.8	2.0 (−11.1)	0.000
Sample_24	0	0.0 (0.0)	100	94.2 (5.8)	0.6	0.6 (0.0)	0.000
Sample_25	50	44.6 (10.9)	50	53.7 (−7.4)	1.2	1.7 (−41.7)	0.000
Sample_26	50	45.0 (9.9)	50	54.3 (−8.5)	0.6	0.7 (−16.7)	0.000

Three samples (Sample_5, Sample _15, and Sample_20) contained EM at higher concentrations compared to the other samples. In fact, those levels of EM were above the higher point of the calibration curve. The analysis of those samples was performed via the appropriate dilution of the sample so that the concentration was within the calibration range. In the literature, there is a lack of information on the safe limit of EM concentration in e-liquids. The study of Omaiye et al. claimed that 46% of the tested e-liquid samples contained EM (0.008–3.13%) in concentrations higher than those added to edible products (up to 0.0142%) and in final products of soap (up to 0.06%), detergents (up to 0.006%), and creams and lotions (up to 0.01%) [15]. The oral LD50 of EM is 1150 mg/Kg for rats, but there is no information on the safe intake in humans. Thus, there is an imperative need to establish the maximum allowed nontoxic concentration for EM in e-liquids.

3. Materials and Methods

3.1. Chemicals and Reagents

Methanol ≥99.9% was purchased from Thermo Fisher Scientific (Waltham, MA, USA), and acetone was from Carlo Erba Reagents (Val de Reuil CEDEX, France). EM, 2,3-Butanedione 97% (DA), 2,3-Pentanedione 97% (AP), N,OBis(trimethylsilyl)trifluorocetamide (BSTFA) 1%–TMCS 99%, and 3-methoxyphenethyl alcohol (internal standard—ISTD) were purchased from Sigma-Aldrich (Steinheim, Germany). The reagent o-Phenylenediamine 98% was purchased from Thermo Fisher Scientific (Waltham, MA, USA). Nicotine, propylene glycol (PG), and vegetable glycol (VG) were provided by NOBACCO (Koropi, Greece). Ultrapure water was produced by a Millipore Direct-Q System (Molsheim, France).

3.2. E-Liquid Samples

Twenty-six e-liquid samples provided by NOBACCO were intended to be screened for nicotine, EM, PG/VG ratio, and the presence of DA and AP. All samples were stored at ambient temperature and protected from light.

3.3. Preparation of Standard Solutions

First method: A stock solution of EM was prepared in methanol at a concentration of 10 mg/mL and was stored at −30 °C to avoid sample degradation. Stock solutions of PG and VG were prepared at ratios of PG/VG—80/20, 70/30, 50/50, and 30/70 mL/mL.

The calibration solutions were obtained by diluting the EM stock solution with 1 mL of PG/VG to the concentration levels of 0.1, 0.2, 0.3, 0.4, and 0.5 mg/mL. Each addition of EM was performed in different ratios of PG/VG so that three calibration curves were obtained in one analytical run. The amounts of EM (mg/mL), PG (mL), and VG (mL) in calibration points (CP) were 1st CP: 0.1, 100, 0; 2nd CP: 0.2, 80, 20; 3rd CP: 0.3, 70, 30; 4th CP: 0.4, 50, 50; 5th CP: 0.5, 30, 70; and 6th CP: 0.5, 0, 100, respectively. The ISTD was added to each sample at a final concentration level of 6 mg/mL.

Derivatization was performed by adding 40 μL of BSTFA + TMCS to 1 μL of the sample. Finally, the solutions were thoroughly mixed and maintained at 55 °C for 30 min.

Second method: Stock solutions of DA and AP were prepared in acetone at concentrations of 1 mg/mL. A stock solution of o-phenylenediamine was prepared in methanol at a concentration of 1 mg/mL.

Calibration standards for nicotine were prepared in PG/VG 50/50 mL/mL at concentration levels of 2, 3, 6, 9, 12, 18, and 20 mg/mL. DA and AP were added to each sample at concentration levels of 5 μg/mL. The ISTD was added to each sample at concentration levels of 6 mg/mL.

Derivatization was performed by adding 20 μL of o-phenylenediamine to 5 μL of the sample. Finally, the solutions were thoroughly mixed and maintained at room temperature for 5 min.

3.4. Sample Preparation

A total of 1 mL of each e-liquid was transferred in 2 mL Eppendorf tube, 6 µL of ISTD were added, and the solution was vortexed for 30 s.

For the measurement of EM, PG, and VG, 40 µL of BSTFA + TMCS were added to 1 µL of the sample, and the solution was maintained at 55 °C for 30 min.

For the measurement of nicotine, DA, and AP, 20 µL of o-phenylenediamine was added to 5 µL of the sample, and the solution was maintained at room temperature for 5 min.

3.5. Instrumentation

The Thermo Trace 2000 series GC (ThermoQuest, Waltham, MA, USA) system coupled with the Q plus (ThermoQuest, Waltham, MA, USA) mass spectrometer was used. The system was equipped with the AS 2000 (ThermoQuest, Waltham, MA, USA) autosampler. Analysis was performed using the Xcalibur version 1.2 (Thermo Fisher Scientific, Waltham, MA, USA) software. An SGE fused silica capillary column BP-5 ms (30 m × 0.25 mm × I.D. 0.25 µm film thickness) (Trajan Scientific and Medical, Victoria, Australia) was used for the chromatographic separation. Helium was used as the carrier gas at a constant flow rate of 0.8 mL/min. The injections were in split mode with a 1:5 ratio, and the injector temperature was 250 °C. The ionization of the compounds was carried out by EI in the positive ion mode at an electron energy of 70 eV with a source temperature of 230 °C, whereas the ion trap temperature was 75 °C. The injection volume was 1 µL. The oven temperature was 50 °C (held for 5 min), programmed to reach 310 at 30 °C/min in the first method and at 60 °C/min in the second method. The final temperature was kept for 1 min. The selected ion monitoring (SIM) mode was selected. In the first method, the ions 205 m/z, 293 m/z, 197 m/z, and 135 m/z were selected for the detection of PG, VG, EM, and ISTD, respectively. In the second method, the ions 84 m/z, 158 m/z, 172 m/z, and 196 m/z were selected for the detection of nicotine, DA, AP, and ISTD, respectively.

3.6. Method Validation

The method was validated by examining linearity, precision (repeatability and intermediate precision), accuracy, reproducibility, stability, robustness, the limit of detection (LOD), and the limit of quantitation (LOQ) according to the ICH Q2(R1) analytical procedure guidelines (https://www.ema.europa.eu/en/documents/scientific-guideline/ich-q-2-r1-validation-analytical-procedures-text-methodology-step-5_en.pdf, accessed on 30 November 2022).

4. Conclusions

A two-stage analytical methodology was developed for the simultaneous determination of nicotine, PG, VG, EM, DA, and AP. The importance of this analytical methodology depends on the fact that DA and AP are referred to as harmful and potentially harmful constituents (HPHCs) of e-liquids, and electronic cigarettes are not totally safe for human health.

The validated methodology is fast and can be applied for the quality control of e-liquids by manufacturers, as it is based on the familiar and commonly available GC-MS instrumentation. As proof of applicability, the validated methods were successfully applied on a small set of EC liquid samples, indicating that this methodology could be used for routine quality control analyses of EC liquids. The quality control of e-liquids is an essential procedure as differences of actual against the labeled concentrations of the PG, VG, and nicotine have been shown. Another advantage of the developed, fast methodology is the possibility for quantitative determination of one of the most used and toxic flavor chemicals, EM, as well as the estimation of toxic DA and AP levels. The analysis of EM, DA, and AP is especially important, as the maximum allowable levels compared to the proposed occupational exposure limits are still under scientific examination.

Finally, the suggested methodology, apart from a quality control routine procedure, can be applied for the examination of the stability of e-liquids under different conditions of storage, as many of the users of these e-liquid refill samples keep them, for instance, at their homes or offices in the direct sunlight. Furthermore, the e-liquid inside the vaporizer undergoes repeated cycles of temperature variations (from vaporing point to room temperature) during vaping, thus making the assessment of the e-liquid content imperative for safeguarding the quality of the refill liquids.

Part of this work was presented at the 2nd Scientific Summit on Tobacco Harm Reduction (Greece, Athens, 29–30 May 2019).

Supplementary Materials: The following supporting information is available online at https://www.mdpi.com/article/10.3390/molecules28041902/s1, Figure S1: Blank chromatogram after addition of derivatization agent and ISTD acquired using the GC-MS methodology developed for the measurement of PG, VG, and EM, Figure S2: Blank chromatogram after addition of derivatization agent and ISTD acquired using the GC-MS methodology developed for the measurement of nicotine, DA, and AP.

Author Contributions: Conceptualization, A.T.; formal analysis, I.D., E.G. and A.T.; investigation, I.D.; methodology, I.D. and E.G.; project administration, A.T.; resources, A.T.; supervision, A.T.; validation, I.D. and E.G.; visualization, E.G. and A.T.; writing—original draft, I.D.; writing—review and editing, E.G. and A.T. All authors have read and agreed to the published version of the manuscript.

Funding: This research received no external funding.

Institutional Review Board Statement: Not applicable.

Informed Consent Statement: Not applicable.

Data Availability Statement: The data presented in this study are available in this article and supplementary material.

Acknowledgments: We especially thank NOBACCO for providing the e-liquid samples.

Conflicts of Interest: The authors declare no conflict of interest.

Sample Availability: Samples of the compounds are not available from the authors.

References

1. Kim, H.; Lim, J.; Buehler, S.S.; Brinkman, M.C.; Johnson, N.M.; Wilson, L.; Cross, K.S.; Clark, P.I. Role of sweet and other flavours in liking and disliking of electronic cigarettes. *Tob. Control.* **2016**, *25*, ii55–ii61. [CrossRef]
2. National Academies of Sciences, Engineering, and Medicine. New Report One of the Most Comprehensive Studies on Health Effects of E-Cigarettes: Finds that Using E-Cigarettes May Lead Youth to Start Smoking, Adults to Stop Smoking—ScienceDaily. Available online: https://www.sciencedaily.com/releases/2018/01/180123121043.htm (accessed on 3 June 2022).
3. Leigh, N.J.; Lawton, R.I.; Hershberger, P.A.; Goniewicz, M.L. Flavorings significantly affect inhalation toxicity of aerosol generated from electronic nicotine delivery systems (ENDS). *Tob. Control.* **2016**, *25*, ii81. [CrossRef]
4. Behar, R.Z.; Luo, W.; Mcwhirter, K.J.; Pankow, J.F.; Talbot, P. Analytical and toxicological evaluation of flavor chemicals in electronic cigarette refill fluids OPEN. *Sci. Rep.* **2018**, *8*, 8288. [CrossRef]
5. Larcombe, A.; Allard, S.; Pringle, P.; Mead-Hunter, R.; Anderson, N.; Mullins, B. Chemical analysis of fresh and aged Australian e-cigarette liquids. *Med. J. Aust.* **2022**, *216*, 27–32. [CrossRef]
6. Page, M.K.; Goniewicz, M.L.; Thornburg, J.; Petters, S.; Kovach, A. New Analytical Method for Quantifying Flavoring Chemicals of Potential Respiratory Health Risk Concerns in e-Cigarette Liquids. *Front. Chem.* **2021**, *9*, 763940. [CrossRef]
7. Bitzer, Z.T.; Goel, R.; Reilly, S.M.; Elias, R.J.; Silakov, A.; Foulds, J.; Muscat, J.; Richie, J.P., Jr. Effect of Flavoring Chemicals on Free Radical Formation in Electronic Cigarette Aerosols. *Free Radic. Biol. Med.* **2018**, *20*, 72–79. [CrossRef]
8. Omaiye, E.E.; Luo, W.; Mcwhirter, K.J.; Pankow, J.F.; Talbot, P. Electronic Cigarette Refill Fluids Sold Worldwide: Flavor Chemical Composition, Toxicity and Hazard Analysis Graphical Abstract. *Chem. Res. Toxicol.* **2020**, *33*, 2972–2987. [CrossRef]
9. Omaiye, E.E.; Mcwhirter, K.J.; Luo, W.; Pankow, J.F.; Talbot, P. High Nicotine Electronic Cigarette Products: Toxicity of JUUL Fluids and Aerosols Correlates Strongly with Nicotine and Some Flavor Chemical Concentrations HHS Public Access. *Chem. Res. Toxicol.* **2019**, *32*, 1058–1069. [CrossRef]
10. Hua, M.; Omaiye, E.E.; Luo, W.; Mcwhirter, K.J.; Pankow, J.F.; Talbot, P. Identification of Cytotoxic Flavor Chemicals in Top-Selling Electronic Cigarette Refill Fluids. *Sci. Rep.* **2019**, *9*, 2782. [CrossRef]

11. Miao, S.; Beach Phd, E.S.; Sommer, T.J.; Zimmerman, J.B.; Jordt, S.-E. High-Intensity Sweeteners in Alternative Tobacco Products. *Nicotine Tob. Res.* **2016**, *18*, 2169–2173. [CrossRef]
12. Duell, A.K.; Pankow, J.F.; Peyton, D.H. Free-Base Nicotine Determination in Electronic Cigarette Liquids by 1H NMR Spectroscopy. *Chem. Res. Toxicol.* **2018**, *31*, 431–434. [CrossRef]
13. Rajapaksha, R.D.; Tehrani, M.W.; Rule, A.M.; Harb, C.C. A Rapid and Sensitive Chemical Screening Method for E-Cigarette Aerosols Based on Runtime Cavity Ringdown Spectroscopy. *Environ. Sci. Technol* **2021**, *55*, 8096. [CrossRef]
14. Miller, D.R.; Buettner-Schmidt, K.; Orr, M.; Rykal, K.; Niewojna, E. A systematic review of refillable e-liquid nicotine content accuracy. *J. Am. Pharm. Assoc.* **2021**, *61*, 20–26. [CrossRef]
15. Omaiye, E.E.; Mcwhirter, K.J.; Luo, W.; Tierney, P.A.; Pankow, J.F.; Talbot, P. High concentrations of flavor chemicals are present in electronic cigarette refill fluids. *Sci. Rep.* **2019**, *9*, 2468. [CrossRef]
16. Klager, S.; Vallarino, J.; Macnaughton, P.; Christiani, D.C.; Lu, Q.; Allen, J.G. Flavoring Chemicals and Aldehydes in E-Cigarette Emissions. *Environ. Sci. Technol.* **2017**, *51*, 10806–10813. [CrossRef]
17. Melvin, M.S.; Avery, K.C.; Ballentine, R.M.; Flora, J.W.; Gardner, W.; Karles, G.D.; Pithawalla, Y.B.; Smith, D.C.; Ehman, K.D.; Wagner, K.A. Formation of Diacetyl and Other α-Dicarbonyl Compounds during the Generation of E-Vapor Product Aerosols. *ACS Omega* **2020**, *5*, 17565–17575. [CrossRef]
18. Kanwal, R.; Kullman, G.; Piacitelli, C.; Boylstein, R.; Sahakian, N.; Martin, S.; Fedan, K.; Kreiss, K. Evaluation of flavorings-related lung disease risk at six microwave popcorn plants. *J. Occup. Environ. Med.* **2006**, *48*, 149–157. [CrossRef] [PubMed]
19. Athleen, K.; Reiss, K.; Omaa, H.G.; Ullman, R.K.; Edan, F.; Duardo, E.; Imoes, J.S.; Aul, P.; Night, L.E. Clinical Bronchiolitis Obliterans in Workers at a Microwave-Popcorn Plant. *N. Engl. J. Med.* **2002**, *347*, 330–338. [CrossRef]
20. Shibamoto, T. Diacetyl: Occurrence, Analysis, and Toxicity. *J. Agric. Food Chem.* **2014**, *62*, 4048–4053. [CrossRef] [PubMed]
21. More, S.; Vartak, A.P.; Vince, R. The butter flavorant, diacetyl, exacerbates β-amyloid cytotoxicity. *Chem. Res. Toxicol.* **2012**, *25*, 2083–2091. [CrossRef] [PubMed]
22. Barhdadi, S.; Canfyn, M.; Courselle, P.; Rogiers, V.; Vanhaecke, T.; Deconinck, E. Development and validation of a HS/GC–MS method for the simultaneous analysis of diacetyl and acetylpropionyl in electronic cigarette refills. *J. Pharm. Biomed. Anal.* **2017**, *142*, 218–224. [CrossRef] [PubMed]
23. Farsalinos, K.E.; Kistler, K.A.; Gillman, G.; Voudris, V. Evaluation of electronic cigarette liquids and aerosol for the presence of selected inhalation toxins. *Nicotine Tob. Res.* **2015**, *17*, 168–174. [CrossRef] [PubMed]
24. Crenshaw, M.D.; Tefft, M.E.; Buehler, S.S.; Brinkman, M.C.; Clark, P.I.; Gordon, S.M. Determination of Nicotine, Glycerol, Propylene Glycol and Water in Electronic Cigarette Fluids Using Quantitative 1 H NMR. *Magn. Reson. Chem.* **2016**, *54*, 901–904. [CrossRef] [PubMed]
25. Beauval, N.; Antherieu, S.; Soyez, M.; Gengler, N.; Grova, N.; Howsam, M.; Hardy, E.M.; Fischer, M.; Appenzeller, B.M.R.; Goossens, J.-F.; et al. Chemical Evaluation of Electronic Cigarettes: Multicomponent Analysis of Liquid Refills and their Corresponding Aerosols. *J. Anal. Toxicol.* **2017**, *41*, 670–678. [CrossRef] [PubMed]
26. Kucharska, M.; Wesołowski, W.; Czerczak, S.; Soćko, R. Testing of the composition of e-cigarette liquids—Manufacturer-declared vs. true contents in a selected series of products. *Med. Pr.* **2016**, *67*, 239–253. [CrossRef]

Disclaimer/Publisher's Note: The statements, opinions and data contained in all publications are solely those of the individual author(s) and contributor(s) and not of MDPI and/or the editor(s). MDPI and/or the editor(s) disclaim responsibility for any injury to people or property resulting from any ideas, methods, instructions or products referred to in the content.

Article

Structure Identification of Adsorbed Anionic–Nonionic Binary Surfactant Layers Based on Interfacial Shear Rheology Studies and Surface Tension Isotherms

Ourania Oikonomidou, Margaritis Kostoglou and Thodoris Karapantsios *

Department of Chemical Technology and Industrial Chemistry, Faculty of Chemistry, Aristotle University of Thessaloniki, University Box 116, 541 24 Thessaloniki, Greece
* Correspondence: karapant@chem.auth.gr; Tel.: +30-2310-997772

Abstract: Mixtures of anionic sodium oleate (NaOl) and nonionic ethoxylated or alkoxylated surfactants improve the selective separation of magnesite particles from mineral ores during the process of flotation. Apart from triggering the hydrophobicity of magnesite particles, these surfactant molecules adsorb to the air–liquid interface of flotation bubbles, changing the interfacial properties and thus affecting the flotation efficiency. The structure of adsorbed surfactants layers at the air–liquid interface depends on the adsorption kinetics of each surfactant and the reformation of intermolecular forces upon mixing. Up to now, researchers use surface tension measurements to understand the nature of intermolecular interactions in such binary surfactant mixtures. Aiming to adapt better to the dynamic character of flotation, the present work explores the interfacial rheology of NaOl mixtures with different nonionic surfactants to study the interfacial arrangement and viscoelastic properties of adsorbed surfactants under the application of shear forces. Interfacial shear viscosity results reveal the tendency on nonionic molecules to displace NaOl molecules from the interface. The critical nonionic surfactant concentration needed to complete NaOl displacement at the interface depends on the length of its hydrophilic part and on the geometry of its hydrophobic chain. The above indications are supported by surface tension isotherms.

Citation: Oikonomidou, O.; Kostoglou, M.; Karapantsios, T. Structure Identification of Adsorbed Anionic–Nonionic Binary Surfactant Layers Based on Interfacial Shear Rheology Studies and Surface Tension Isotherms. *Molecules* **2023**, *28*, 2276. https://doi.org/10.3390/molecules28052276

Academic Editors: Paraskevas D. Tzanavaras and Victoria Samanidou

Received: 6 February 2023
Revised: 23 February 2023
Accepted: 23 February 2023
Published: 28 February 2023

Copyright: © 2023 by the authors. Licensee MDPI, Basel, Switzerland. This article is an open access article distributed under the terms and conditions of the Creative Commons Attribution (CC BY) license (https://creativecommons.org/licenses/by/4.0/).

Keywords: sodium oleate; surfactant binary mixtures; adsorption; synergism; interfacial shear rheology; viscoelasticity; surface tension isotherms

1. Introduction

Vulnerable multiphase systems such as foams and emulsions exist at the majority of chemical products, such as processed food, beverages, cleaning detergents, medicine and cosmetics [1–3]. The stability of these systems strongly depends on mass transport phenomena and chemical interactions ongoing at the region of the involving interfaces. Thus, understanding the interactions between bubbles and droplets, so as to control the quality of foam and emulsions in products, requires a deep understanding of the ongoing dynamic and equilibrium interfacial properties [3–5]. In this manner, a plethora of research works examine the interfacial properties of bubbles and droplets in the presence of surface active molecules (i.e., surfactants) [1,4,6–11]. The amphiphilic nature of these molecules makes them adsorb to the incorporating gas/liquid interfaces and prevent them from collapse [12]. The technical and economic advantages of using surface active molecules as mixtures and not as individuals have been reported several times in the literature. Proper surfactant combinations can achieve better stabilizing properties at much lower concentrations; therefore, in practice, binary surfactant mixtures are applied quite often in chemical processes [1,6,7,10]. Mixing molecules of different chemical structure introduces the complexity of competitive adsorption and intermolecular interactions to the resulting interfacial arrangement [4].

Interaction parameter 'β' can reveal the nature and magnitude of surfactant interactions at the interface of a binary solution resulting from the reformation of intermolecular forces (i.e., electrostatic and steric repulsive forces, ion-dipole and Van der Waals attraction forces) upon mixing. Negative β values indicate synergistic interactions between the mixed surfactants, as the repulsive forces weaken and attractive forces develop instead. On the contrary, positive β values correspond to antagonistic interactions between mixed surfactants, as in this case, the repulsive forces between two different surfactants are stronger compared to the self-repulsive forces between two identical surfactant molecules [6–13]. To calculate the interaction parameter of a system, surface tension isotherms of the binary surfactant solution and the solutions of individual surfactant components need to be employed. Up to now, the interactions between numerous ionic–ionic and ionic–nonionic surfactant combinations have been tested for the purpose of generating formulas of better surface activity [6–11,14]. However, surface tension results are restricted to equilibrium conditions and lack the ability to provide dynamic information on the interfacial rearrangement of surfactant molecules under the application of stresses, as encountered in the majority of chemical processes [3].

To understand the stability and mobility response of adsorbed molecular layers under external forces, researchers perform interfacial rheology tests. Interfacial rheology is studied under both dilatational and shear deformation, since both of them are present in processes involving such multiphase systems [15]. Dilatational deformation changes the area but not the shape of the tested interface and depends on the nature of the molecules adsorbed on the interface [16]. Interfacial dilatational rheology of systems is extensively examined in the literature [1,17–19]. Liggieri and Miller., 2010, resume some critical studies on the dilatational rheology of adsorbing surfactant layers at liquid interfaces, while Miller et al., 1996, report the different instrument designs available for these studies [20,21]. On the other hand, shear deformation changes the shape but not the area of the tested interface and depends on intermolecular interactions between the different species [16,22].

Interfacial shear rheology is less exploited; however, there are some critical works that use it to identify the structure of adsorbed mixed layers as a result of different adsorption kinetics and intermolecular forces [14,23]. The majority of systems tested under shear deformation are surfactant–protein mixtures [24–26]. Contrary to small surfactant molecules, large protein molecules impose some significant resistance towards motion of the interface and allow such kind of measurements. In practice, small surfactant molecules co-exist with biodegradable large protein molecules in many food- and medicine-related systems. Some indicative examples are as follows: Kragel et al., 2008, review the interfacial shear rheology behavior of various protein–surfactant mixtures used as stabilizers in foams and emulsions employing several experimental techniques [4]. As they report, surfactant molecules tend to displace the adsorbed protein molecules from the interface, turning the elastic proteins properties into viscous. The displacement is dictated by the nature of the surfactant and the protein/protein interactions. Dwyer et al., 2012, studied the interaction of designed small unstructured peptides with large-structured protein molecules at the interface of air/water systems [2]. Synergistic interaction is reported as the adsorption kinetics of the mixture is higher than that of individual molecules. Erni et al., 2003, present the results from steady shear and oscillatory experiments as well as creep recovery and stress relaxation tests for ovalbumin protein adsorbed films and sorbitan tristearate spread films on both oil/water and air/water interfaces [27]. Torcello-Gómez et al., 2011, studied the surface rheology of sorbitan tristearate (food emulsifier and stabilizer) and b-lactoglobulin (protein present in cow milk) mixtures with both shear and dilatational deformation tests [16]. Amongst the few surfactant–surfactant mixtures tested as a matter of interfacial rheology is sodium oleate—C12(EO)6, aiming to understand the effect of surfactants concentration on foam drainage during the process of flotation deinking [28]. However, in this case, the resulting surface properties are restricted as derived by dynamic surface tension measurements.

In the field of flotation, sodium oleate (NaOl) is a rather valuable surfactant as it is considered a very compatible collector for the recovery of magnesite particles from mineral ores [29,30]. The addition of nonionic cocollectors to the primary anionic NaOl is believed to enhance further the recovery rates [31]. The efficiency of flotation depends on the hydrodynamics at the interface of bubbles that may affect the motion of the bubble through the liquid [15,32]. It is worth noticing that shear interfacial viscosity has no effect on the motion of a bubble due to an external force field (e.g., gravity). This has been proved by solving the corresponding set of the fluid dynamics equations [33,34]. The axial symmetry of the resulting flow field prevents the influence of interfacial shear viscosity. However, in case of three-dimensional turbulent flow conditions met in flotation, there may be an effect of interfacial shear velocity on altering the flow field around the bubble. The interfacial shear viscosity is of paramount importance for froth drainage through Plateau borders [35,36]. Up to now, researchers study the collector–cocollector interactions through surface tension isotherms [37]. Going a step further, the present work attempts to examine the structure and viscoleasticity of adsorbing mixed surfactant layers at the air/liquid interface of flotation systems, by studying the interfacial rheology of different NaOl and ethoxylated/alkoxylated nonionic surfactant mixtures under the application of shear forces. Surface tension isotherms are performed as supporting indications of the adsorbing surfactant layers' structure. The structure of this work is the following: the Materials and Methods section presents the employed surfactant reagents and experimental devices. Experimental measurements are presented and extensively discussed in the Results and Discussion section. Finally, the conclusions are presented in the last section of this work.

2. Results and Discussion

Table 1 reports the pH values of 1CMC NaOl solution and anionic-nonionic binary mixtures at the maximum tested nonionic concentration. Measurements show that all pH values are close to 10 since the addition of nonionic surfactants in 1CMC NaOl solution does not much affect the pH. At this level of pH, the flotation process is efficient, so no further pH regulation is needed to consider the following intermolecular results as appropriate for making conclusions [38].

Table 1. Measured pH of 1CMC NaOl solution and ionic–nonionic surfactant mixtures at 50:50 mass ratio.

Solution	pH
300 ppmNaOl	9.8
300 ppmNaOl + 300 ppm Iso Eth 10	9.6
300 ppmNaOl + 300 ppm Iso Eth 03	9.8
300 ppmNaOl + 300 ppm Dod Eth 03	9.8
300 ppmNaOl + 300 ppm Dod Alk 54	9.2
300 ppmNaOl + 300 ppm Iso Alk 52	9.2

Figure 1a shows the shear viscosity of anionic NaOl collector molecules adsorbing on the air/liquid interface of two different 1CMC (300 ppm) NaOl solutions at 15 °C. The results are quite repeatable. The artificial shear viscosity of a 'surfactants-free' ultrapure water surface is reported as a supporting control measurement. Pure water viscosity numbers indicate the lower measuring limits of the rheometer and have no physical meaning. This explains the missing measuring points below 1/s shear rate that are discarded by the rheometer as not trusted, as they correspond to extremely low viscosity values. Comparing the interfacial shear viscosity values of NaOl solutions to those of pure water, it can be seen that the adsorbing NaOl molecules attain some measurable viscosity at low shear rates, below 1/s. Higher shear rates seem to break the interfacial NaOl layer leaving the liquid surface free of surfactant molecules. This experimental conclusion is verified by visual inspection of the liquid surface before and after each run (continuous and accelerating bicone rotation). Figure 1b shows NaOl layers on the liquid surface, perimetrically to the walls of the test cell, before the application of shear stresses. After each experimental run, these layers are no longer detected. The formation of NaOl layers on the liquid surface is an indication that NaOl is not soluble to water at 15 °C. Therefore, these layers are considered as aggregates of non-dissolved NaOl solid particles that precipitate on the liquid surface. Experiments show that the present working temperature is not applicable to the studied flotation systems. Thereafter, all experiments are conducted at 30 °C.

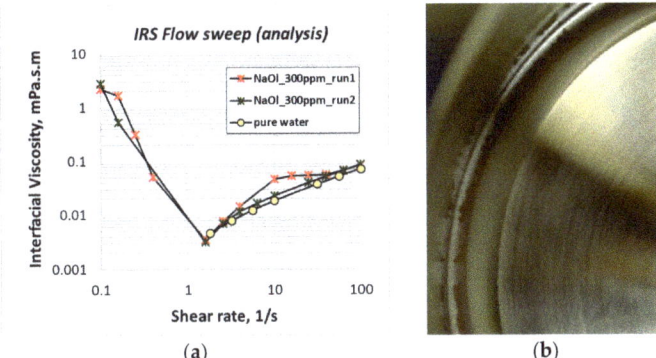

Figure 1. (a) Interfacial viscosity of undissolved sodium oleate (NaOl) particles precipitates under various shear rates, at 15 °C working temperature; (b) layer of NaOl solid particles precipitating at the solution surface before rotation of the bicone geometry.

Figure 2a shows the shear viscosity of NaOl molecules adsorbed on the air/liquid interface for 1CMC (300 ppm) NaOl solution at 30 °C. At this temperature, NaOl is soluble to water. However, NaOl is a frother, and when the solution is poured into the test cell, it creates foam at the perimeter of the air/liquid interface (see Figure 2b). Runs 1 and 2 show the interfacial shear viscosity of two different 1CMC NaOl solutions. These measurements are not considered repeatable; however, they have a similar trend. Interfacial viscosity decreases with the increase in shear rate up to 1/s and has an unstable behavior at shear rates above 1/s. To investigate this behavior further, additional 'reverse' runs are performed by decreasing the shear rate from 100/s to 0.1/s with a logarithmic ramp. The measuring duration per shear rate does not change. The resulting reverse runs 1 and 2 are repeatable, indicating the following conclusions: under low shear rates (<10/s), the air trapped in the foam lowers the measured interfacial resistance, resulting in lower interfacial viscosity values than the real ones. Under high shear rates (>10/s), the fast bicone rotation breaks the foam, resulting in realistic viscosity measurements. In the case of reverse measurements that start under high shear rates, the foam breaks at the beginning of each run and allows us to attain proper viscosity measurements during the whole run. To overcome this problem, NaOl solution is poured slowly into the test cell to avoid foaming and manage proper (repeatable) interfacial viscosity measurements, as shown in run 3. Discussing on the measurement itself, it seems that the anionic NaOl molecules adsorbing on the air/liquid surface acquire some interfacial shear viscosity (all measuring values are above those of pure water). Furthermore, the adsorbed NaOl layer has a non-Newtonian shear thinning behavior, as its viscosity decreases with shear rate.

Figure 3 shows the variation of NaOl solution interfacial viscosity at the region of pre-micellar concentrations. As expected, the interfacial viscosity decreases with the decrease in NaOl concertation; however, the qualitative behavior of interfacial viscosity with shear rate remains the same. This indication is critical as in flotation applications the concentration of the collector is in the order of 100 ppm. Therefore, it seems that the interfacial rheology findings at 1CMC NaOl can also stand for the real flotation concentrations.

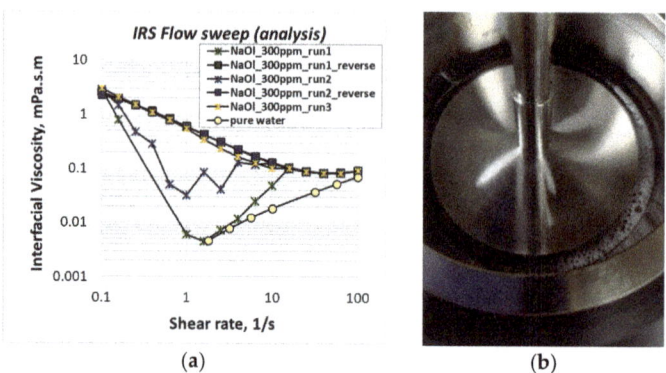

Figure 2. (a) Shear viscosity of NaOl molecules adsorbed at the interface of 1CMC NaOl solution at 30 °C; (b) NaOl solution foaming at the walls of the IRS test cell.

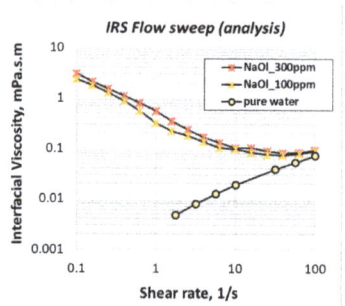

Figure 3. The effect of NaOl solution concentration at the interfacial viscosity of adsorbed NaOl molecules under various shear rates.

In case flotation liquids contain binary collector mixtures, both anionic and nonionic surfactant molecules adsorb on the air–liquid interface, forming arrays based on the on-going intermolecular forces. The structure of adsorbing binary surfactant monolayers is examined through interfacial rheology studies. Different cocollectors are tested at different anionic:nonionic mass ratios varying from 95:05 to 50:50. NaOl concentration at the test solutions is always 1CMC (300 ppm). Interfacial shear viscosity results are presented in Figure 4, accompanied by the corresponding control measurements of 1CMC NaOl, nonionic cocollector solution at the same concentration and pure water. The interfacial shear viscosity of nonionic cocollector solutions is insignificant and equal to that of pure water. This means that the adsorbing nonionic molecular layers do not impose any extra interfacial viscosity under shear stress. The gradual addition of a nonionic cocollector at 1CMC NaOl solution decreases the resulting shear viscosity of the surfactants binary layer adsorbed at the air/liquid interface. Above some critical nonionic mass ratio, the interfacial shear viscosity is reduced to that of the nonionic cocollector molecules. This trend exists for all cocollector cases, and it is a strong indication that by increasing the cocollector concentration in the binary solution, the cocollector molecules displace the adsorbed anionic NaOl collector molecules at the air/liquid interface. The proposed displacement mechanism is illustrated in Figure 5. Above the critical mass ratio, the air/liquid interface is occupied to the utmost by the nonionic surfactant molecules. The above observation is quite similar to that of Kragel et al., 2008, and Bosa and van Vlieta, 2001, reporting that the displacement of large protein molecules by surfactant molecules of low molecular weight changes the shear viscoelasticity of the adsorbed layer [4,39].

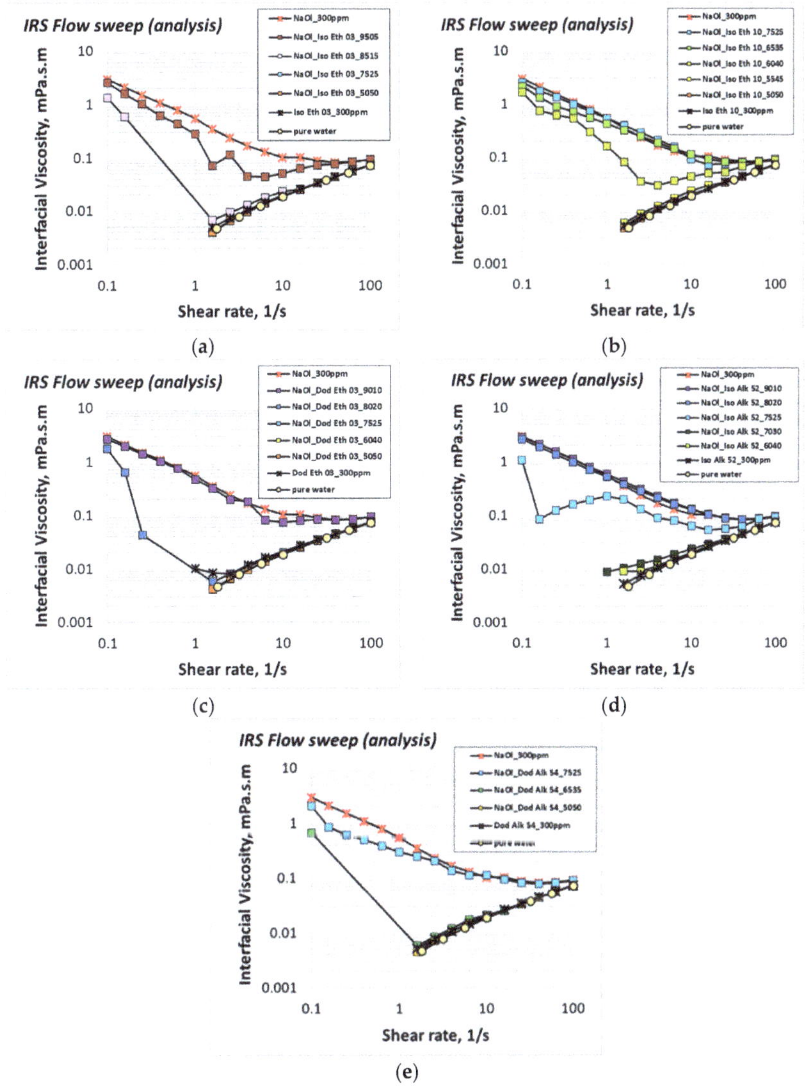

Figure 4. Shear viscosity of adsorbed surfactants layers at the interface of binary solutions of 1CMC NaOl collector and (**a**) Iso Eth 03; (**b**) Iso Eth 10; (**c**) Dod Eth 03; (**d**) Iso Alk 52; (**e**) Dod Alk 54; nonionic cocollector, at 30 °C.

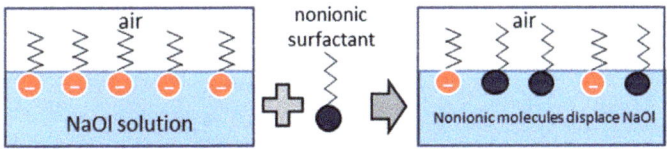

Figure 5. Nonionic cocollector molecules displace ionic primary collector molecules from the air/liquid interface of the binary solution.

The critical nonionic cocollector mass ratio varies with the molecular structure of the surfactant. Table 2 summarizes the critical mass ratios of all tested nonionic cocollectors. Critical mass ratio increases for nonionic molecules with longer hydrophilic heads, either due to more ethoxylated groups (compare values of Isotridecyl Ethoxylate 03 and Isotridecyl Ethoxylate 10) or due to alkoxylated chains (compare values of Isotridecyl Ethoxylate 03 and Isotridecyl Alkoxylate 52, or values of Dodecyl Ethoxylate 03 and Dodecyl Alkoxylate 54). Molecules with an enhanced hydrophilic part are more soluble in the liquid phase, and thus, a greater concentration of these molecules is needed to cover the air/liquid interface. On the other hand, the critical mass ratio is lower for nonionic molecules with a branched hydrophobic chain (compare values of Isotridecyl Ethoxylate 03 and Dodecyl Ethoxylate 03). The Isotridecyl branched chain occupies larger space on the interface due to the strong stearic repulsive forces between the hydrophobic heads of nonionic molecules, and thus, fewer molecules are needed to cover the entire interface [8].

Table 2. Regions of critical noninonic cocollector mass ratios resulting from the interfacial shear viscosity measurements of the tested ionic–nonionic binary mixtures.

Binary System	Critical Nonionic Concentration
NaOl—Iso Eth 03	15–25%
NaOl—Iso Eth 10	40–45%
NaOl—Dod Eth 03	20–25%
NaOl—Dod Alk 54	40–50%
NaOl—Iso Alk 52	25–30%

Additional conventional surface tension isotherms at 30 °C verify the interfacial structure of flotation bubbles as resulting from the present rheological study. The results are presented in Figure 6. Increasing the concentration of each nonionic cocollector in water decreases the surface tension of the solution (dark-colored points of each subfigure). Surface tension becomes stable at 1CMC condition of each cocollector. Surface tension of the corresponding anionic–nonionic binary mixture increases by increasing the nonionic cocollector concentration in the solution (light-colored points of each subfigure). Surface tension at zero nonionic concentration corresponds to surface tension of 1CMC NaOl solution (marked with a red star). Measurements show that the CMC of all tested nonionic surfactant solutions is below CMC of NaOl (300 ppm). Thus, all tested nonionic cocollectors are more surface active than NaOl and dominate on the solution surface. This explains the fact that the surface tension of binary mixtures does not decrease with the increase in surfactants concentration, but it increases towards surface tension at 1CMC of nonionic surfactant solutions. The above conclusion agrees with the rheological finding that nonionic cocollector molecules displace NaOl molecules from the air/liquid interface. The nonionic cocollector concentration that makes surface tension of the binary solution equal to that of 1CMC nonionic surfactant solution, is the concentration needed to displace all NaOl molecules from the interface. This concentration is always higher than the corresponding critical mass ratio deriving from the interfacial viscosity measurements (Table 3 shows the nonionic surfactant concentration that corresponds to each mass ratio of the binary mixture). This means that even in the presence of some NaOl molecules on the interface, the shear viscosity of the adsorbed binary layer is insignificant. For the cases of Isotridecyl Ethoxylate 10, Isotridecyl Alkoxylate 52 and Dodecyl Alkoxylate 54, the maximum tested concentration (300 ppm) is not adequate for the full displacement of NaOl from the interface, since the surface tension of the binary mixture never reaches surface tension at 1CMC of each nonionic surfactant. The aforementioned observation is in line with the molecular structure of the surfactants. The long hydrophilic head of these surfactants makes them water soluble and increases the concentration needed to compete NaOl molecules on surface adsorption. This phenomenon is even stronger for the case of Dodecyl Alkoxylate

54 as its straight Dodecyl chain is less space demanding due to the weak steric repulsive forces between the neighboring hydrophobic heads.

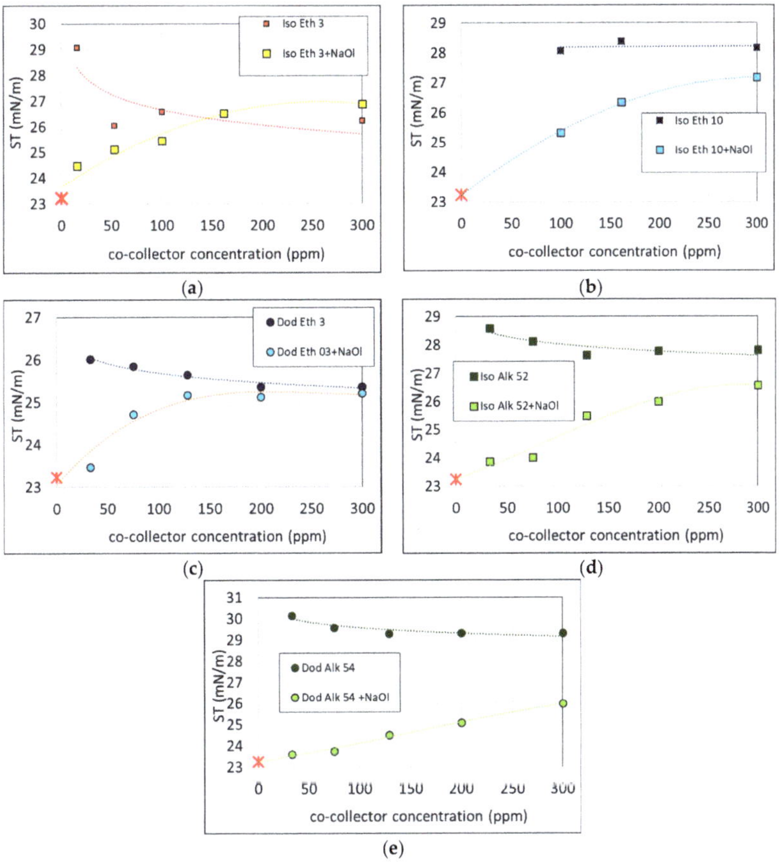

Figure 6. Surface tension (ST) isotherms of nonionic surfactant solutions: (**a**) Iso Eth 03; (**b**) Iso Eth 10; (**c**) Dod Eth 03; (**d**) Iso Alk 52; (**e**) Dod Alk 54; and anionic–nonionic binary mixtures at 30 °C, for 300 ppm NaOl (1CMC) and different nonionic surfactant concentrations at 30 °C. The red star symbol shows the surface tension of 1CMC NaOl solution.

Table 3. The nonionic cocollector concentration corresponding to each anionic:nonionic binary collectors solution mass ratio. Anionic concentration is 300 ppm for all cases.

Nonionic Cocollector Concentration (ppm)	Anionic:Nonionic Binary Mixture Mass Ratio
15.7	95:05
33.3	90:10
52.9	85:15
75	80:20
100	75:25
128.5	70:30
161.5	65:35
200	60:40
270	55:45
300	50:50

Based on the literature, the synergistic efficiency between ionic and nonionic molecules is defined as follows: the total concentration of the mixed surfactant required to reduce the surface tension of the solvent to a given value is less than that of either individual surfactants. This condition refers to the concentrations of surfactants that adsorb to the interface of solvents and not to surfactants concentrations in the solutions bulk. In the literature, all works that result in synergistic ionic–nonionic interactions report the exact same trend of the surface tension isotherms variation; the surface tension isotherm of the binary mixture is always between those of the individual surfactants and very close to the nonionic one [6,9,11].

Surface tension isotherms of Figure 7 aim to examine qualitatively the nature of intermolecular interactions at the indicative binary mixture of NaOl- Isotridecyl Ethoxylate 03. More specifically, Figure 7 presents the surface tension variation of Isotridecyl Ethoxylate 03 solution (collector 1), NaOl solution (collector 2) and their binary mixture (12), with the total collectors concentration. The mass ratio of the binary mixture is 40:60 (as at this mass ratio and for 1CMC NaOl concentration, the surface tension of the solution reaches surface tension of 1CMC Isotridecyl Ethoxylate 03 solution as shown in Figure 6a). Solid lines connect the surface tension measuring points to show the overall trend. Points at which trend lines become horizontal correspond to 1CMC values of the solutions. Expanding the surface tension trend lines of Figure 7 towards lower surfactant concentrations, shows that the present system has the same behavior with those reported in the literature [6,9,11]; the trend line of the mixture (in black) is between those of the individual surfactants (in blue and red). This finding indicates that for low-surfactant concentrations that correspond to surface tension values above 30 mN/m, NaOl interacts synergistically with Isotridecyl Ethoxylate 03. Calculation of interaction parameter β is not feasible for the present case, as the tested concentrations are above the pre-micellar regions of surfactant solutions. For the purpose of flotation research, it is meaningless to focus on lower surfactant concentrations; thus, the interaction between ionic and nonionic surfactants is only qualitatively estimated. In general, Zhou and Rosen., 2003, claim that repulsive (antagonistic) interactions are found only in mixtures of hydrocarbon chain and fluorocarbon chain surfactants of the same sign [8]. However, they straighten that interaction between an anionic surfactant and a nonionic surfactant with a polyoxyethylene chain of many oxyethylene units, acquires a weak positive charge that indicates a weak repulsive interaction nature [8,40–42]. NaOl—Isotridecyl Ethoxylate 10 system is an example with an expected repulsive behavior. Moreover, nonionic surfactants with branched hydrophobic chains affect the ionic–nonionic interactions, resulting in lower values of β parameter (stronger synergism); however, this effect considers mostly surfactants micellization in the bulk and not the surfactants interaction at the interface [10,12].

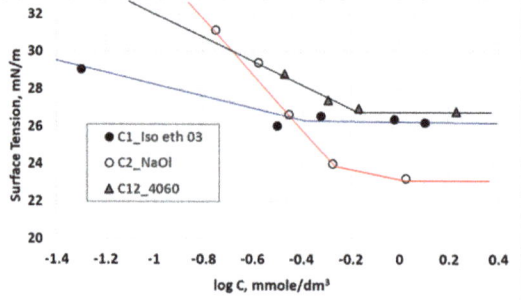

Figure 7. Surface tension variation with collectors concentration for NaOl solution, Iso Eth 03 solution and their binary mixture at 60:40 mass ratio.

As a next step, shear viscoelasticity studies indicate the mobility and stability of surfactant layers adsorbed on the surface of flotation bubbles. The viscoelasticity of each binary system is tested at the critical mass ratio that results from the interfacial shear viscosity measurements. The reason is quite simple. Below critical mass ratios, the viscoelasticity of binary layers will be equal to that of individual NaOl molecules, and above the critical mass ratios, their viscoelasticity will resemble that of individual nonionic cocollector or pure water. The above conclusions derive from the interfacial shear viscosity measurements of Figure 4. Therefore, shear viscoelasticity of binary mixtures with a mass ratio diverging from the critical ones, are expected to coincide with one of the two aforementioned extreme tested conditions.

At first, Figure 8 presents the amplitude sweep measurements that are performed to identify the linear viscoelastic region of the adsorbed binary surfactant layers. Amplitude sweep runs of 1CMC NaOl solution and pure water are reported as control measurements. The resulting storage modulus (G') and loss modulus (G'') are presented in separate subfigures (Figure 8a,b in respect) so as to avoid data overcrowding. For the ease of comparison, the two subfigures have the same scaling. The results show that both storage and loss moduli of all binary mixtures at critical mass ratios are in between the corresponding moduli of 1CMC NaOl solution and the artificial measurements of pure water, meaning that the addition of nonionic cocollectors in the NaOl solution suppresses the magnitude of both elastic and viscous nature of the initial layer. This behavior is in consistency with the interfacial shear viscosity measurements of Figure 4. Moreover, the results show that for all tested binary mixtures and the anionic NaOl solution, and for the whole range of tested amplitudes, the loss modulus is approximately one order of magnitude higher than storage modulus. Since there is no intersection of the corresponding storage and loss moduli curves, the yield point cannot be identified. This clear domination of loss modulus elucidates the pure viscous behavior of all adsorbed surfactant layers, meaning that upon the application of some stress, the layers do not retain any memory of their initial condition. However, the absence of yield point shows that the structure of these layers does not break even under intense shear deformation. Another crucial point is that in both moduli measurements, it is not easy to detect the linear viscoelastic region of surfactant layers. The linear viscoelastic region is a very short part of the tested amplitude: 0.1–0.3%. The measuring sensitivity of the rheometer does not allow us to perform tests at lower amplitude values and obtain more clearly the linear viscoelastic region of the surfactant layers.

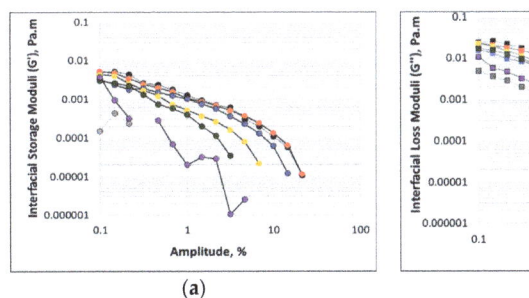 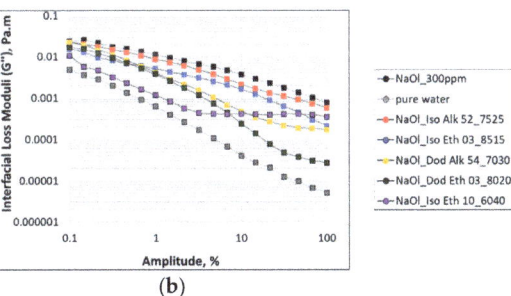

(a) (b)

Figure 8. Oscillatory amplitude sweep runs under 1 rad/s constant frequency, to measure: (**a**) Interfacial storage modulus, G', and (**b**) Interfacial loss modulus, G'', of 1CMC NaOl solution and NaOl-nonionic collector mixtures at their critical mass ratios.

The interfacial shear stresses applied to the binary layers during amplitude sweep oscillations are reported in Figure 9. Deviations between the applied shear stresses become clear at amplitudes above 10%. Measuring data show that shear stress depends on the molecular structure of nonionic cocollectors. The most important factor that results in high shear stresses is the branched hydrophobic chain of cocollector. Comparing separately the shear stresses of brunched and straight cocollectors, it appears that the length of their hydrophilic chain is the second factor that results in high shear stress values [17].

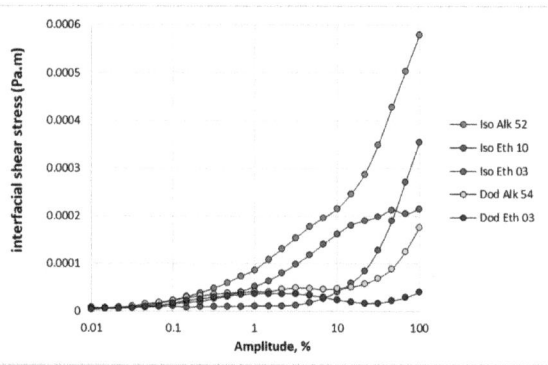

Figure 9. Interfacial shear stress applied on the interface of ionic–nonionic collector solutions under oscillations of different amplitude.

Oscillatory frequency sweep runs should be performed for an amplitude in the linear viscoelastic region. For such low amplitudes, the resulting disturbance is not adequate to provide trustful measurements. Thus, the tested amplitude condition is a bit higher: 1%. Figure 10 shows the interfacial shear moduli of binary collector mixtures at the critical mass ratios. Measurements of 1CMC NaOl solution and pure water are reported as control conditions. The results show that both storage and loss moduli of binary mixtures coincide with those of pure water, indicating that the structure of layers break since the beginning of rotation. For oscillation conditions beyond the linear viscoelastic region, such a result is expected. Commenting on NaOl curves, the viscous character prevails along the whole frequency range, denoting the fluidic interface of flotation bubbles in the anionic collector solution.

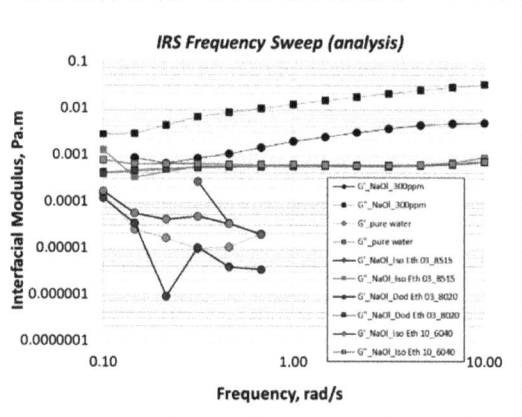

Figure 10. Oscillatory frequency sweep runs under 1% amplitude to measure the interfacial storage (G′) and loss (G″) moduli of 1CMC NaOl solution and NaOl-nonionic collector mixtures at their critical mass ratios.

3. Materials and Methods

Sodium Oleate (Mw = 304.44 g/mol, ≥82%) is supplied from Sigma-Aldrich company. BASF company provided Isotridecyl Ethoxylate 03, Isotridecyl Ethoxylate 10, Dodecyl Ethoxylate 03, Dodecyl Alkoxylate 54 and Isotridecyl Alkoxylate 52 nonionic surfactants to be tested as cocollectors for flotation. For simplicity, from now on, the aforementioned nonionic cocollectors are named using the following abbreviations: Iso Eth 03, Iso Eth

10, Dod Eth 03, Dod Alk 54 and Iso Alk 52, respectively. These cocollectors are either water soluble, miscible or non-soluble in water. The chemical structure of all primary and cocollectors used in this work is illustrated in Figure 11.

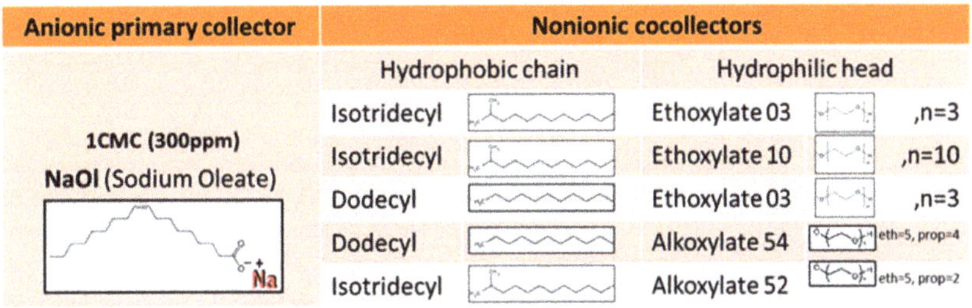

Anionic primary collector	Nonionic cocollectors			
	Hydrophobic chain		Hydrophilic head	
1CMC (300ppm) NaOl (Sodium Oleate)	Isotridecyl		Ethoxylate 03	,n=3
	Isotridecyl		Ethoxylate 10	,n=10
	Dodecyl		Ethoxylate 03	,n=3
	Dodecyl		Alkoxylate 54	eth=5, prop=4
	Isotridecyl		Alkoxylate 52	eth=5, prop=2

Figure 11. Molecular structure of anionic and nonionic surfactants that are tested as flotation collector and cocollectors in this work.

A Millipore Direct-Q 3 UV water purification system is used to produce ultrapure water (Type 1) as a solvent for the preparation of test solutions. NaOl concentration in all tested surfactant solutions is 300 ppm (1CMC), as this concentration ensures the maximum accumulation of surfactant molecules on the solution surface without the formation of micelles in the bulk. The tested nonionic cocollector concentrations are calculated as mass ratios to the primary NaOl collector, ranging from 5:95 to 50:50. The dissolution of surfactants in water is performed under mild heating at 30 °C and mixing on a hot plate magnetic stirrer for 10 min. Additional ultrasonic mixing for 20 min helps to achieve the homogeneous dispersion of non-water-soluble surfactants in water bulk. Characterization of the resulting flotation liquids is performed at two working temperatures, 15 °C and 30 °C, that simulate ambient conditions during winter and summer months. The PH is a parameter that affects the intermolecular interactions between the different surfactants of a solution [43–45]. The pH of test solutions is measured using a HANNA INSTRUMENTS HI2020-edge Multiparameter pH meter equipped with a HI-10430 glass probe.

The interactions between NaOl and nonionic cocollectors on the surface of flotation liquids indicate the structure of surfactant layers at the interface of the liquid and flotation bubbles. These interactions are examined through interfacial rheology and surface tension isotherms. Physica MCR 301 rheometer (Anton Paar) equipped with an Interfacial Rheology System (IRS) and a bicone geometry is used to study viscoelasticity at the interface between the test liquid solution and air under shear deformation. The diameter of the bicone is 60 mm, the diameter of the cylindrical test cell is 80 mm and the volume of the bottom phase (test solution) within the test cell is 110 mL. The test cell is covered with an evaporation trap to avoid disturbance of the examined interface. A Peltier element is placed at the bottom of the test cell to regulate temperature of the test solution at the desired level. Continuous air supply and Fisher Scientific Isotemp 3013 circulating chiller water bath are used to cool down the Peltier element. The application of a Controlled Shear Rate (CSR) measuring mode gives the interfacial viscosity of surfactant solutions in a wide range of shear rates (0.1–100/s). Five measuring points are recorded per decade of shear rate (logarithmic scale). The measuring duration of each point decreases from 8 min to 0.5 min with a logarithmic scale ramp while moving from low to high shear rate values. In case of low shear deformation, the measuring points are not considered trusted, and the rheometer does not transform the bulk data to interfacial data. Therefore, in some cases, the number of measuring points per decade of shear deformation can be less than five. Oscillatory motion is applied to study the interfacial viscoelasticity of surfactant solutions. The interfacial storage (G') and loss (G'') moduli measurements resulting from amplitude sweep runs show the linear viscoelastic region of surfactants layer adsorbing on the surface of test solutions. During these measurements, bicone geometry oscillates with 1 rad/s constant frequency and amplitude escalating from 0.1 to 100% with a logarithmic time ramp from

3 min to 30 s. The applied frequency is low in order to avoid violent disturbances that would immediately break the thin interfacial layer structure. Oscillatory frequency sweep runs give the interfacial storage and loss moduli under a constant amplitude in the range of the resulting linear viscoelastic region and a frequency increasing from 0.1 rad/s to 10 rad/s, with a logarithmic time ramp from 3 min to 0.5 min. Both interfacial viscosity and interfacial viscoelastic moduli values result from a numerical hydrodynamic analysis on the flow field of the bulk phase [27].

A LAUDA TE2 tensiometer is used to measure the surface tension of test liquids with Wilhelmy plate method. The tensiometer is connected to a Brookfield TC-102 water bath for thermalization of test liquids at the desired working temperature. Upon determination of liquid surface location, surface tension measurements are recorded every 1 s.

4. Conclusions

The present work studies the individual and interactive performance of sodium oleate (NaOl) and nonionic collectors on the air–water interface related to flotation system reagents. Rheological and surface tension measurements illustrate the structure and viscoelasticity of interfacially adsorbed binary surfactant layers. Interfacial shear viscosity measurements under controlled shear rate indicate interactions between different types of adsorbed surfactant molecules on air–liquid interface. The adsorption of nonionic cocollectors on air–water interface does not change the interfacial viscosity of pure water surface. On the other hand, the adsorption of NaOl anionic collector molecules on the surface of 1CMC aqueous solution gives a significant interfacial viscosity that decreases with the applied shear rate. To obtain reliable results, the working temperature of the liquid must be above the room temperature so that NaOl is fully dissolved in water. The interfacial shear viscosity of anionic–nonionic binary collector mixtures shows that NaOl molecules adsorbed on the bubble surface are gradually displaced by nonionic cocollector molecules. NaOl molecules displacement is completed at lower anionic:nonionic w/w mass ratios for nonionic molecules of long hydrophilic head and branched hydrophobic chain. Surface tension isotherms show the synergistic interaction between NaOl and the tested nonionic cocollector molecules. Interfacial shear oscillatory runs show the pure viscous behavior of all binary surfactant layers under the whole range of tested amplitudes. The emerging linear viscoelastic region is quite short. The implementation of interfacial dilatational rheology studies in future will complete the examination of dynamic surfactant interactions for the present systems.

Author Contributions: Conceptualization, O.O., M.K. and T.K.; methodology, O.O., M.K. and T.K.; validation, M.K. and T.K.; investigation, O.O.; methodology, O.O., M.K. and T.K.; writing—original draft preparation, O.O.; writing—review and editing, M.K. and T.K.; supervision, M.K. and T.K. All authors have read and agreed to the published version of the manuscript.

Funding: This research was funded by European Union's Horizon 2020 research and innovation programme, grant number 821265.

Institutional Review Board Statement: Not applicable.

Informed Consent Statement: Not applicable.

Data Availability Statement: Data are available on request, due to restrictions.

Acknowledgments: The authors would like to acknowledge BASF SE located in Ludwigshafen of Germany for donating the nonionic surfactants tested in the present work.

Conflicts of Interest: The authors declare no conflict of interest. The funders had no role in the design of the study; in the collection, analyses, or interpretation of data; in the writing of the manuscript; or in the decision to publish the results.

Sample Availability: Not applicable.

References

1. Aono, K.; Suzuki, F.; Yomogida, Y.; Hasumi, M.; Kado, S.; Nakahara, Y.; Yajima, S. Effects of Polypropylene Glycol at Very Low Concentrations on Rheological Properties at the Air–Water Interface and Foam Stability of Sodium Bis(2-ethylhexyl)sulfosuccinate Aqueous Solutions. *Langmuir* **2020**, *36*, 10043–10050. [CrossRef] [PubMed]
2. Dwyer, M.D.; He, L.; James, M.; Nelson, A.; Middelberg, A.P.J. Insights into the role of protein molecule size and structure on interfacial properties using designed sequences. *J. R. Soc.* **2012**, *10*, 20120987. [CrossRef]
3. Erni, P.; Fischer, P.; Windhab, E.J. Rheology of surfactant assemblies at the air/liquid and liquid/liquid interface. In *Proceedings of the 3rd International Symposium on Food Rheology and Structure, Zürich, Switzerland, 9–13 February 2003*; Fischer, P., Marti, I., Windhab, E.J., Eds.; Laboratory of food process engineering, Institute of Food Science and Nutrition, Swiss Federal Institute of Technology: Zürich, Switzerland; p. 8092.
4. Krägel, J.; Derkatch, S.R.; Miller, R. Interfacial shear rheology of protein–surfactant layers. *Adv. Colloid Interface Sci.* **2008**, *144*, 38–53. [CrossRef]
5. Ferrari, M.; Navarini, L.; Liggieri, L.; Ravera, F.; Liverani, F.S. Interfacial properties of coffee-based beverages. *Food Hydrocoll.* **2007**, *21*, 1374–1378. [CrossRef]
6. Bagheri, A.; Khalili, P. Synergism between non-ionic and cationic surfactants in a concentration range of mixed monolayers at an air–water interface. *RSC Adv.* **2017**, *7*, 18151–18161. [CrossRef]
7. Zawala, J.; Wiertel-Pochopien, A.; Kowalczuk, P.B. Critical Synergistic Concentration of Binary Surfactant Mixtures. *Minerals* **2020**, *10*, 192. [CrossRef]
8. Zhou, Q.; Rosen, M.J. Molecular Interactions of Surfactants in Mixed Monolayers at the Air/Aqueous Solution Interface and in Mixed Micelles in Aqueous Media: The Regular Solution Approach. *Langmuir* **2003**, *19*, 4555–4562. [CrossRef]
9. El-Aila, H.J.Y. Interaction of Nonionic Surfactant Triton-X-100 with Ionic Surfactants. *J. Dispers. Sci. Technol.* **2009**, *30*, 1277–1280. [CrossRef]
10. Rosen, M.J.; Zhou, Q. Surfactant-Surfactant Interactions in Mixed Monolayer and Mixed Micelle Formation. *Langmuir* **2001**, *17*, 3532–3537. [CrossRef]
11. Sis, H.; Chander, G.; Chander, S. Synergism in Sodium Oleate/Ethoxylated Nonylphenol Mixtures. *J. Dispers. Sci. Technol.* **2005**, *26*, 605–614. [CrossRef]
12. Bagheri, A.; Abolhasani, A. Binary mixtures of cationic surfactants with triton X-100 and the studies of physicochemical parameters of the mixed micelles. *Korean J. Chem. Eng.* **2015**, *32*, 308–315. [CrossRef]
13. Bosa, M.A.; van Vliet, T. Interfacial rheological properties of adsorbed protein layers and surfactants: A review. *Adv. Colloid Interface Sci.* **2001**, *91*, 437–471. [CrossRef] [PubMed]
14. Franck, A. TA Instruments Germany. In *Interfacial Rheometry and the Stability of Foams and Emulsions*; TA Instruments Germany: Hüllhorst, Germany, 2005.
15. Edwards, D.A.; Brenner, H.; Wasan, D.T. *Interfacial Transport Processes and Rheology*; Butterworth-Heinemann: Boston, MA, USA, 2013; pp. 21–33.
16. Torcello-Gómez, A.; Maldonado-Valderrama, J.; Gálvez-Ruiz, M.J.; Martín-Rodríguez, A.; Cabrerizo-Vílchez, M.A.; de Vicente, J. Surface rheology of sorbitan tristearate and b-lactoglobulin: Shear and dilatational behavior. *J. Nonnewton Fluid. Mech.* **2011**, *166*, 713–722. [CrossRef]
17. Erni, P.; Fischer, P.; Windhab, E.J. Sorbitan Tristearate Layers at the Air/Water Interface Studied by Shear and Dilatational Interfacial Rheology. *Langmuir* **2005**, *21*, 10555–10563. [CrossRef] [PubMed]
18. Ravera, F.; Loglio, G.; Kovalchuk, V.I. Interfacial dilational rheology by oscillating bubble/drop methods. *Curr. Opin. Colloid Interface Sci.* **2010**, *15*, 217–228. [CrossRef]
19. Moradi, N.; Zakrevskyy, Y.; Javadi, A.; Aksenenko, E.V.; Fainerman, V.B.; Lomadze, N.; Santer, S.; Miller, R. Surface tension and dilation rheology of DNA solutions in mixtures with azobenzene-containing cationic surfactant. *Colloids Surf. A Physicochem. Eng. Asp.* **2016**, *505*, 186–192. [CrossRef]
20. Liggieri, L.; Miller, R. Relaxation of surfactants adsorption layers at liquid interfaces. *Curr. Opin. Colloid Interface Sci.* **2010**, *15*, 256–263. [CrossRef]
21. Miller, R.; Wüstneck, R.; Krägel, J.; Kretzschmar, G. Dilational and shear rheology of adsorption layers at liquid interfaces. *Colloids Surf. A Physicochem. Eng. Asp.* **1996**, *111*, 75–118. [CrossRef]
22. Krägel, J.; Clark, D.; Wilde, P.; Krägel, J.; Miller, R. Studies of adsorption and surface shear rheology of mixed β-lactoglobulin/surfactant systems. In *Trends in Colloid and Interface Science IX. Progress in Colloid & Polymer Science*; Appell, J., Porte, G., Eds.; Springer: Berlin/Heidelberg, Germany, 1995; p. 98.
23. Ruhs, P.A.; Boni, L.; Fuller, G.G.; Inglis, R.F.; Fischer, P. In-Situ Quantification of the Interfacial Rheological Response of Bacterial Biofilms to Environmental Stimuli. *PLoS ONE* **2013**, *8*, 78524. [CrossRef]
24. Kragel, J.; Siegel, S.; Miller, R.; Born, M.; Schano, K.H. Measurement of interfacial shear rheological properties: An automated apparatus. *Colloids Surf. A Physicochem. Eng. Asp.* **1994**, *91*, 169–180. [CrossRef]
25. Kotsmar, C.; Pradines, V.; Alahverdjieva, V.S.; Aksenenko, E.V.; Fainerman, V.B.; Kovalchuk, V.I.; Krägel, J.; Leser, M.E.; Miller, R. Thermodynamics, adsorption kinetics and rheology of mixed protein-surfactant interfacial layers. *Adv. Colloid Interface Sci.* **2009**, *150*, 41–54. [CrossRef] [PubMed]

26. Dan, A.; Gochev, G.; Kragel, J.; Aksenenko, E.V.; Fainerman, V.B.; Miller, R. Interfacial rheology of mixed layers of food proteins and surfactant. *Curr. Opin. Colloid Interface Sci.* **2013**, *18*, 302–310. [CrossRef]
27. Erni, P.; Fischer, P.; Windhab, E.J. Stress- and strain-controlled measurements of interfacial shear viscosity and viscoelasticity at liquid/liquid and gas/liquid interfaces. *Rev. Sci. Instrum.* **2003**, *74*, 4916. [CrossRef]
28. Beneventi, D.; Pugh, R.J.; Carré, B.; Gandini, A. Surface rheology and foaming properties of sodium oleate and C12(EO)6 aqueous solutions. *J. Colloid Interface Sci.* **2003**, *268*, 221–229. [CrossRef] [PubMed]
29. Basarová, P.; Bartovská, L.; Korínek, K.; Horn, D. The influence of flotation agent concentration on the wettability and flotability of polystyrene. *J. Colloid Interface Sci.* **2005**, *286*, 333–338. [CrossRef]
30. Zhang, H.; Liu, W.; Han, C.; Hao, H. Effects of monohydric alcohols on the flotation of magnesite and dolomite by sodium oleate. *J. Mol. Liq.* **2018**, *249*, 1060–1067. [CrossRef]
31. Sis, H.; Chander, S. Improving froth characteristics and flotation recovery of phosphate ores with nonionic surfactants. *Min. Eng.* **2003**, *16*, 587–595. [CrossRef]
32. Brabcová, Z.; Karapantsios, T.; Kostoglou, M.; Basarová, P.; Matis, K. Bubble–particle collision interaction in flotation systems. *Colloids Surf. A Physicochem Eng. Asp.* **2015**, *473*, 95–103. [CrossRef]
33. Narsimhan, V. Letter: The effect of surface viscosity on the translational speed of droplets. *Phys. Fluids* **2018**, *30*, 081703. [CrossRef]
34. Levan, M.D. Motion of a Droplet with a Newtonian Interface. *J. Colloid Interface Sci.* **1981**, *83*, 11–17. [CrossRef]
35. Langevin, D. Influence of interfacial rheology on foams and emulsion properties. *Adv. Colloid Interface Sci.* **2000**, *88*, 209–222. [CrossRef] [PubMed]
36. Nguyen, A.V.; Schulze, H.J. *Colloidal Science of Flotation*; Marcel Dekker: New York, NY, USA, 2004.
37. Chen, C.; Zhu, H.; Sun, W.; Hu, Y.; Qin, W.; Liu, R. Synergetic Effect of the Mixed Anionic/Non-Ionic Collectors in Low Temperature Flotation of Scheelite. *Minerals* **2017**, *7*, 87. [CrossRef]
38. Wonyen, D.G.; Kromah, V.; Gibson, B.; Nah, S.; Chelgani, S.C. A Review of Flotation Separation of Mg Carbonates (Dolomite and Magnesite). *Minerals* **2018**, *8*, 354. [CrossRef]
39. Rosen, M.J.; Yuan Hua, X. Surface concentrations and molecular interactions in binary mixtures of surfactants. *J. Colloid Interface Sci.* **1982**, *86*, 164–172. [CrossRef]
40. Rosen, M.J.; Zhao, F.J. Binary mixtures of surfactants. The effect of structural and microenvironmental factors on molecular interaction at the aqueous solution/air interface. *J. Colloid Interface Sci.* **1983**, *95*, 443. [CrossRef]
41. Nagarajan, R. *New Horizons: Detergents for the New Millennium Conference Invited Papers*; American Oil Chemists Society and Consumer Specialty Products Association: Fort Myers, FL, USA, 2001.
42. Nagarajan, R.; Kalpakci, B. Visometric investigation of complexes between polyethyleneoxide and surfactant micelles. *Polym. Prepr. Am. Chem. Soc.* **1982**, *23*, 41.
43. Żamojć, K.; Wyrzykowski, D.; Chmurzyński, L. On the Effect of pH, Temperature, and Surfactant Structure on Bovine Serum Albumin–Cationic/Anionic/Nonionic Surfactants Interactions in Cacodylate Buffer–Fluorescence Quenching Studies Supported by UV Spectrophotometry and CD Spectroscopy. *Int. J. Mol. Sci.* **2022**, *23*, 41. [CrossRef]
44. Li, R.; Wu, Z.; Wangb, Y.; Ding, L.; Wang, Y. Role of pH-induced structural change in protein aggregation in foam fractionation of bovine serum albumin. *Proc. Biotechnol. Rep.* **2016**, *9*, 46–52. [CrossRef]
45. Roberts, S.A.; Kellaway, I.W.; Taylor, K.M.; Warburton, B.; Peters, K. Combined surface pressure-interfacial shear rheology study of the effect of pH on the adsorption of proteins at the air-water interface. *Langmuir* **2005**, *21*, 7342–7348. [CrossRef]

Disclaimer/Publisher's Note: The statements, opinions and data contained in all publications are solely those of the individual author(s) and contributor(s) and not of MDPI and/or the editor(s). MDPI and/or the editor(s) disclaim responsibility for any injury to people or property resulting from any ideas, methods, instructions or products referred to in the content.

Article

Lipidomic Analysis of Liver and Adipose Tissue in a High-Fat Diet-Induced Non-Alcoholic Fatty Liver Disease Mice Model Reveals Alterations in Lipid Metabolism by Weight Loss and Aerobic Exercise

Thomai Mouskeftara [1,2], Olga Deda [1,2], Grigorios Papadopoulos [3], Antonios Chatzigeorgiou [3] and Helen Gika [1,2,*]

1. Laboratory of Forensic Medicine & Toxicology, Department of Medicine, Aristotle University of Thessaloniki, 54124 Thessaloniki, Greece; mousthom@auth.gr (T.M.); oliadmy@gmail.com (O.D.)
2. Biomic AUTh, Center for Interdisciplinary Research and Innovation (CIRI-AUTH), Balkan Center B1.4, 10th km Thessaloniki-Thermi Rd, P.O. Box 8318, 57001 Thessaloniki, Greece
3. Department of Physiology, Medical School, National and Kapodistrian University of Athens, 75 Mikras Asias Str., 11527 Athens, Greece; grpapad@hotmail.com (G.P.); achatzig@med.uoa.gr (A.C.)
* Correspondence: gkikae@auth.gr

Citation: Mouskeftara, T.; Deda, O.; Papadopoulos, G.; Chatzigeorgiou, A.; Gika, H. Lipidomic Analysis of Liver and Adipose Tissue in a High-Fat Diet-Induced Non-Alcoholic Fatty Liver Disease Mice Model Reveals Alterations in Lipid Metabolism by Weight Loss and Aerobic Exercise. *Molecules* 2024, 29, 1494. https://doi.org/10.3390/molecules29071494

Academic Editor: Susy Piovesana

Received: 5 January 2024
Revised: 11 March 2024
Accepted: 12 March 2024
Published: 27 March 2024

Copyright: © 2024 by the authors. Licensee MDPI, Basel, Switzerland. This article is an open access article distributed under the terms and conditions of the Creative Commons Attribution (CC BY) license (https://creativecommons.org/licenses/by/4.0/).

Abstract: Detailed investigation of the lipidome remodeling upon normal weight conditions, obesity, or weight loss, as well as the influence of physical activity, can help to understand the mechanisms underlying dyslipidemia in metabolic conditions correlated to the emergence and progression of non-alcoholic fatty liver disease (NAFLD). C57BL/6 male mice were fed a normal diet (ND) or a high-fat diet (HFD) for 20 weeks. Subgroups within the high-fat diet (HFD) group underwent different interventions: some engaged in exercise (HFDex), others were subjected to weight loss (WL) by changing from the HFD to ND, and some underwent a combination of weight loss and exercise (WLex) during the final 8 weeks of the 20-week feeding period. To support our understanding, not only tissue-specific lipid remodeling mechanisms but also the cross-talk between different tissues and their impact on the systemic regulation of lipid metabolism are essential. Exercise and weight loss-induced specific adaptations in the liver and visceral adipose tissue lipidomes of mice were explored by the UPLC–TOF–MS/MS untargeted lipidomics methodology. Lipidomic signatures of ND and HFD-fed mice undergoing weight loss were compared with animals with and without physical exercise. Several lipid classes were identified as contributing factors in the discrimination of the groups by multivariate analysis models, such as glycerolipids, glycerophospholipids, sphingolipids, and fatty acids, with respect to liver samples, whereas triglycerides were the only lipid class identified in visceral adipose tissue. Lipids found to be dysregulated in HFD animals are related to well-established pathways involved in the biosynthesis of PC, PE, and TG metabolism. These show a reversing trend back to basic levels of ND when animals change to a normal diet after 12 weeks, whereas the impact of exercise, though in some cases it slightly enhances the reversing trend, is not clear.

Keywords: non-alcoholic fatty liver disease; exercise; de novo lipogenesis; lipidomics; liver; adipose tissue

1. Introduction

Non-alcoholic fatty liver disease (NAFLD) has emerged as one of the most common forms of chronic liver disease and has evolved into a significant health issue worldwide [1]. Pathophysiologic hallmarks of the disease include triglyceride deposition in the liver, observed in the cytoplasm of at least 5% of hepatocytes, necrosis of hepatocytes, and inflammation [2]. According to its histological characteristics, NAFLD can be categorized as non-alcoholic fatty liver (NAFL), described by simple steatosis, and non-alcoholic steatohepatitis (NASH), which is characterized by the coexistence of steatosis, inflammation, and hepatocellular ballooning and can progress to cirrhosis, liver failure, and hepatocellular carcinoma [3].

Reviews and meta-analyses have amply demonstrated that type 2 diabetes mellitus, hyperlipidemia, metabolic syndrome, and other factors related to an unhealthy high-caloric diet are comorbid conditions commonly associated with NAFLD, or, according to the new nomenclature, metabolic dysfunction-associated steatotic liver disease (MASLD) [4–10]. Among all these contributors, obesity appears to play a significant role in the initial stages that lead to simple steatosis and in the progression to NASH. Obesity-induced NAFLD has been investigated in animal models applying high-fat diets (HFD) in combination with various sugars, such as fructose, which is commonly consumed in the Western world, aiming to evaluate its role in the development of NAFLD. Notably, dietary sugar intake is considered a significant mediator of hepatic steatosis and plays an essential role in the progression of NAFLD [11]. Currently, high-fructose corn syrup (HFCS), which mainly contains glucose and fructose, is the primary source of added sugars to beverages and fat-enriched diets [12]. In dysregulated liver metabolism, fructose has been proposed to affect de novo lipogenesis (DNL) by increasing the levels of enzymes involved in this process while decreasing insulin efficiency [13]. Importantly, HFCS significantly contributes to steatosis deterioration in obesity-related NAFLD, probably due to DNL upregulation in combination with tricarboxylic acid (TCA) cycle overactivation and further impairment in hepatic insulin resistance [14].

The improvement of NAFLD in obese patients can be achieved with weight loss and, more significantly, with weight loss maintenance. Studies have demonstrated that greater weight loss can lead to improvements in steatohepatitis and fibrosis [15]. In overweight patients, a 3–5% weight loss has shown a reduction in steatosis [16], while a weight loss >10% has been associated with a reduction in both steatosis and fibrosis [17]. Exercise is another effective approach for weight loss and can be utilized as part of the treatment regimen for NAFLD [18]. Notably, combining weight loss with regular exercise has been associated with improvements in transaminase levels, including a significant decrease in alanine aminotransferase (ALT) levels in patients with NASH [19]. In general, enhanced peripheral insulin sensitivity leads to a decrease in the excessive supply of free fatty acids and glucose for the synthesis of free fatty acids in the liver. Elevation in fatty acid oxidation and reduction in lipogenesis in the liver are associated with protective effects against damage to mitochondria and hepatocytes by reducing the release of damage-associated molecular patterns during exercise [5].

However, the mechanisms underlying these effects in both liver and adipose tissue have not been investigated in depth. Lipidomic studies examine the contributors to the lipid profile of the liver, but none so far have compared the changes in the hepatic and adipose tissue lipid signature triggered by weight loss and/or exercise during obesity and NAFLD. Therefore, it is important to examine the potential unidentified roles of dietary components and exercise in NAFLD pathology and its progression. In this research, we utilized a lipidomic workflow to investigate the weight loss and the effect of exercise in an obesity-induced NAFLD mice model in both hepatic and visceral adipose tissues. Our primary objective was to identify potential pathways and lipids species that might be responsible for the escalation of steatosis severity in obesity-related NAFLD mice models, particularly within the context of HFCS consumption, and explore the effect of weight loss in combination with exercise in NAFLD.

2. Results

2.1. Animal Study

Many research findings have amply demonstrated that NAFLD is closely linked to metabolic imbalances associated with obesity. The examination of animal weight records indicated significant alterations primarily in relation to changes in diet, followed by exercise. Finally, body weights of weight loss (WL) and weight loss in combination with exercise

(WLex) groups at the end of the 8-week intervention period were similar to the control (ND) group, whereas the HFDex group showed a decrease compared to HFD, as demonstrated in Figure 1a, where the average weekly body weight (g) (±SD) for all mice is provided. The *p*-values for the comparisons of body weights between HFD-HFDex, HFD-WL, HFD-WLex, ND-WL, and ND-WLex are listed in Table S1. According to the histological examination performed on the mouse liver samples, increased fat deposition was observed in the HFD group compared to the ND group. A reduction in hepatic steatosis was also observed in the HFDex group relative to the HFD group; however, the greatest difference was observed in the WL and WLex groups, where histological images showed remission of NAFLD, as shown in Figure 1b.

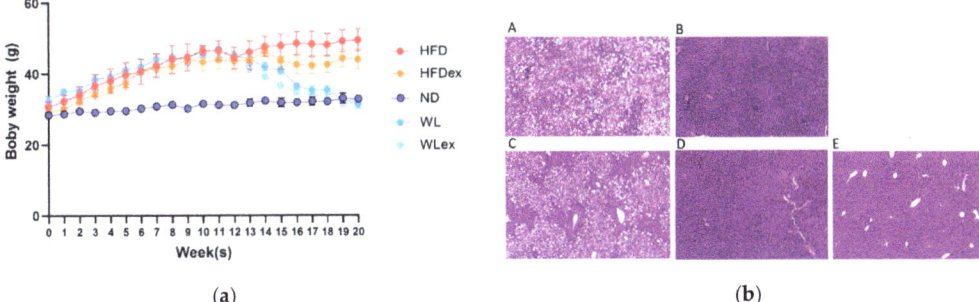

(a) (b)

Figure 1. (a) Average (±SD) weekly body weight (g) in all mice (b). Representative images from Hematoxylin–Eosin (H&E)-stained liver tissue sections of (**A**) HFD, (**B**) ND, (**C**) HFDex, (**D**) WL, and (**E**) WLex mice. The *p*-values for the comparisons of body weights between HFD-HFDex, HFD-WL, HFD-WLex, ND-WL, and ND-WLex are listed in Table S1.

2.2. LC–TOF–MS Lipidomics Data

By applying the methodology for comprehensive lipid profiling, a wealth of lipid species could be detected; in liver tissue, 5298 ion signals in positive mode and 3037 in negative mode were considered after quality control filtering. Data from visceral adipose tissue were acquired only in positive ionization mode, aiming to investigate the content of glycerolipids, which ionize more effectively under this condition. The total number of corresponding ion signals reached 1092 after filtering. These data were studied by multivariate statistical methods to unravel patterns expressed by the lipidomic phenotype in the different groups. To evaluate the analytical quality of the data, in the first place, QC samples were examined (see Figures S1–S3).

Based on the projection of the samples in the principal components analysis (PCA) models, it can be observed that a variation of 31% (positive ionization) and 36% (negative ionization) can be explained in the t1 axis (first component) between the hepatic tissue of mice fed with a high-fat diet (plotted in red and orange), irrespective of the parameter of exercise, and those fed with a normal diet (plotted in different color scales of blue) (Figure 2a,b). It is also clear that the two groups induced to a normal diet (WL and WLex) after the 12th week of the study exhibit greater similarity to the liver lipidomic profile of the ND mice. This finding suggests that after 8 weeks on a normal diet, lipid metabolism in the liver tends to return close to ND. This observation, however, is not reflected in the visceral adipose tissue lipidomic profile of the same mice. As can be seen in Figure 2c, weight loss groups are projected closely to HFD groups, whereas it can be concluded that ND mice have a more distinct visceral adipose tissue lipidomic profile—projected in a different quantile. This is an interesting finding that could be explained by the critical role of adipose tissue in the regulation of the systemic energy homeostasis of the organism. By using supervised models, a clearer distinction between the different groups could be observed (see PLS score plot in Figure S4), and lipid species contributing to this differentiation could be identified.

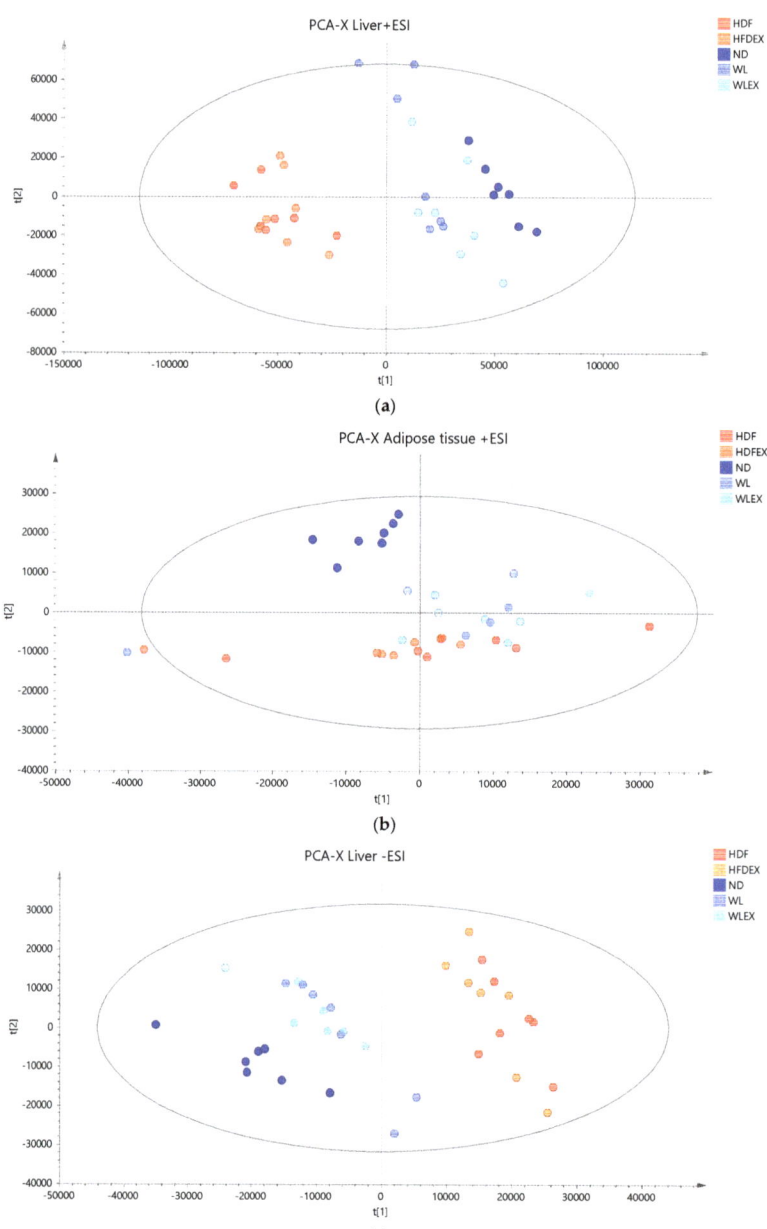

Figure 2. PCA score plots of mice tissue samples from the five studied groups: ND, HFD, HFDex, WL, and WLex. (**a**) Liver tissue samples projection based on positive ionization mode analysis data, (**b**) liver tissue based on negative ionization mode data, and (**c**) visceral adipose tissue samples based on positive ionization mode data. Logarithmic transformation of the data and Pareto scaling were used in all liver models in negative ionization, whereas only Pareto scaling was used in liver models in positive ionization. Logarithmic transformation of the data and Pareto scaling were used in all visceral adipose tissue models.

2.3. Hepatic Lipids Profile Reveal Alterations in Major Lipid Classes with Diet

To allow for a thorough understanding of the molecular characteristics of NAFLD associated with a high-fat diet in obesity, weight loss, and the effect of exercise, a comprehensive analysis of liver lipids was conducted. Pairwise OPLS-DA analysis between the groups revealed significant alterations in lipids between high-fat diet groups and all three normal diet mice groups (either remained—ND, either returned to normal diet—WL, WLex) independently of exercise, while no differentiation was observed when exercise was considered as the only differentiating factor (e.g., HFD vs. HFDex or WL vs. WLex). Glycerolipids, glycerophospholipids, fatty acids, and sphingolipids were found to be statistically differentiated when HFD vs. ND, HFD vs. WL, and HFD vs. WLex were compared. The quality metrics of the constructed models are presented in Table S2, whereas the detailed annotations of the statistically significant lipid species revealed from these comparisons can be found in Table S3. Table 1 summarizes all identified lipids found to be significantly altered in the liver of high-fat diet mice after weight loss or after weight loss in combination with exercise. The statistical parameters of Log2FC, p-values, VIP, and CV% based on the pairwise comparisons are also provided, indicating the impact on the lipids' levels and their contributing effect on the distinct profiles that were observed.

Table 1. Hepatic tissue lipids were found to be significant in the binary group comparisons between HFD-ND, HFD-WL, and HFD-WLEX. p-values, Log2FCs, CV% values, and VIP scores are provided for each lipid after univariate Kruskal–Wallis, followed by a post hoc Bonferroni's test and multivariate analysis. Values highlighted in red correspond to $p \leq 0.05$ (bold $p < 0.01$) or $|\text{Log2FC}| \geq 1.5$. Lipids highlighted in green were found in negative ionization mode.

	HFD-ND			HFD-WL			HFD-WLEX			
Lipids	p-Value	Log2FC	VIP	p-Value	Log2FC	VIP	p-Value	Log2FC	VIP	CV%
DG 36:2	8.03×10^{-5}	−3.61	4.2	9.75×10^{-3}	−1.43	4.1	2.44×10^{-1}	−0.96	3.4	1.53
DG 36:3	2.55×10^{-2}	0.27	6.8	1.00×10^{0}	−0.04	3.6	7.42×10^{-2}	0.24	7.2	0.89
DG 36:4	3.78×10^{-5}	2.13	11	1.00×10^{0}	0.94	5.9	3.81×10^{-2}	1.56	9.9	1.03
DG 38:2	3.24×10^{-5}	−1.81	4.8	3.12×10^{-2}	−1.31	5.2	1.63×10^{-1}	−1.17	4.7	1.79
DG 38:3	3.45×10^{-2}	−0.96	4.2	3.24×10^{-5}	−1.58	5.9	1.50×10^{-1}	−0.88	4.5	6.21
DG 38:4	5.07×10^{-4}	−1.66	4.3	3.44×10^{-3}	−1.58	5.1	1.77×10^{-1}	−0.98	4.1	2.06
DG 40:7	1.51×10^{-2}	−0.51	5.3	1.00×10^{0}	−0.11	2.9	7.61×10^{-1}	−0.29	8.4	18.7
DG 40:8	2.37×10^{-5}	5.03	8.0	3.07×10^{-1}	3.50	4.5	5.61×10^{-2}	4.12	6.2	18.3
FA 16:0	1.92×10^{-1}	0.18	4.6	2.64×10^{-1}	0.17	5.4	2.11×10^{-3}	0.31	7.2	4.68
FA 16:1	5.85×10^{-1}	−0.23	3.3	2.55×10^{-2}	−0.60	6.6	1.00×10^{0}	−0.02	1.1	3.43
FA 18:1	2.55×10^{-2}	−0.32	9.5	9.17×10^{-1}	−0.18	8.8	1.00×10^{0}	−0.01	5.4	5.26
FA 18:2	8.64×10^{-4}	0.82	12	2.85×10^{-1}	0.53	11	4.37×10^{-3}	0.77	14	8.73
FA 18:3	1.08×10^{-4}	2.38	8.1	4.76×10^{-1}	1.32	5.7	2.70×10^{-3}	2.13	8.4	3.38
FA 18:4	1.92×10^{-4}	3.26	2.5	4.76×10^{-1}	2.11	1.8	1.64×10^{-3}	3.18	2.7	3.51
FA 20:1	1.64×10^{-3}	−1.06	4.2	6.95×10^{-3}	−1.05	5.2	9.72×10^{-2}	−0.81	4.3	12.4
FA 20:2	7.57×10^{-4}	−1.27	3.7	2.30×10^{-2}	−1.06	4.2	6.16×10^{-2}	−0.94	3.8	2.78
FA 20:3	2.55×10^{-2}	−1.36	3.7	1.38×10^{-1}	−0.85	3.8	1.92×10^{-1}	−0.69	3.3	3.99
FA 20:5	5.96×10^{-5}	2.95	11	1.77×10^{-1}	2.07	9.1	2.82×10^{-2}	2.56	11	2.58
FA 22:3	3.78×10^{-5}	−2.86	2.4	1.16×10^{-1}	−1.79	2.7	4.20×10^{-2}	−2.08	2.6	5.01
FA 22:4	9.75×10^{-3}	−1.03	3.7	5.61×10^{-2}	−0.96	4.4	2.30×10^{-2}	−0.95	4.4	3.68
FA 22:5	3.05×10^{-3}	1.54	6.5	5.61×10^{-2}	1.31	6.4	3.05×10^{-3}	1.54	7.6	3.86
FA 22:6	7.57×10^{-4}	0.87	10	6.76×10^{-2}	0.63	10	6.76×10^{-2}	0.65	9.9	2.43
LPC 18:0	1.00×10^{0}	0.60	2.2	3.12×10^{-2}	1.07	5.0	1.00×10^{0}	0.34	2.2	4.23
LPC 18:2	8.12×10^{-2}	1.25	5.2	6.20×10^{-3}	1.41	7.2	2.39×10^{-3}	1.44	6.5	3.09
LPC 20:4	1.64×10^{-3}	−1.06	4.2	3.12×10^{-2}	−0.94	4.5	1.00×10^{0}	−0.44	2.8	4.30
LPE 18:2	6.20×10^{-3}	2.13	2.9	2.30×10^{-2}	1.85	3.2	5.61×10^{-2}	1.77	2.7	2.68
PC 32:0	1.64×10^{-3}	0.81	5.5	3.45×10^{-2}	0.65	5.8	1.50×10^{-1}	0.44	3.3	12.0
PC 32:1	2.85×10^{-1}	0.22	2.5	1.87×10^{-2}	0.37	4.4	4.63×10^{-2}	0.29	3.3	4.38

Table 1. Cont.

Lipids	HFD-ND			HFD-WL			HFD-WLEX			CV%
	p-Value	Log2FC	VIP	p-Value	Log2FC	VIP	p-Value	Log2FC	VIP	
PC 34:1	6.20×10^{-3}	−0.45	3.8	1.00×10^{0}	−0.01	0.5	6.26×10^{-1}	−0.23	2.9	1.98
PC 34:2	1.26×10^{-5}	0.88	5.5	7.42×10^{-2}	0.64	5.5	1.26×10^{-1}	0.61	4.9	2.66
PC 34:3	3.45×10^{-2}	0.64	2.9	1.00×10^{0}	−0.04	0.5	1.00×10^{0}	0.29	1.3	3.60
PC 36:2	1.12×10^{-3}	0.87	4.7	4.91×10^{-3}	0.85	5.7	4.13×10^{-1}	0.55	3.7	1.63
PC 36:3	1.09×10^{-2}	0.91	3.3	1.00×10^{0}	0.46	1.6	8.71×10^{-3}	0.96	3.9	2.63
PC 36:4	2.03×10^{-5}	1.99	4.7	3.45×10^{-2}	1.30	3.8	3.57×10^{-1}	0.98	2.8	4.42
PC 36:5	1.07×10^{-5}	3.58	2.7	1.26×10^{-1}	2.00	2.0	8.12×10^{-2}	2.43	2.0	2.69
PC 38:3	2.93×10^{-4}	−1.58	8.0	1.00×10^{0}	−0.24	3.1	4.20×10^{-2}	−0.89	6.9	1.50
PC 38:5	5.07×10^{-4}	−0.94	5.5	3.81×10^{-2}	−0.46	5.4	2.30×10^{-2}	−0.62	5.3	3.27
PC 40:5	6.76×10^{-2}	1.25	2.8	8.71×10^{-3}	1.33	3.3	1.35×10^{-2}	1.36	3.4	3.35
PC 40:6	1.00×10^{0}	0.07	3.6	9.75×10^{-3}	0.20	7.1	1.00×10^{0}	0.09	3.9	2.97
PC 40:8	3.12×10^{-2}	0.96	5.5	3.05×10^{-3}	1.09	7.4	1.51×10^{-2}	0.99	6.3	1.27
PE 34:2	5.80×10^{-4}	1.57	5.1	8.89×10^{-2}	1.19	5.4	4.44×10^{-1}	0.95	3.8	1.39
PE 36:2	3.78×10^{-5}	1.34	5.2	1.87×10^{-2}	1.07	5.4	2.26×10^{-1}	0.94	4.7	1.57
PE 36:3	9.31×10^{-5}	1.35	4.6	3.57×10^{-1}	0.81	3.9	1.26×10^{-1}	0.80	3.1	1.25
PE 36:4	6.92×10^{-5}	2.26	11	9.72×10^{-2}	2.06	12	3.45×10^{-2}	2.07	11	0.95
PE 36:5	1.21×10^{-2}	2.23	2.8	2.39×10^{-3}	2.33	3.6	2.07×10^{-2}	2.16	3.2	3.64
PE 38:5	6.68×10^{-1}	1.05	2.6	3.86×10^{-4}	1.64	4.9	1.35×10^{-2}	1.32	3.8	1.08
PE 38:6	3.86×10^{-4}	0.79	5.9	1.77×10^{-1}	0.42	4.8	1.63×10^{-1}	0.42	4.4	1.83
PE 38:7	9.31×10^{-5}	1.22	2.8	3.31×10^{-1}	0.67	2.1	1.38×10^{-1}	0.74	2.1	1.30
PE 40:6	3.45×10^{-2}	0.60	2.9	2.30×10^{-2}	0.50	3.1	1.16×10^{-1}	0.43	2.5	1.43
PE O-38:5	1.00×10^{0}	−0.36	1.5	1.51×10^{-2}	−0.93	2.7	1.35×10^{-2}	−0.83	2.5	2.50
PI 34:2	2.78×10^{-5}	1.81	3.5	9.72×10^{-2}	1.30	3.4	6.16×10^{-2}	1.29	3.2	4.91
PI 38:5	1.87×10^{-2}	2.84	2.1	7.57×10^{-4}	3.08	2.8	3.45×10^{-2}	2.65	2.2	2.58
PS 38:6	4.43×10^{-4}	1.70	3.3	7.13×10^{-1}	0.77	2.0	1.00×10^{0}	0.35	1.2	2.88
SM 40:1;O2	2.55×10^{-4}	1.48	4.8	5.10×10^{-1}	0.79	3.4	1.00×10^{0}	0.50	2.5	4.90
SM 41:1;O2	3.44×10^{-3}	1.20	4.3	2.70×10^{-3}	1.25	5.2	1.63×10^{-1}	0.98	3.7	1.00
SM 42:1;O2	2.78×10^{-5}	1.23	6.7	2.85×10^{-1}	0.98	6.6	1.68×10^{-2}	1.07	6.6	6.28
TG 52:2	4.43×10^{-4}	−1.35	4.1	2.85×10^{-1}	−0.71	3.6	4.44×10^{-1}	−0.69	2.7	12.6
TG 52:4	6.20×10^{-3}	0.53	4.3	1.00×10^{0}	0.23	2.3	2.08×10^{-1}	0.45	3.9	1.65
TG 56:8	2.08×10^{-1}	0.92	3.2	7.79×10^{-3}	1.12	4.2	6.76×10^{-2}	0.90	3.3	5.10
TG 56:9	3.24×10^{-5}	1.63	3.9	4.13×10^{-1}	1.00	2.6	1.38×10^{-1}	1.24	3.4	3.78

2.3.1. Hepatic Phospholipids

In the context of NAFLD, phosphatidylcholines (PCs) and phosphatidylethanolamines (PEs) have been linked to liver damage. In our study, we have noted statistically significant alterations in PCs, even though fold changes were relatively low between HFD mice and the other groups, as demonstrated in Table 1. On the contrary, all identified PEs, except for PE-O (38:5), exhibited substantial downregulation in the HFD group when compared to the ND, WL, and WLex groups. The ratio PC/PE was statistically significantly higher in livers from HFD compared to ND, WL, and WLex mice (Figure 3), a finding that is reported in the literature as an imbalance in the PC/PE ratio, which is linked to both fat accumulation in the liver and progression to NASH [20]. This is an indication of the reversing trend in fat accumulation in both WL groups, though no significantly enhanced impact of exercise was noted (WLex). Total phosphatidylinositol (PI) content was found to be decreased in the HFD group compared to ND, WL, and WLex mice. Lower levels of phosphatidylserine (PS) and specifically PS 38:6 were observed in the HFD group. Lysophosphatidylcholines (LPC) total intensities were also found to decrease in the HFD group. However, individual species such as LPC 18:0 and LPC 20:4 were upregulated in the HFD group when they were compared to ND mice. Total lysophosphatidylethanolamines (LPE) were, in addition, downregulated in the HFD group.

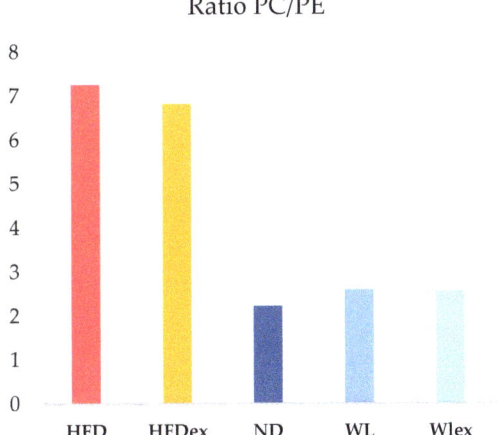

Figure 3. Hepatic PC/PE ratio in the five groups. A reversing trend is shown in the WL and WLex groups.

2.3.2. Fatty Acids Dysregulation in Hepatic Tissue

Fatty acids constitute the fundamental structural components of complex lipids and can enter the system through dietary intake, be released from visceral adipose tissue during lipolysis, or be synthesized within the liver through DNL. Fourteen (14) liver fatty acids were found statistically significant between the three binary comparisons, as presented in Table 1. More specifically, FA 16:0, FA 18:2, FA 18:3, FA 18:4, FA 20:5, FA 22:5, FA 22:6 were decreased in HFD mice compared to ND, WL, and WLex mice, while FA 16:1, FA 18:1, FA 20:1, FA 20:2, FA 20:3, FA 22:3, FA 22:4 exhibited an increase in the HFD group compared to all other groups. Considering the consistent association between hepatic steatosis and saturated, monounsaturated, and polyunsaturated fatty acids, an investigation of their levels across the studied groups was performed.

The combined levels of saturated fatty acids (SFA, FA 16:0, FA 18:0) in the WL group were found to be similar or even slightly higher in WLex mice compared to ND, whereas in HFD, they were significantly suppressed. The latter has been reported in previous studies of NAFLD animal models [21]. Similar levels are also observed in polyunsaturated fatty acids (PUFA, FA 18:2, FA 18:3, FA 18:4, FA 20:5, FA 22:5, FA 22:6, FA 20:2, FA 20:3, FA 22:3, FA 22:4) of ND, WL, and WLex, whereas their levels in HFD are decreased. However, a reverse pattern was observed in monounsaturated fatty acids (MUFA, FA 16:1, FA 18:1, FA 20:1), which were found to increase in HFD mice compared to all other groups. With regards to non-essential fatty acids (NEFA, FA 16:0, FA 16:1, FA 18:0, FA 18:1), an increase was observed in their levels in HFD groups in comparison to the ND mice and remained at higher levels even after weight loss and exercise. Contrastingly, essential fatty acids (EFA, FA 18:2, FA 18:3) exhibited a suppression in the HFD group, with a trend towards levels closer to those of the ND after weight loss and exercise.

The lipogenic index derived from the ratio of palmitic acid (FA 16:0) to the essential ω-6 linoleic acid (FA 18:2), which reflects rates of DNL in HFD animals, has shown a reversing trend in WL and WLex. A similar trend was also observed for the desaturation indices Δ6-desaturase (FA 18:2/FA 18:3) and Δ9-desaturase (FA 16:1/FA 16:0 or FA 18:1/FA 18:0), which showed a higher activity of the enzymes in HFD. Interestingly, Δ5-desaturase (FA 18:2/FA 20:4) activity seems to show a decrease in HFD. Nevertheless, the impact of weight loss or exercise seems to reverse this trend, approaching ND levels. Similar findings were observed for the elongation index (FA 18:0/FA 16:0) and the ratio of ω6/ω3 fatty acids. These findings are shown graphically in the box plots in Figure 4. In Table S4, p-values and Log2FC values are summarized for all those parameters.

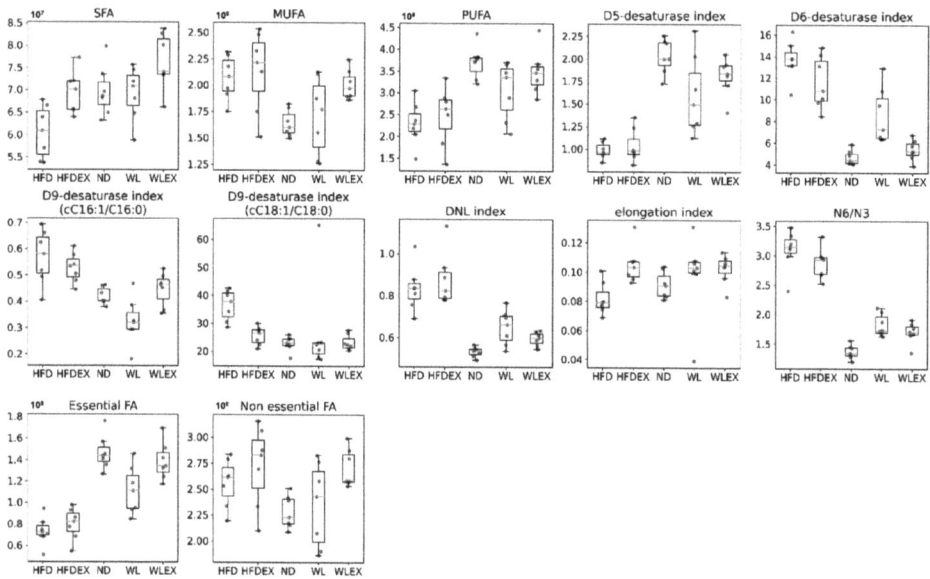

Figure 4. Boxplots showing the distribution of the SFA, MUFA, PUFA, NEFA, EFA, Δ5-desaturase, Δ6-desaturase, Δ9-desaturase, DNL, and elongation indices studied in the studied groups HFD, HFDex, ND, WL, WLex. *p*-values can be found in Table S4.

2.3.3. Glycerolipid and Sphingomyelins Dysregulation

Glycerolipids, particularly TGs, are closely linked to the transition from NAFL to NASH. This connection is partially attributed to changes in liver DNL, the rate of lipolysis, and VLDL metabolism. Four (4) TGs with bulk numbers TG(52:2), TG(52:4), TG(56:8), and TG(56:9) were found to be statistically significant between the three binary comparisons. Only TG(52:2) was elevated in the HFD group, a fact that is probably attributed to the higher monounsaturated fatty content (TG 16:1_18:0_18:1). Eight (8) diglycerides were identified as statistically significant among the four groups that were studied, namely DG(36:2), DG(36:3), DG(36:4), DG(38:2), DG(38:3), DG(38:4), DG(40:7), DG(40:8). Only three out of eight DG species, DG(36:3), DG(36:4) and DG(40:8) exhibited lower levels in HFD mice, while all others were found to increase in the HFD group, as demonstrated in Table 1. This trend seems to get reversed with the levels upturning back closer to the ND with exercise and diet. This finding is likely linked to the presence of polyunsaturated fatty acids, particularly FA 18:2 and FA 22:6, in the structure of these three DGs. Concerning SMs, all of them were found to be downregulated in the HFD group compared to ND, WL, and WLex mice.

2.4. Visceral Adipose Tissue Triglyceride Profile

Insulin resistance in visceral adipose tissue is a key factor leading to increased lipolysis and the release of non-esterified fatty acids into the bloodstream, which is considered the primary metabolic dysfunction in individuals with NAFLD [22]. In this study, the analysis of visceral adipose tissue by OPLS-DA revealed notable differences between HFD mice and those with normal diets, as well as between mice induced to a normal diet for 8 weeks with or without exercise. Similarly, in hepatic tissue, no distinct discrimination was observed between HFD and HFDex mice, nor between the WL and WLex groups, indicating the absence of an impact of the exercise protocol applied on alternating lipid levels in adipose tissue. The validation parameters of the OPLS-DA models can be found in Table S2. Variations among the studied groups were specifically detected in TGs, as they constitute the predominant lipid class in adipose tissue. In the comparison between the

HFD and ND, 20 TGs exhibited alterations. When comparing HFD to WL, 5 TGs were altered, and regarding HFD to WLex discrimination, 2 TGs showed changes. Interestingly, the majority of identified TGs exhibited lower concentration levels in HFD compared to ND, WL, and WLex mice. However, an increase in the levels of molecular species of TGs containing saturated fatty acids or/and fatty acids with a low number of double bonds was observed in HFD mice. Specifically, saturated fatty acids such as FA 16:0, FA 18:0, and FA 20:0 and monounsaturated fatty acids, including FA 16:1, FA 18:1, and FA 20:1, were the main fatty acids composed of TGs that were elevated in the HFD group. Additionally, TGs with a low number of carbons (TG 38:1, TG 38:2, TG 40:1, TG 40:3, TG 42:2) were found elevated in the weight loss group. The detailed tables with TG isomer annotations for each comparison are provided in Table S5. Table 2 summarizes all identified lipids that were found significant between HFD-ND, HFD-WL, and HFD-WLex comparisons in visceral adipose tissue along with their statistical parameters, p-values, Log2FC, VIP, and CV%, while in Figure 5, significant TGs are illustrated for the different groups in box plots.

Table 2. Visceral adipose tissue lipids were identified as significant in the group comparisons between HFD-ND, HFD-WL, and HFD-WLex. p-values, Log2FCs, QC CV values, and VIP scores are provided for each lipid after univariate Kruskal–Wallis, followed by a post hoc Bonferroni's test and multivariate analysis. Values highlighted in red correspond to $p \leq 0.05$ or $|\text{Log2FC}| \geq 1.5$.

	HFD-ND			HFD-WL			HFD-WLex			
Compounds	p-Values	Log2FC	VIP	p-Values	Log2FC	VIP	p-Values	Log2FC	VIP	CV%
TG 38:1	1.00×10^0	−0.01	0.3	7.44×10^{-3}	2.87	2.6	2.60×10^{-1}	1.19	1.0	5.13
TG 38:2	2.82×10^{-1}	1.52	0.6	5.94×10^{-4}	3.57	3.0	5.71×10^{-2}	2.00	1.3	4.75
TG 40:1	1.00×10^0	−0.01	0.2	5.17×10^{-3}	3.04	2.8	1.87×10^{-1}	1.42	0.9	5.77
TG 40:3	1.32×10^{-1}	2.01	0.7	6.79×10^{-4}	3.33	3.0	1.17×10^{-2}	2.42	1.3	3.92
TG 42:2	1.00×10^0	0.43	0.2	2.74×10^{-3}	2.93	2.8	2.27×10^{-2}	1.89	1.0	6.83
TG 50:4	2.40×10^{-4}	1.72	3.1	2.19×10^{-1}	0.91	1.6	4.14×10^{-1}	0.82	2.4	4.12
TG 50:5	4.51×10^{-5}	3.23	1.2	1.21×10^{-1}	1.95	2.6	2.60×10^{-1}	1.68	0.8	6.50
TG 51:3	5.63×10^{-4}	1.20	1.8	1.00×10^0	0.49	1.2	5.94×10^{-1}	0.55	1.4	5.03
TG 52:1	1.11×10^{-3}	−1.65	3.7	1.00×10^0	−0.26	0.9	9.44×10^{-1}	−0.56	3.6	3.97
TG 52:2	1.66×10^{-3}	−0.90	8.1	1.00×10^0	−0.17	0.8	1.00×10^0	−0.14	5.7	2.40
TG 52:4	3.20×10^{-4}	1.42	8.0	8.20×10^{-1}	0.49	1.2	7.78×10^{-1}	0.57	5.7	4.88
TG 52:5	3.29×10^{-5}	2.33	4.6	1.45×10^{-1}	1.07	1.7	3.56×10^{-1}	0.95	2.8	4.21
TG 52:6	1.14×10^{-4}	3.48	1.6	1.91×10^{-1}	1.78	2.5	3.10×10^{-1}	1.63	1.1	9.23
TG 53:4	8.51×10^{-4}	1.28	1.6	1.00×10^0	0.34	1.1	6.36×10^{-1}	0.59	1.2	4.42
TG 54:2	1.66×10^{-3}	−2.04	5.3	1.00×10^0	−0.50	1.0	1.00×10^0	−0.33	4.3	4.06
TG 54:3	3.12×10^{-2}	−1.42	8.9	1.00×10^0	−0.13	0.7	1.00×10^0	0.21	5.6	2.94
TG 54:4	2.02×10^{-5}	2.22	1.6	2.43×10^{-1}	1.09	1.7	1.11×10^{-1}	1.19	1.4	3.68
TG 54:5	1.14×10^{-4}	1.84	7.2	1.00×10^0	0.48	1.1	2.03×10^{-1}	0.80	5.2	3.41
TG 54:6	2.02×10^{-5}	2.98	4.5	1.84×10^{-1}	1.28	1.8	1.44×10^{-1}	1.32	3.0	4.10
TG 54:7	7.18×10^{-5}	4.84	1.8	2.09×10^{-1}	2.61	2.9	2.62×10^{-1}	2.44	1.1	10.2
TG 56:2	2.81×10^{-2}	−1.05	1.9	1.00×10^0	−0.06	0.6	1.00×10^0	0.07	1.1	3.19
TG 56:3	2.38×10^{-5}	2.63	2.2	1.99×10^{-1}	1.22	1.8	1.72×10^{-1}	1.24	1.6	3.54
TG 56:4	1.66×10^{-3}	0.95	1.6	6.35×10^{-1}	0.55	1.1	1.21×10^{-1}	0.70	1.9	3.21
TG 56:5	2.44×10^{-3}	1.81	1.5	8.91×10^{-1}	1.06	1.4	1.72×10^{-1}	1.48	1.9	6.52
TG 56:6	2.38×10^{-5}	2.63	2.2	1.99×10^{-1}	1.22	1.8	1.72×10^{-1}	1.24	1.6	3.54

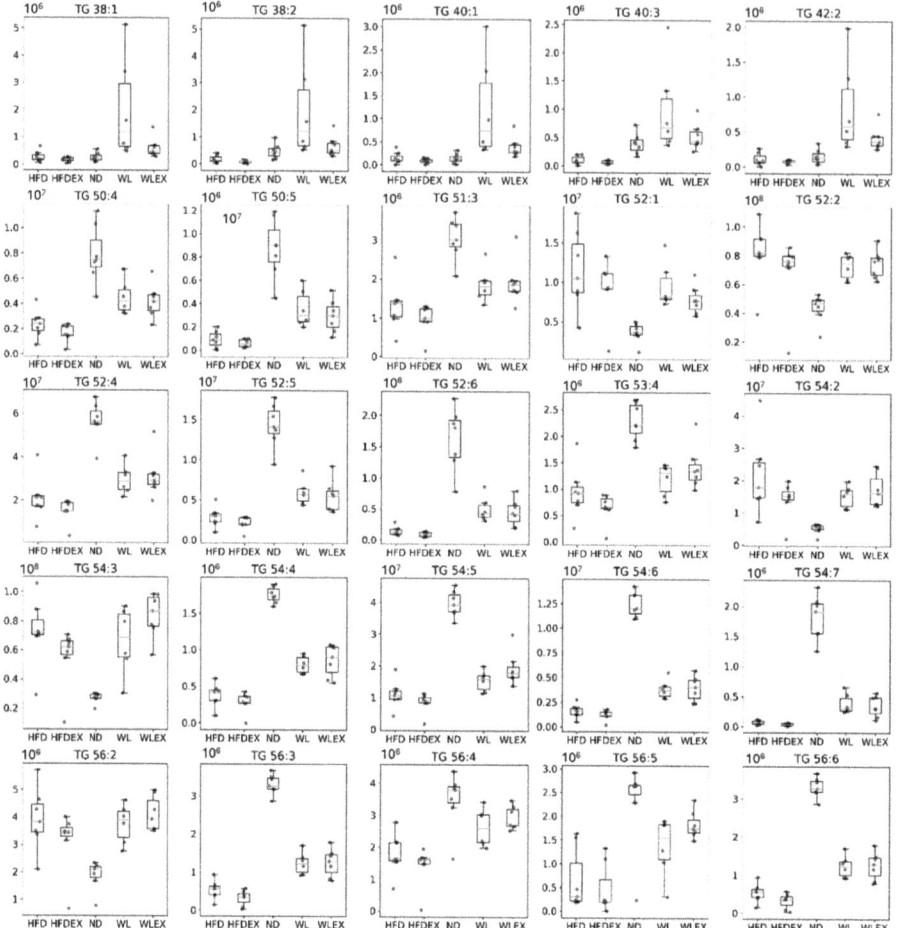

Figure 5. Box plots showing the distribution of TG intensities in the five groups HFD, HFDex, ND, WL, and WLex examined in the study. p-values can be found in Table 2.

3. Discussion

In this study, an animal model of NAFLD induced by an HF diet and 5% HFCS was studied based on its hepatic and visceral adipose tissue lipidomic phenotype. The primary focus was directed towards investigating the initial two hallmarks of disease management: the transition to a normal diet and the incorporation of exercise, specifically examining their effects on hepatic and visceral adipose tissue for the first time. Clinical parameters, such as weight loss and histological findings, including steatosis, were examined in conjunction with lipidomic analysis results. The integration of these assessments allowed for a comprehensive conclusion regarding the effectiveness of the animal model in elucidating the mechanisms underlying steatosis.

The analysis of animal body weight records showed significant changes primarily in response to dietary interventions, with exercise playing a secondary role. Both the WL and WLex groups exhibited similar results in terms of weight loss, with a noticeable difference in body weight emerging after the third week of the dietary intervention. The HFDex group demonstrated lower weight loss compared to the dietary intervention groups, suggesting that the influence of diet has a higher impact on body weight in comparison to exercise alone.

When considering the results from the untargeted lipidomic analysis, it also becomes evident that the impact of dietary modification is more influential in shaping hepatic lipid profiles than exercise. This aligns well with our expectations based on the typical NAFLD pathophysiology. As illustrated in the PCA score plots, samples are grouped based on dietary constitution independently of exercise, with samples from the ND group lying close to the WL and WLex groups, suggesting that an 8-week dietary modification, even after 12 weeks of a high-fat diet, could restore mice's hepatic tissue lipid metabolism to the basal level. When pairwise comparisons between the HFD and HFDex groups, as well as the WL and WLex groups, were considered, discernible discriminations were not identified. Despite the well-established evidence supporting the beneficial effects of exercise on NAFLD, these effects were not clearly apparent on the basis of the lipidome in the studied model.

It is important to note that in a prior study employing an animal model featuring Apolipoprotein E knockout mice exposed to an HFD [23], the authors did not observe a discernible impact of exercise intervention on various biochemical parameters, apart from hepatic transaminases. However, a notable reduction in lipid species, including TG 56:8, TG 52:4, TG 52:2, and epoxy-eicosadiene, was observed when they compared the exercise group to the matched HFD group. In our study, we observed a reduction in TG 56:8 and TG 52:4, along with higher levels of TG 52:2 as well. This alignment with our results suggests a consistent impact of exercise on specific lipid species. One plausible explanation for the observed variations may be the exercise protocol's duration and frequency, as in the mentioned study [23], which spanned over 12 weeks with 30 to 40 min swimming sessions conducted five days per week [24–26]. A very recent comprehensive review [27], encompassing 43 animal studies and 14 randomized clinical trials, systematically investigates the impact of physical activity protocols on the management of NAFLD. The review underscores the significance of standardized exercise durations, highlighting that in animal studies focused on NAFLD management, the exercise intervention ranged from 8 to 12 weeks.

PCs and PEs constitute the predominant phospholipids in mammalian cell membranes. Although no significant variations were observed in PC levels, a notable reduction in PEs was noted. Their relevance in NAFLD is acknowledged, particularly the significance of their ratio, PC/PE. This specific ratio holds crucial importance in various tissues, with both low and high PC/PE ratios correlating with elevated NAS scores in the liver [28]. The balance between PCs and PEs on lipid droplet surfaces is crucial for their dynamics. Inhibiting PC biosynthesis during triglyceride storage conditions enlarges droplets due to altered surface area-to-volume ratios. Additionally, an increased PE presence promotes the fusion of smaller droplets. Adding PC to expanding lipid droplets reduces PE abundance, preventing droplet coalescence [28]. Moreover, diminished concentrations of PCs abundant in PUFAs, specifically those containing FA 22:6, are also evident in NAFLD. The observed decline in the conversion of PE to PC likely signifies a reduction in production via the methionine cycle, impacting the synthesis of S-adenosylmethionine (SAM), which is essential for the enzymatic conversion of PE to PC [29].

The decrease in PI species indicates a connection between disrupted lipid metabolism, inflammation, and hepatic steatosis in NAFLD. In a previous study by Ščupáková [30], the importance of PI and arachidonic acid metabolism in non-steatotic tissue areas was highlighted while linking LDL and VLDL metabolism specifically to steatotic tissue.

Regarding lysophospholipids, our results show significant changes in both LPC and LPE lipid species. Interestingly, changes in LPC in the context of NAFLD/NASH might be related to LPC acyltransferase activity. In the study of Béland-Bonenfant et al., the circulating lipid profiles of 679 patients were examined, analyzing over 400 lipid species to predict hepatic fat content and NAFLD, concluding to lower LPC levels, especially C16:0 and C18:0 for NAFLD patients [29,31]. This observation aligns with our study, where we found significantly lower hepatic levels of LPC 18:0, highlighting the potential relevance of this LPC as a biomarker for NAFLD.

Elevated liver fat content in the TGs and PLs of NAFLD patients is linked to increased levels of TGs containing saturated or monounsaturated fatty acids and reduced levels of phospholipids containing PUFAs [31–34]. Interestingly, our observations revealed a slight decrease in SFAs, likely attributed to only two identified lipids within this class. We observed an increase in MUFAs and a reduction in PUFAs. The rise in palmitoleic acid (FA 16:1n7), coupled with lower stearic acid (FA 18:0), suggests enhanced $\Delta 9$ stearoyl-CoA desaturase activity, consistent with our findings. The increase in palmitoleic acid might represent an adaptive anti-inflammatory reaction to counteract the pro-inflammatory effects of fatty acid overload [29]. Increased dietary intake of ω-6 polyunsaturated fatty acids has raised the ω-6 to ω-3 ratio, contributing to NAFLD. Animal studies support a beneficial reduction in this ratio to address steatosis. Both ω-6 and ω-3 fatty acids undergo oxidation, potentially generating oxylipins through enzymatic or non-enzymatic pathways [32]. Excessive ω-6 may lead to mitochondrial dysfunction, causing cell death. In our model, the estimated ω-6/ω-3 ratio likely increased, aligning with findings in NASH patients by Puri et al. [33].

In exploring additional indicators to unravel the complexities of disrupted metabolism in NAFLD, our investigation unveiled an elevated $\Delta 6$ desaturated index and a concurrent reduction in the $\Delta 5$ desaturated index associated with the HFD. Significantly, the liver's desaturation of PUFAs is orchestrated by key enzymes, Δ-6D and Δ-5D, critical for synthesizing essential highly unsaturated FAs like 20:4, n-6, and 22:6, n-3 [35]. Crucially, our findings point to a decline in these specific FAs, mirroring the observed decrease in Δ-6D activity in our study. Importantly, emerging evidence consistently indicates a marked reduction in the activities of both Δ-6D and Δ-5D within the livers of obese NAFLD patients compared to their non-NAFLD counterparts [36]. These insights highlight potential metabolic shifts in NAFLD and underscore the central roles these enzymatic pathways play in the liver's PUFA desaturation mechanism [37].

In our examination of adipose tissue, the focus shifted to TG content. Utilizing multivariate statistical analysis on profiling data, we observed a pronounced impact of the HFD and, to a lesser extent, exercise on liver outcomes. Prolonged overnutrition and obesity induce a state of reduced metabolic flexibility in adipose tissue, impairing the efficient storage and mobilization of lipids. This metabolic inflexibility, exacerbated by a persistent surplus of circulating free fatty acids (FFAs), extends its systemic effects, influencing the liver and muscle and ultimately contributing to insulin resistance (IR) [38].

Notably, adipose tissue insulin resistance emerges as a crucial factor in driving hepatic fat accumulation and the progression of NAFLD. Insulin's principal role in adipose tissue involves suppressing lipolysis and facilitating the uptake of fatty acids. This resistance disrupts the regulatory actions of lipases, including adipose tissue triglyceride lipase (ATGL) and hormone-sensitive lipase (HSL), resulting in the release of diacylglycerol and free fatty acids [39,40]. Our findings specifically identified elevated levels of TGs composed of certain fatty acids, such as FA 16:0, FA 18:0, and FA 20:0, alongside MUFAs like FA 16:1, FA 18:1, and FA 20:1 under the HFD. SFAs, particularly 16:0, are recognized as lipotoxic lipids, significantly contributing to chronic inflammation and organ damage in both the liver and adipose tissue [40,41].

We have summarized clinical and microscopic findings, highlighting lipids' abnormalities, such as shifts in free fatty acid distribution, fluctuations in glycerolipids and sphingolipids, and a diminished PC/PE ratio associated with NAFLD. These alterations likely contribute to the mechanisms driving excessive hepatic triglyceride accumulation, linked to an increased supply of free fatty acids from peripheral adipose tissue to the liver and enhanced de novo lipid synthesis via the lipogenic pathway. This is speculated to potentially result in less liver disposal through β-oxidation and VLDL export. Irrespective of the NAFLD stage, the organism endeavors to maintain the compositional integrity of the hepatocyte membrane lipidome, and thus, triglyceride synthesis may be an adaptive, beneficial response when hepatocytes are exposed to potentially toxic triglyceride metabolites. Therefore, advancing our understanding is crucial, not only in tissue-specific mechanisms but also in the cross-talk between different tissues and its systemic impact on the regulation of lipid metabolism [42], considering the vast structural complexity of approximately 40,000 distinct lipids identified to date [43]. Nevertheless, the investigation was limited to specific variations such as the type of high-fat diet and exercise as well as training durations in animal models. These differences in the experimental design of the various studies could influence the pathogenesis and mitigation of NAFLD and consequently be reflected in the related findings [24,27].

4. Materials and Methods

4.1. Chemicals and Materials

Methanol (MeOH), acetonitrile (ACN), methyl-tert-butyl-ether (MTBE; ≥99%), chloroform ($CHCl_3$), and formic acid (all ULC/MS-CC/SFC grade) were obtained from CHEM-LAB NV (Zedelgem, Belgium). Isopropanol (IPA) was purchased from Fisher Scientific (International Inc., Hampton, NH, USA). Ammonium formate (NH_4HCO_2; MS grade) was obtained from Sigma-Aldrich (Merck, Darmstadt, Germany). Deionized water (ddH_2O) was ultrapurified by a Millipore (Bedford, MA, USA) instrument delivering water quality with a resistivity of ≥ 18.2 MΩ·cm.

4.2. Animal Study

For this study, male mice of the C57BL/6 strain, obtained from the Hellenic Pasteur Institute, were utilized. These animals were housed within the EZEFIS animal facility (Department of Pharmacology, Medical School, NKUA), where the environmental conditions were strictly controlled to maintain a temperature of 20–22 °C, continuous air renewal, and a 12 h light-dark cycle. The mice were provided with unrestricted access to both food and water supplies (ad libitum). Initially, the animal model included two groups. The first group (n = 7) was provided with a normal diet composed 10% kcal from fat (D12450B Research Diets, New Brunswick, NJ, USA) and had access to tap water for a period of 12 weeks. This diet was referred to as a normal diet (ND). The second group (n = 28) consumed a high-fat diet containing 60% kcal from fat (D12492 Research Diets, New Brunswick, NJ, USA) and was supplied with water containing 5% HFCS (Best Flavors, Orange, CA, USA) for 12 weeks. The combination of the high-fat diet and the addition of 5% HFCS to the water was classified as a high-fat diet (HFD). After the end of 12 weeks, the mice previously subjected to HFD were divided randomly into four subgroups: Group a, mice that continued on the same HFD (n = 7); Group b, mice that continued to consume the same HFD and underwent weekly sessions of supervised aerobic exercise (HFD exercise, HFDex, n = 7); Group c, mice that changed diet to ND without the addition of aerobic exercise (Weight Loss, WL, n = 7); and Group d, mice that switched diet to ND and simultaneously placed in weekly sessions of supervised aerobic exercise (Weight Loss exercise, WLex, n = 7). The four subgroups, along with the original ND group, were maintained on their respective diets for an additional eight-week period, and they continued to consume food as previously described. The mice were weighed to monitor their progress every week. In Figure 6, a schematic illustration of the experimental design, including the respective timelines, is provided. The HFDex and WLex feeding groups underwent intense aerobic exercise training, consisting of three 30 min running exercise sessions on a weekly basis, on a treadmill specialized for studies pertinent to exercise and metabolism (PanLab LE8700 Treadmill). Animals were acclimatized to trial exercise sessions on the treadmill. For experiments, mice were urged to achieve a running speed of 20 cm/s. Once the feeding period was complete, the animals were euthanized upon systemic perfusion with phosphate-buffered saline (PBS). Blood, as well as liver and visceral adipose tissues (VAT), were isolated. Animal experiments performed for this study were in accordance with the regulations of the European Union, and the protocol was approved by the Region of Attica, Greece.

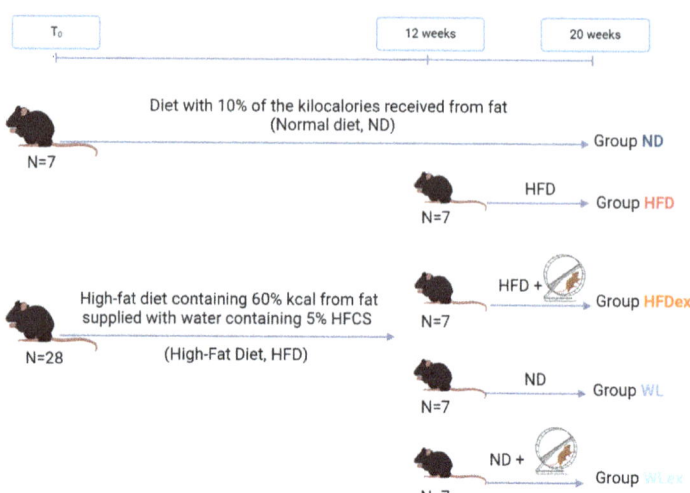

Figure 6. Graphical illustration of the NAFLD mouse model induced by a high-fat diet in combination with 5% HCFS. In the 12 weeks of the experiment, the HFD group was divided into four subgroups: HFD, HFDex, WL, and WLex.

4.3. Lipidomic Analysis

4.3.1. Liver and Visceral Adipose Tissue Extraction

Liver tissues were transferred to 2.0 mL Eppendorf tubes containing 1.0 mm ceramic beads. An organic solvent mixture of MTBE-MeOH 3:1 (v/v) was added to the weighed tissue proportionally to weight and up to 1200 µL for the maximum weight. The exact extraction solvent's volumes are provided in Table S6. Homogenization, followed by performing 4 cycles with a 30 s duration and speed set at 6.00 m/s using a Bead mill Homogenizer (BEAD RUPTOR ELITE, Omni International, Kennesaw, Georgia). The mixture was then centrifuged at 4 °C for 30 min at 10,000 rpm. One hundred and fifty (150) µL of the supernatant were transferred to a 1.5 mL Eppendorf tube and evaporated to dryness in vacuo (SpeedVac, Eppendorf Austria GmbH, Wien, Austria), followed by reconstitution with 150 µL of H_2O-ACN-IPA in a 1:1:3 (v/v) ratio.

For visceral adipose tissue lipid extraction, a modified Folch protocol was used. The weighted tissue was transferred to a 2.0 mL Eppendorf tube containing 1.0 mm ceramic beads. The volumes of the extraction solvents were adjusted proportionally to the weighted tissue, as described in Table S6. In 10 mg of adipose tissue, 200 µL of MeOH were added. Homogenization was followed by four cycles of 30 s at 6.00 m/s speed, and the homogenates were transferred to new 2.0 mL Eppendorf tubes, where 640 µL of $CHCl_3$-MeOH 7:1 (v/v) were added. After 30 min of vortexing at room temperature, 360 µL of $CHCl_3$ were added and vortexed for another 30 min. One hundred eighty (180) µL of H_2O were added to enhance phase separation, and the samples were centrifuged for 30 min at 10,000 rpm at 4 °C. Two hundred (200) µL of the lower phase were collected in a new 1.5 mL Eppendorf tube. The extraction process was repeated, and the organic phases were combined and evaporated to dryness, as described above. The dry residues were reconstituted in 200 µL of IPA and diluted 40 times with the same solvent.

For the quality control of the analyses, a pooled sample (Quality Control Sample, QC) was prepared by mixing equal volumes of each supernatant. Phenotypic QC samples were prepared for each of the five groups by mixing equal volumes of the supernatants from the samples of the same group. Diluted QCs (1:2, 1:4, 1:6, 1:8) were also prepared for the evaluation of the dilution integrity of the detected features.

4.3.2. Analytical Instrumentation and Conditions

A UHPLC Elute system equipped with an Elute autosampler operating at 8 °C was used. The separation was performed with an Acquity UPLC CSH C18, 2.1 × 100 mm, 1.7 μm column (Waters Ltd., Elstree, UK) equipped with a pre-column Acquity UPLC CSH C18Van-Guard (Waters Ltd., Elstree, UK), maintained at 55 °C. A 20 min gradient mobile phase system was employed using an HPLC binary solvent manager. The mobile phase A consisted of ACN/H_2O (60:40), 10 mM ammonium formate, and 0.1% formic acid, and the mobile phase B consisted of IPA/ACN (90:10) and 0.1% formic acid. The gradient was as follows: 60–57% A (0.0–2.0 min), 57–50% A (2.0–2.1 min; curve 1), 50–46% A (2.1–12.0 min), 46–30% A (12.0–12.1 min; curve 1), 30–1% A (12.1–18 min), 1–60% A (18.0–18.1 min), and 60% A (18.1–20.0 min). The flow rate was set at 0.4 mL/min. The injection volume was set at 5 μL in positive mode and at 10 μL in negative mode. The needle was initially washed with the strong wash solvent IPA/ACN (90:10) (1000 μL), followed by the weak wash solvent ACN/H_2O (60/40) (1000 μL) before and after each injection.

The MS data were acquired using a TIMS TOF mass spectrometer (Bruker, Billerica, MA, USA) in positive and negative ionization modes in liver tissue and only in positive mode for adipose tissue, performing data-dependent acquisition (DDA) for MS/MS analyses. The settings in ESI were as follows: capillary ±4.2 kV, dry temperature 200 °C, dry gas 10 L/min, and nebulizer gas 2 Bar. Auto MS/MS was applied using dynamic MS/MS spectra acquisition with 6 and 10 Hz as minimum and maximum spectra rates, respectively. Collision energy was set at 20 V for precursor ions below 100 m/z, 30 V for precursor ions with m/z ranging from 100 to 1000, and 40 V for precursor ions with m/z ranging from 1000 to 2000 m/z. Calibrant (sodium formate, 10 mM) was infused into MS at a 10 μL/h flow rate in the first 0.2 min of each analysis.

4.3.3. Data Analysis

Raw data from TIMS-TOF were recalibrated using sodium formate clusters by data analysis (version 5.3, Bruker, Bremen, Germany) and converted to mzML by MSConvert (ProteoWizard 3.0.11567). Retention time alignment and feature grouping are performed by XCMS (version 3.2.0) in R programming prior to chromatographic peak detection. Variables containing empty/zero/missing values >50% in each sample group and those with QC CV values >30% were removed. Raw data were normalized using QC samples [44]. SIMCA 13.0.3 (UMETRICS AB Sweden) software was used for unsupervised principal component analysis (PCA), and the data were further processed by partial and orthogonal-partial least squares discriminant analysis (PLS, OPLS-DA). "S-plot" was used, applying absolute p and p(corr) cutoff values of >|0.05| and |0.5|, respectively, for identifying significant features. Parameters that demonstrate the quality of the models, including goodness of fit in the X (R2X) and Y (R2Y) variables and predictability (Q2YCV), were determined by permutation and CV ANOVA analysis. Univariate statistical analysis was performed in the Python programming language to assess the differences between the study groups. The Kruskal–Wallis test was performed, followed by post hoc Bonferroni's test for multiple comparisons, and statistical significance was defined as a value of $p \leq 0.05$. The analysis of body weights was conducted using GraphPad Prism v8.0.1 software.

4.3.4. Lipids' Annotation

Identification of the statistically significant lipid species was performed in Lipostar2 (version 2.0.2, Molecular Discovery Ltd., Hertfordshire, UK) equipped with the LIPID MAPS structure database (version September 2021) [45]. The raw files were imported directly and aligned using the default settings. Automatic peak picking was performed with the Savitzky–Golay algorithm using the following parameters: window size set to 7, degree to 2, multi-pass iterations to 1, and minimum S/N ratio of 3. Mass tolerance settings were set to 10 ppm with an RT tolerance of 0.2 min. Filters "Retain lipids with isotopic pattern" and "Retain lipids with MS/MS" were applied to keep only features with isotopic patterns and MS/MS spectra for identification. The following parameters were used for lipid identification: 5 ppm precursor ion mass tolerance and 20 ppm product ion mass tolerance. The automatic approval was performed to keep structures with a quality of 3–4 stars.

5. Conclusions

In the present study, an untargeted lipidomic analysis was used to investigate the effect of weight loss, exercise, and their combination on NAFLD, both in the liver and visceral adipose tissue, for the first time. The results showed disturbed lipid metabolism and variations in important classes such as glycerolipids, glycerophospholipids, and sphingolipids in the HFD animals, which seem to alleviate mainly upon dietary intervention. This finding was not observed with the effects of physical exercise alone. Thus, nutritional intervention has been proven based on the observed lipidomic phenotype as the strongest differentiating factor between the groups, which is in agreement with the histological findings where a remission of NAFLD is observed in these cases.

Supplementary Materials: The following supporting information can be downloaded at: https://www.mdpi.com/article/10.3390/molecules29071494/s1, Figures S1–S3. PCA score plots of mice hepatic and adipose tissue samples of the five studied groups and QC samples. All groups (HFD, HFDex, ND, WL, WLex) are illustrated with grey color while QC samples are presented with purple color and clustered together. Figure S4. PLS score plot of mice adipose tissue samples of the five studied groups. HFD and HFDex groups are clustered together on the negative part of the y-axis, ND group is in the center of the eclipse, while WL and WLex mice are grouped together on the positive side of the same axis; Table S1: Comparisons of the mice body weights among the groups HFD-HFDex, HFD-WL, HFD-WLex, ND-WL, ND-WLex during diet and exercise innervations. p value < 0.05 were indicated as significant and highlighted; Table S2: Characteristics of the constructed unsupervised and supervised models, log transformation and pareto (PAR) scale were used in all models in Liver -ESI and adipose tissue +ESI, while only pareto (PAR) scale was used in Liver +ESI models; Table S3: Summary of all statistically identified lipids in the liver of mice for the three binary comparisons, HFD-ND, HFD-WL, HFD-WLex. Information is provided regarding the structural formula and fatty acid chains of the lipids, molecular structure, monoisotopic mass, detected derivatives, retention time, and mass accuracy; Table S4: P values and Log2FC for all sums of fatty acids and indices studied in the comparisons HFD-ND, HFD-WL, HFD-WLex; Table S5: Summary of all statistically significant identified lipids in adipose tissue of mice for the three binary comparisons, HFD-ND, HFD-WL, HFD-WLex. Information is provided regarding the structural formula and fatty acid chains of the lipids, molecular structure, monoisotopic mass, detected derivatives, retention time, and mass accuracy; Table S6: Description of the extractions performed on the mice liver and adipose tissue samples.

Author Contributions: Conceptualization, A.C. and H.G.; methodology, T.M., G.P., A.C. and H.G.; animal experiments, G.P. and A.C.; visualization, T.M. and G.P.; investigation, T.M. and G.P.; data curation, T.M.; G.P. and O.D.; writing—original draft preparation, T.M. and O.D.; writing—review and editing, H.G., G.P., A.C. and O.D.; supervision, H.G. and A.C.; project administration, H.G. and A.C. All authors have read and agreed to the published version of the manuscript.

Funding: This research received no external funding.

Institutional Review Board Statement: Mouse experiments were performed according to a protocol approved by the Region of Attica, Greece (5847, approved 1 November 2018).

Informed Consent Statement: Not applicable.

Data Availability Statement: Data are contained within the article and Supplementary Materials.

Acknowledgments: We would like to thank Theodoros Liapikos for creating the box plots.

Conflicts of Interest: The authors declare no conflicts of interest.

References

1. Lazarus, J.V.; Mark, H.E.; Anstee, Q.M.; Arab, J.P.; Batterham, R.L.; Castera, L.; Cortez-Pinto, H.; Crespo, J.; Cusi, K.; Dirac, M.A.; et al. Advancing the Global Public Health Agenda for NAFLD: A Consensus Statement. *Nat. Rev. Gastroenterol. Hepatol.* **2022**, *19*, 60–78. [CrossRef] [PubMed]
2. Huby, T.; Gautier, E.L. Immune Cell-Mediated Features of Non-Alcoholic Steatohepatitis. *Nat. Rev. Immunol.* **2022**, *22*, 429–443. [CrossRef] [PubMed]
3. Han, S.K.; Baik, S.K.; Kim, M.Y. Non-Alcoholic Fatty Liver Disease: Definition and Subtypes. *Clin. Mol. Hepatol.* **2023**, *29*, S5–S16. [CrossRef] [PubMed]
4. Nseir, W.; Hellou, E.; Assy, N. Role of Diet and Lifestyle Changes in Nonalcoholic Fatty Liver Disease. *World J. Gastroenterol.* **2014**, *20*, 9338–9344. [PubMed]
5. van der Windt, D.J.; Sud, V.; Zhang, H.; Tsung, A.; Huang, H. The Effects of Physical Exercise on Fatty Liver Disease. *Gene Expr.* **2018**, *18*, 89–101. [CrossRef] [PubMed]
6. Zarghamravanbakhsh, P.; Frenkel, M.; Poretsky, L. Metabolic Causes and Consequences of Nonalcoholic Fatty Liver Disease (NAFLD). *Metab. Open* **2021**, *12*, 100149. [CrossRef] [PubMed]
7. Martin, A.; Lang, S.; Goeser, T.; Demir, M.; Steffen, H.-M.; Kasper, P. Management of Dyslipidemia in Patients with Non-Alcoholic Fatty Liver Disease. *Curr. Atheroscler. Rep.* **2022**, *24*, 533–546. [CrossRef] [PubMed]
8. Lee, C.; Lui, D.T.; Lam, K.S. Non-alcoholic Fatty Liver Disease and Type 2 Diabetes: An Update. *J. Diabetes Investig.* **2022**, *13*, 930–940. [CrossRef] [PubMed]
9. Francque, S.M.A.; Dirinck, E. NAFLD Prevalence and Severity in Overweight and Obese Populations. *Lancet Gastroenterol. Hepatol.* **2023**, *8*, 2–3. [CrossRef] [PubMed]
10. Hsu, C.L.; Loomba, R. From NAFLD to MASLD: Implications of the New Nomenclature for Preclinical and Clinical Research. *Nat. Metab.* **2024**, 1–3. [CrossRef]
11. Eng, J.M.; Estall, J.L. Diet-Induced Models of Non-Alcoholic Fatty Liver Disease: Food for Thought on Sugar, Fat, and Cholesterol. *Cells* **2021**, *10*, 1805. [CrossRef] [PubMed]
12. Sigala, D.M.; Hieronimus, B.; Medici, V.; Lee, V.; Nunez, M.V.; Bremer, A.A.; Cox, C.L.; Price, C.A.; Benyam, Y.; Chaudhari, A.J.; et al. Consuming Sucrose- or HFCS-Sweetened Beverages Increases Hepatic Lipid and Decreases Insulin Sensitivity in Adults. *J. Clin. Endocrinol. Metab.* **2021**, *106*, 3248–3264. [CrossRef]
13. Jensen, T.; Abdelmalek, M.F.; Sullivan, S.; Nadeau, K.J.; Green, M.; Roncal, C.; Nakagawa, T.; Kuwabara, M.; Sato, Y.; Kang, D.-H.; et al. Fructose and Sugar: A Major Mediator of Nonalcoholic Fatty Liver Disease. *J. Hepatol.* **2018**, *68*, 1063–1075. [CrossRef] [PubMed]
14. Papadopoulos, G.; Legaki, A.-I.; Georgila, K.; Vorkas, P.; Giannousi, E.; Stamatakis, G.; Moustakas, I.I.; Petrocheilou, M.; Pyrina, I.; Gercken, B.; et al. Integrated Omics Analysis for Characterization of the Contribution of High Fructose Corn Syrup to Non-Alcoholic Fatty Liver Disease in Obesity. *Metabolism* **2023**, *144*, 155552. [CrossRef] [PubMed]
15. Finer, N. Weight Loss Interventions and Nonalcoholic Fatty Liver Disease: Optimizing Liver Outcomes. *Diabetes Obes. Metab.* **2022**, *24*, 44–54. [CrossRef]
16. Wong, V.W.-S.; Wong, G.L.-H.; Chan, R.S.-M.; Shu, S.S.-T.; Cheung, B.H.-K.; Li, L.S.; Chim, A.M.-L.; Chan, C.K.-M.; Leung, J.K.-Y.; Chu, W.C.-W.; et al. Beneficial Effects of Lifestyle Intervention in Non-Obese Patients with Non-Alcoholic Fatty Liver Disease. *J. Hepatol.* **2018**, *69*, 1349–1356. [CrossRef] [PubMed]
17. Vilar-Gomez, E.; Martinez-Perez, Y.; Calzadilla-Bertot, L.; Torres-Gonzalez, A.; Gra-Oramas, B.; Gonzalez-Fabian, L.; Friedman, S.L.; Diago, M.; Romero-Gomez, M. Weight Loss Through Lifestyle Modification Significantly Reduces Features of Nonalcoholic Steatohepatitis. *Gastroenterology* **2015**, *149*, 367–378.e5, quiz e14–15. [CrossRef] [PubMed]
18. Suzuki, A.; Lindor, K.; St Saver, J.; Lymp, J.; Mendes, F.; Muto, A.; Okada, T.; Angulo, P. Effect of Changes on Body Weight and Lifestyle in Nonalcoholic Fatty Liver Disease. *J. Hepatol.* **2005**, *43*, 1060–1066. [CrossRef] [PubMed]
19. Hickman, I.J.; Jonsson, J.R.; Prins, J.B.; Ash, S.; Purdie, D.M.; Clouston, A.D.; Powell, E.E. Modest Weight Loss and Physical Activity in Overweight Patients with Chronic Liver Disease Results in Sustained Improvements in Alanine Aminotransferase, Fasting Insulin, and Quality of Life. *Gut* **2004**, *53*, 413–419. [CrossRef] [PubMed]
20. Shama, S.; Jang, H.; Wang, X.; Zhang, Y.; Shahin, N.N.; Motawi, T.K.; Kim, S.; Gawrieh, S.; Liu, W. Phosphatidylethanolamines Are Associated with Nonalcoholic Fatty Liver Disease (NAFLD) in Obese Adults and Induce Liver Cell Metabolic Perturbations and Hepatic Stellate Cell Activation. *Int. J. Mol. Sci.* **2023**, *24*, 1034. [CrossRef] [PubMed]
21. Knebel, B.; Fahlbusch, P.; Dille, M.; Wahlers, N.; Hartwig, S.; Jacob, S.; Kettel, U.; Schiller, M.; Herebian, D.; Koellmer, C.; et al. Fatty Liver Due to Increased de Novo Lipogenesis: Alterations in the Hepatic Peroxisomal Proteome. *Front. Cell Dev. Biol.* **2019**, *7*, 248. [CrossRef] [PubMed]
22. Ziolkowska, S.; Binienda, A.; Jabłkowski, M.; Szemraj, J.; Czarny, P. The Interplay between Insulin Resistance, Inflammation, Oxidative Stress, Base Excision Repair and Metabolic Syndrome in Nonalcoholic Fatty Liver Disease. *Int. J. Mol. Sci.* **2021**, *22*, 11128. [CrossRef] [PubMed]
23. Huang, W.-C.; Xu, J.-W.; Li, S.; Ng, X.E.; Tung, Y.-T. Effects of Exercise on High-Fat Diet-Induced Non-Alcoholic Fatty Liver Disease and Lipid Metabolism in ApoE Knockout Mice. *Nutr. Metab.* **2022**, *19*, 10. [CrossRef]

24. Poole, D.C.; Copp, S.W.; Colburn, T.D.; Craig, J.C.; Allen, D.L.; Sturek, M.; O'Leary, D.S.; Zucker, I.H.; Musch, T.I. Guidelines for Animal Exercise and Training Protocols for Cardiovascular Studies. *Am. J. Physiol. Heart Circ. Physiol.* **2020**, *318*, H1100–H1138. [CrossRef] [PubMed]
25. Adamovich, Y.; Ezagouri, S.; Dandavate, V.; Asher, G. Monitoring Daytime Differences in Moderate Intensity Exercise Capacity Using Treadmill Test and Muscle Dissection. *STAR Protoc.* **2021**, *2*, 100331. [CrossRef] [PubMed]
26. Hastings, M.H.; Herrera, J.J.; Guseh, J.S.; Atlason, B.; Houstis, N.E.; Abdul Kadir, A.; Li, H.; Sheffield, C.; Singh, A.P.; Roh, J.D.; et al. Animal Models of Exercise from Rodents to Pythons. *Circ. Res.* **2022**, *130*, 1994–2014. [CrossRef] [PubMed]
27. PubMed. Physical Activity Protocols in Non-Alcoholic Fatty Liver Disease Management: A Systematic Review of Randomized Clinical Trials and Animal Models. Available online: https://pubmed.ncbi.nlm.nih.gov/37510432/ (accessed on 19 November 2023).
28. van der Veen, J.N.; Kennelly, J.P.; Wan, S.; Vance, J.E.; Vance, D.E.; Jacobs, R.L. The Critical Role of Phosphatidylcholine and Phosphatidylethanolamine Metabolism in Health and Disease. *Biochim. Biophys. Acta BBA Biomembr.* **2017**, *1859*, 1558–1572. [CrossRef]
29. Béland-Bonenfant, S.; Rouland, A.; Petit, J.-M.; Vergès, B. Concise Review of Lipidomics in Nonalcoholic Fatty Liver Disease. *Diabetes Metab.* **2023**, *49*, 101432. [CrossRef] [PubMed]
30. Ščupáková, K.; Soons, Z.; Ertaylan, G.; Pierzchalski, K.A.; Eijkel, G.B.; Ellis, S.R.; Greve, J.W.; Driessen, A.; Verheij, J.; De Kok, T.M.; et al. Spatial Systems Lipidomics Reveals Nonalcoholic Fatty Liver Disease Heterogeneity. *Anal. Chem.* **2018**, *90*, 5130–5138. [CrossRef]
31. Orešič, M.; Hyötyläinen, T.; Kotronen, A.; Gopalacharyulu, P.; Nygren, H.; Arola, J.; Castillo, S.; Mattila, I.; Hakkarainen, A.; Borra, R.J.H.; et al. Prediction of Non-Alcoholic Fatty-Liver Disease and Liver Fat Content by Serum Molecular Lipids. *Diabetologia* **2013**, *56*, 2266–2274. [CrossRef]
32. Masoodi, M.; Gastaldelli, A.; Hyötyläinen, T.; Arretxe, E.; Alonso, C.; Gaggini, M.; Brosnan, J.; Anstee, Q.M.; Millet, O.; Ortiz, P.; et al. Metabolomics and Lipidomics in NAFLD: Biomarkers and Non-Invasive Diagnostic Tests. *Nat. Rev. Gastroenterol. Hepatol.* **2021**, *18*, 835–856. [CrossRef]
33. Puri, P.; Wiest, M.M.; Cheung, O.; Mirshahi, F.; Sargeant, C.; Min, H.-K.; Contos, M.J.; Sterling, R.K.; Fuchs, M.; Zhou, H.; et al. The Plasma Lipidomic Signature of Nonalcoholic Steatohepatitis. *Hepatology* **2009**, *50*, 1827–1838. [CrossRef] [PubMed]
34. Gorden, D.L.; Myers, D.S.; Ivanova, P.T.; Fahy, E.; Maurya, M.R.; Gupta, S.; Min, J.; Spann, N.J.; McDonald, J.G.; Kelly, S.L.; et al. Biomarkers of NAFLD Progression: A Lipidomics Approach to an Epidemic1[S]. *J. Lipid Res.* **2015**, *56*, 722–736. [CrossRef] [PubMed]
35. Catalá, A. Five Decades with Polyunsaturated Fatty Acids: Chemical Synthesis, Enzymatic Formation, Lipid Peroxidation and Its Biological Effects. *J. Lipids* **2013**, *2013*, e710290. [CrossRef] [PubMed]
36. Araya, J.; Rodrigo, R.; Pettinelli, P.; Araya, A.V.; Poniachik, J.; Videla, L.A. Decreased Liver Fatty Acid Delta-6 and Delta-5 Desaturase Activity in Obese Patients. *Obesity* **2010**, *18*, 1460–1463. [CrossRef]
37. Mäkelä, T.N.K.; Tuomainen, T.-P.; Hantunen, S.; Virtanen, J.K. Associations of Serum N-3 and n-6 Polyunsaturated Fatty Acids with Prevalence and Incidence of Nonalcoholic Fatty Liver Disease. *Am. J. Clin. Nutr.* **2022**, *116*, 759–770. [CrossRef] [PubMed]
38. Fuchs, A.; Samovski, D.; Smith, G.I.; Cifarelli, V.; Farabi, S.S.; Yoshino, J.; Pietka, T.; Chang, S.-W.; Ghosh, S.; Myckatyn, T.M.; et al. Associations Among Adipose Tissue Immunology, Inflammation, Exosomes and Insulin Sensitivity in People with Obesity and Nonalcoholic Fatty Liver Disease. *Gastroenterology* **2021**, *161*, 968–981.e12. [CrossRef]
39. Guerra, S.; Mocciaro, G.; Gastaldelli, A. Adipose Tissue Insulin Resistance and Lipidome Alterations as the Characterizing Factors of Non-alcoholic Steatohepatitis. *Eur. J. Clin. Investig.* **2022**, *52*, e13695. [CrossRef] [PubMed]
40. Rosso, C.; Kazankov, K.; Younes, R.; Esmaili, S.; Marietti, M.; Sacco, M.; Carli, F.; Gaggini, M.; Salomone, F.; Møller, H.J.; et al. Crosstalk between Adipose Tissue Insulin Resistance and Liver Macrophages in Non-Alcoholic Fatty Liver Disease. *J. Hepatol.* **2019**, *71*, 1012–1021. [CrossRef]
41. Lange, M.; Angelidou, G.; Ni, Z.; Criscuolo, A.; Schiller, J.; Blüher, M.; Fedorova, M. AdipoAtlas: A Reference Lipidome for Human White Adipose Tissue. *Cell Rep. Med.* **2021**, *2*, 100407. [CrossRef]
42. Cockcroft, S. Mammalian Lipids: Structure, Synthesis and Function. *Essays Biochem.* **2021**, *65*, 813–845. [CrossRef] [PubMed]
43. Balakrishnan, M.; Patel, P.; Dunn-Valadez, S.; Dao, C.; Khan, V.; Ali, H.; El-Serag, L.; Hernaez, R.; Sisson, A.; Thrift, A.P.; et al. Women Have Lower Risk of Nonalcoholic Fatty Liver Disease but Higher Risk of Progression vs Men: A Systematic Review and Meta-Analysis. *Clin. Gastroenterol. Hepatol. Off. Clin. Pract. J. Am. Gastroenterol. Assoc.* **2021**, *19*, 61. [CrossRef] [PubMed]
44. Sánchez-Illana, Á.; Piñeiro-Ramos, J.D.; Sanjuan-Herráez, J.D.; Vento, M.; Quintás, G.; Kuligowski, J. Evaluation of Batch Effect Elimination Using Quality Control Replicates in LC-MS Metabolite Profiling. *Anal. Chim. Acta* **2018**, *1019*, 38–48. [CrossRef] [PubMed]
45. Goracci, L.; Tortorella, S.; Tiberi, P.; Pellegrino, R.M.; Di Veroli, A.; Valeri, A.; Cruciani, G. Lipostar, a Comprehensive Platform-Neutral Cheminformatics Tool for Lipidomics. *Anal. Chem.* **2017**, *89*, 6257–6264. [CrossRef] [PubMed]

Disclaimer/Publisher's Note: The statements, opinions and data contained in all publications are solely those of the individual author(s) and contributor(s) and not of MDPI and/or the editor(s). MDPI and/or the editor(s) disclaim responsibility for any injury to people or property resulting from any ideas, methods, instructions or products referred to in the content.

Article

Detection of 26 Drugs of Abuse and Metabolites in Quantitative Dried Blood Spots by Liquid Chromatography–Mass Spectrometry

Thomas Meikopoulos [1,2], Helen Gika [2,3], Georgios Theodoridis [1,2,4] and Olga Begou [2,4,*]

1. Laboratory of Analytical Chemistry, Department of Chemistry, Aristotle University of Thessaloniki, 54124 Thessaloniki, Greece; meikthom@chem.auth.gr (T.M.); gtheodor@chem.auth.gr (G.T.)
2. BIOMIC_Auth, Center for Interdisciplinary Research, and Innovation (CIRI-AUTH), 57001 Thessaloniki, Greece; gkikae@auth.gr
3. Laboratory of Forensic Medicine & Toxicology, School of Medicine, Aristotle University of Thessaloniki, 54124 Thessaloniki, Greece
4. ThetaBiomarkers, Center for Interdisciplinary Research, and Innovation (CIRI-AUTH), Balkan Center, 10th Km Thessaloniki-Thermi Rd., P.O. Box 8318, 57001 Thessaloniki, Greece
* Correspondence: olgabegou@thetabiomarkers.com

Citation: Meikopoulos, T.; Gika, H.; Theodoridis, G.; Begou, O. Detection of 26 Drugs of Abuse and Metabolites in Quantitative Dried Blood Spots by Liquid Chromatography–Mass Spectrometry. *Molecules* **2024**, *29*, 975. https://doi.org/10.3390/molecules29050975

Academic Editors: Rosa Herráez Hernández and James Barker

Received: 23 January 2024
Revised: 11 February 2024
Accepted: 16 February 2024
Published: 23 February 2024

Copyright: © 2024 by the authors. Licensee MDPI, Basel, Switzerland. This article is an open access article distributed under the terms and conditions of the Creative Commons Attribution (CC BY) license (https://creativecommons.org/licenses/by/4.0/).

Abstract: A method was developed for the determination of 26 drugs of abuse from different classes, including illicit drugs in quantitative dried blood spots (qDBSs), with the aim to provide a convenient method for drug testing by using only 10 μL of capillary blood. A satisfactory limit of quantification (LOQ) of 2.5 ng/mL for 9 of the compounds and 5 ng/mL for 17 of the compounds and a limit of detection (LOD) of 0.75 ng/mL for 9 of the compounds and 1.5 ng/mL for 17 of the compounds were achieved for all analytes. Reversed-phase liquid chromatography was applied on a C18 column coupled to MS, providing selective detections with both +ESI and -ESI modes. Extraction from the qDBS was performed using AcN-MeOH, 1:1 (v/v), with recovery ranging from 84.6% to 106%, while no significant effect of the hematocrit was observed. The studied drugs of abuse were found to be stable over five days under three different storage conditions (at ambient temperature 21 °C, at −20 °C, and at 35 °C), thus offering a highly attractive approach for drug screening by minimally invasive sampling for individuals that could find application in forensic toxicology analysis.

Keywords: LC–MS/MS analysis; drugs of abuse; quantitative dried blood spot; blood micro sampling; drug screening

1. Introduction

Dried blood spots (DBSs) represent a minimally invasive sample collection approach that can find applications in various fields. From its introduction in bioanalysis a century ago [1–4], applications have been increasing over the years, including newborn screening, analysis of small molecules, DNA, proteins for diagnostics [5], therapeutic drug monitoring [6], preclinical drug development, and forensic toxicology [7] among others. The cost-effectiveness, facilitated storage, and shipment conditions of the DBSs in combination with decreased biohazard risk support their potential in the area of toxicological analysis.

Whole blood, plasma, and urine are the typical specimens used for drug analysis in forensic toxicology [8–11]. However, shipment, sample storage, and handling of these types of biological samples may bring limitations. DBSs offer numerous benefits, the most important being the ability to collect blood without venipuncture. Reduced invasiveness and low biohazard risk, for instance, for HIV or other infectious pathogens, during sample shipment, are two additional advantages of DBS sampling [12,13]. Moreover, many reports have already focused on the enhanced stability of the analytes in these types of samples at room temperature, without the need for refrigeration [14,15].

A crucial challenge, however, is the need for highly sensitive instrumentation and methodologies capable of detecting low levels of drugs in such small blood volumes collected in a DBS. With regard to quantification, the accurate and reproducible collection of blood is often difficult; thus, valid and accurate quantitative data are not feasible. Quantitative dried blood spot (qDBS) analysis overcomes this limitation, offering the advantage of collecting small but precise predefined blood volume, e.g., 10, 20, 50 µL. Thus, it can be used for accurate determination [16,17] besides the screening applications.

To date, many analytical protocols have been developed to determine/quantify drugs of abuse in DBSs by modern analytical techniques, including liquid chromatography tandem–mass spectrometry (LC–MS/MS) [16,18–26] mainly for quantification purposes; liquid chromatography quadrupole time-of-flight mass spectrometry (LC-QTOF-MS) [17,27] and gas chromatography–mass spectrometry (GC–MS) [11,28] for drug screening. Nowadays, various methodologies have been developed for DBSs for both quantification and screening purposes, offering precise analyte level measurements and efficiency of detection in biological samples. Most of these focus on the detection or/and quantification of a specific category of drugs, such as cocaine and its metabolites, benzodiazepines, new psychoactive substances (NPSs), amphetamines, and cannabinoids [14,16,29]. There are few protocols that enable the simultaneous determination of a plethora of illicit drugs belonging to various categories in DBSs but require a minimum blood volume of 15 µL [17,20,21,30–33]. In all these cases, spots in a simple filter paper, such as Whatman protein saver cards, are used, which presents various limitations, especially in terms of accurate quantification.

The current study aimed to highlight the opportunities arising from the implementation of qDBSs in drug screening for forensic toxicology purposes. The focus was set on the development of a valuable method for the detection of 26 drugs of abuse and metabolites (both illicit drugs and others) in a 10 µL qDBS that offers high precision in sampling [34–38] and has the potential to be applied for quantitative purposes. The studied substances are those frequently abused. To represent these, we chose the following drugs: benzodiazepines and metabolites (diazepam, bromazepam, temazepam, oxazepam, alprazolam, and 7-aminflunitrazepam), synthetic opioids (methadone and AH-7921), cannabinoids (tetrahydrocannabinol (THC), cannabinol (CBN), cannabidiol (CBD), and synthetic cannabinoids (JWH-018)), cocaine and metabolites (benzoylecgonine, methylecgonine, and cocaethylene), amphetamines analogues (3,4-methylenedioxymethamphetamine (MDMA), 25B-NB2OMe, 25C-NB2Ome, and 25I-NB2OMe), and other stimulants (cathine, mescaline, mephedrone, methylone, 3,4-methylenedioxypyrovalerone (MDPV), and 1-benzylpiperazine). A sensitive qualitative method was developed and evaluated for the detection of the target analytes from qDBSs using liquid chromatography coupled to high tandem mass spectrometry.

To our knowledge, this is the first method developed for drug abuse screening utilizing a commercially available qDBS device. The proposed protocol delivers efficient and reliable results by using only 10 µL of blood, obtained in a minimally invasive way by finger pricking.

2. Results

2.1. LC–MS/MS Optimization

For the efficient separation of twenty-six (26) drugs and metabolites, some of which have similar structures, various chromatographic systems were tested. Two different C18 columns, an Intensity Solo 2 C18 column (2.1 mm × 100 mm, 2.0 µm) and an Acquity BEH C18 (2.1 mm × 100 mm, 1.7 µm), were used. Mobile phase systems applied included the following: (a) A: H_2O-MeOH, 90:10 (v/v), 0.01% FA and B: MeOH, 0.01% FA and (b) A: 5 mM AF in H_2O-MeOH, 90:10 (v/v) acidified with 0.01% FA and B: 5 mM AF in MeOH acidified with 0.01% FA. Different gradient elution profiles were tested, starting with up to 95% aqueous phase. Based on the literature, retention of more hydrophilic analytes on RPLC assays requires initial mobile phase conditions of 80–90% water [39]. The optimum selected conditions were finally obtained with A: 5 mM aq. AF-MeOH, 90:10 (v/v), 0.01% FA and B: 5 mM AF in MeOH, 0.01% FA on an Intensity Solo 2 C18 column. The system demonstrated satisfactory chromatographic performance under all test conditions. However, further improved peak shapes and signals of higher intensity were achieved under these conditions. The chromatographic traces of the analytes obtained on the two columns can be seen in Figure 1.

Furthermore, to achieve the highest level of sensitivity, detection parameters in the mass spectrometer were optimized. Different MS parameters were tested to achieve the highest intensities and signals for all compounds. Specifically, ion spray voltage for +ESI was set at 5000, while for -ESI at −4500. Cone temperature was set at 300 °C, heated probe temperature at 250 °C, while curtain gas was set at 20 psi. Multiple reaction monitoring (MRM) mode was employed to monitor selective transitions for the target analytes. Two daughter ions for each analyte were selected for detection confirmation. The MRM transitions, retention times, and collision energies for every analyte are listed in Table 1.

Table 1. MRM transitions, retention times, detection parameters, LOD, and LOQ for all analytes.

Analyte	Molecular Weight	Parent Ion	Daughter Ions	Collision Energy (eV)	Retention Time (min)	LOQ (ng/mL)	LOD (ng/mL)
Bromazepam	316.2	317.0	**228.0** | 209.0	40	6.69	5.00	1.50
Temazepam	300.7	301.3	**255.1** | 282.9	35	8.42	5.00	1.50
Oxazepam	286.7	287.3	**241.1** | 269.1	35	8.02	5.00	1.50
Alprazolam	308.8	309.8	**281.2** | 274.1	25	8.13	5.00	1.50
7-aminoflunitrazepam (7-AF)	283.3	284.0	**226.9** | 135.1	32	4.48	5.00	1.50
Methadone	309.4	310.0	**265.0** | 105.0	14	7.84	5.00	1.50
Cathine	151.2	152.0	**134.0** | 117.0	10	2.86	5.00	1.50
Mescaline	211.3	212.0	**195.1** | 180.1	22	2.93	5.00	1.50
25B-NB2OMe	380.3	381.0	**121.3** | 91.2	25	6.69	5.00	1.50
25C-NB2OMe	335.8	336.8	**121.3** | 91.2	20	7.28	5.00	1.50
25I-NB2OMe	427.3	428.3	**121.3** | 91.2	20	7.78	5.00	1.50
Mephedrone	177.2	178.1	**160.3** | 147.3	10	3.56	5.00	1.50
AH-7921	329.3	329.2	**173.2** | 95.3	30	6.56	5.00	1.50
1-Benzylpiperazine	176.3	177.0	**91.2** | 85.2	20	3.33	5.00	1.50
Methylone	207.2	208.1	**132.2** | 160.3	25	2.91	5.00	1.50
3,4-Methylenedioxypyrovalerone (MDPV)	275.3	276.2	**135.2** | 126.3	25	4.35	5.00	1.50
JWH-018	341.5	342.3	**214.4** | 155.3	25	11.5	5.00	1.50
Diazepam	284.7	285.5	**154.0** | 193.0	28	9.45	2.50	0.75
Cocaine	303.4	304.0	**182.0** | 82.0	20	3.99	2.50	0.75
Benzoylecgonine	289.3	290.0	**105.0** | 168.0	18	3.58	2.50	0.75
Methylecgonine	199.3	200.0	**82.0** | 182.0	18	1.98	2.50	0.75
Cocaethylene	317.4	318.3	**196.2** | 82.0	18	4.77	2.50	0.75
3,4-methylenedioxymethamphetamine (MDMA)	193.3	194.0	**163.0** | 135.0	12	3.17	2.50	0.75
Tetrahydrocannabinol (THC)	314.4	313.0	**245.2** | 191.0	25	11.7	2.50	0.75
Cannabinol (CBN)	310.4	311.2	**223.2** | 240.0	20	11.6	2.50	0.75
Cannabidiol (CBD)	314.5	315.0	**193.0** | 92.7	25	11.5	2.50	0.75

Daughter ion in bold indicates the quantifier ion.

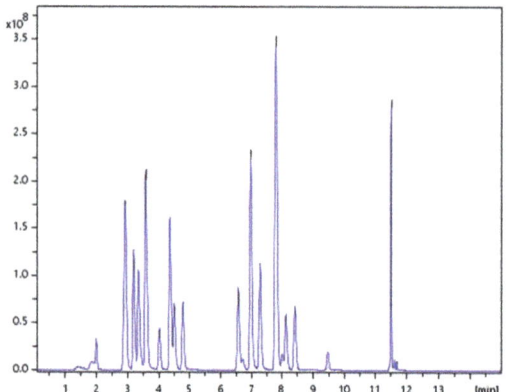
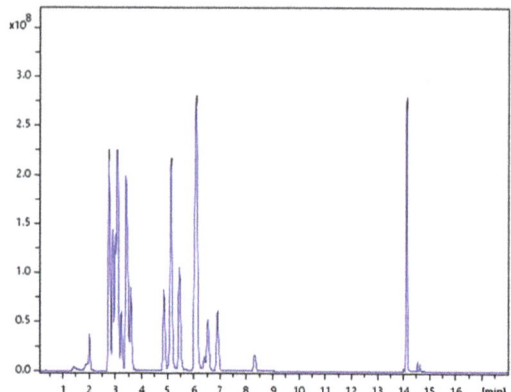

Figure 1. Total ion chromatograms (TICs) of the elution of all 26 illicit drugs in both stationary phases under the same mobile phases and gradient program. The left chromatogram corresponds to Intensity Solo 2 C18 column and the right chromatogram corresponds to Acquity BEH C18.

2.2. Optimization of qDBS Sample Treatment

In the analysis of DBS samples, the analytes extraction is a crucial step. The applied extraction protocol should be carefully designed to achieve the maximum recovery and sufficient sensitivity levels [40], as this can be challenging due to the small sample volume. During this process, several factors should be considered, including the type of paper used in the device, as there is a possibility of substances being released from its materials during sample extraction [41]. Also, blood hematocrit should be considered, as it has multiple effects on the analyte's extraction from DBSs. Blood hematocrit determines blood viscosity, which can cause varying DBS homogeneity on the filter paper [42]. In addition, in the field of toxicological analyses and drug screening, several other factors should be taken into account, such as the characteristics of the blood (postmortem or in vivo) and the age of the bloodstains, as they might have a substantial impact on the extraction efficiency [21]. Hence, the validity of the extraction system should be assessed in relation to the aforementioned factors [6,7,12].

Despite the fact that there are numerous methods reported in the literature for the quantitative measuring of drugs in DBS, no data exist yet on the extraction of drugs of abuse from qDBSs. Different paper substrates have been tested, aiming to determine illicit drugs, with Whatman 903 protein saver card being the most commonly used [16–19,22–24,26]. Moreover, Whatman BFC 180 [21], Bond Elute Dried matrix spotting cards (Agilent) [25], FTA DMPK cards [43], and Sartorius Stedim Biotech Sample carrier paper [20] have been examined and proved to be adequately efficient for drug extraction purposes. However, these approaches do not offer the possibility for accurate collection of a specific volume of blood; thus, they have some accuracy limitations in drug analysis.

The Capitainer qDBS device allows for the accurate collection of an exact volume of blood as it is transferred through a capillary to a precut 6 mm paper disc. A more detailed description of the device can be found in a previous work of the authors [38].

In the present study, a Capitainer qDBS with two collecting discs (10 µL each) was used. In previously reported studies on the determination of drugs, a variety of samples volume have been used on DBS, in all cases larger than 10 µL. More specifically, either 25 µL [16,17], 30 µL [23,24,43], 50 µL [21,22,26], or 85 µL [18,19] of blood were spotted on paper cards. Hence, the Capitainer qDBS (10 µL of whole blood spotted) is the smallest volume of blood used for such analyses.

The initial step was to design a comprehensive experimental approach to identify the optimum extraction protocol for the analytes of interest from the qDBS sample. Based on our literature findings, several studies in DBS determining either benzodiazepines or cocaine and metabolites, or both, performed SPE [14,18,19]. Extraction solvents tested include MeOH-AcN 3:1 (v/v) [21,22], pure H_2O [23], 0.1% FA in MeOH [25], 1% FA in H_2O [26], MeOH [24], AcN-H_2O 8:2 (v/v) [17], and AcN-MeOH 1:1 (v/v) [43]. Herein, AcN, MeOH, and a mixture of AcN-MeOH, 1:1 (v/v) were evaluated. An MQC sample (fortified with 50 ng/mL for bromazepam, temazepam, oxazepam, alprazolam, 7-AF, methadone, cathine, mescaline, 25B-NB2OMe, 25C-NB2OMe, 25I-NB2OMe, mephedrone, AH-7921, 1-benzylpiperazine, methylone, MDPV, JWH-018, and with 25 ng/mL for diazepam, cocaine, benzoylecgonine, methylecgonine, cocaethylene, MDMA, THC, CBN, CBD) was used to investigate which would be the most efficient extraction system for all the studied analytes from qDBSs. As can be seen in Figure 2, with the exception of methylone and mephedrone, which were extracted more efficiently with AcN, pure MeOH or mixture of MeOH with AcN provided better extraction recoveries. AcN-MeOH, 1:1 (v/v) was selected as the optimal qDBS extraction solvent given the fact that 14 out of the 24 drugs had higher intensity using this solvent (in comparison to the other test solvents). Acidification of the AcN-MeOH, 1:1 (v/v) mixture by adding 0.1% FA, as previously reported [25,26], did not improve recovery and thus was not considered further. One milliliter of solvent provided satisfactory results; higher volumes were tested with no enhanced recoveries.

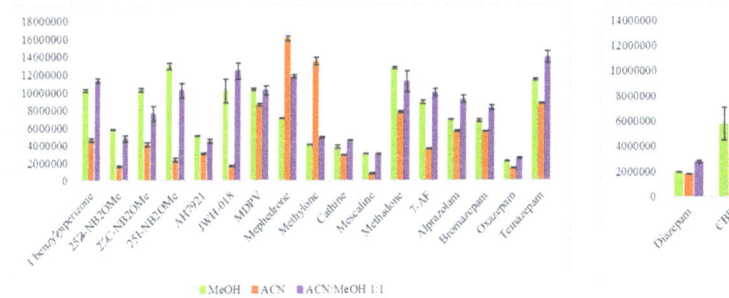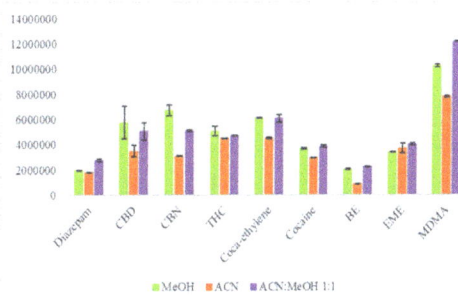

Figure 2. Bar blots illustrating the efficiency of three different extraction solvents for all analytes (x-axes: analyte's peak area; y-axes: analyte).

2.3. Drugs Screening

2.3.1. Sensitivity, LOD, and LOQ

The method developed here aims for the application of a minimally invasive sample collection approach for drug screening in blood. Thus, the study was focused on the detection of the drugs of abuse. For this, LOQ and LOD were estimated, as described in Section 3.6.1. Further parameters, such as intra- and interday accuracy and precision related to quantification, were not studied; however, it will be the goal of a future study. For bromazepam, temazepam, oxazepam, alprazolam, 7-AF, methadone, cathine, mescaline, 25B-NB2OMe, 25C-NB2OMe, 25I-NB2OMe, mephedrone, AH-7921, 1-benzylpiperazine, methylone, MDPV, and JWH 018, LOQ was accessed at 5 ng/mL, while for diazepam, cocaine, benzoylecgonine, methylecgonine, cocaethylene, MDMA, THC, CBN, and CBD, at 2.5 ng/mL. Details for LODs are demonstrated in Table 1.

2.3.2. Extraction Recovery (ER%), Hematocrit Effect

Sample volume and hematocrit (Hct) have proved to be two major factors, affecting spot formulation, homogeneity, drying time, and analyte recovery, and are, thus, studied in DBS applications [42,44,45]. In a qDBS device, a precisely measured sample volume is collected on the disc. Nonetheless, the Hct may have an impact on the accuracy of the sampling or, even more so, on the success of the analyte extraction, and, as previously reported [46,47], an independent hematocrit response bias is likely to be observed.

Herein, the impact of Hct on the extraction recovery of all analytes of interest was investigated by estimating the percentage recovery in three different Hct levels. Results of extraction recovery at LH (35%), FH (40%), and HH (50%) in two different fortified levels are illustrated in Table 2A,B. Based on the results, it was concluded that no effect of Hct was observed, given the fact that ER% spans to similar levels (ranging from 84.6% to 106%) and was within the acceptable criteria [48].

Table 2. (**A**,**B**) Extraction recoveries ± sd in two fortified levels, in three hematocrit levels for the listed analytes.

	(A)					
	Fortified Concentration					
Analyte	**5 ng/mL**			**50 ng/mL**		
	LH (ER% ± sd)	FH (ER% ± sd)	HH (ER% ± sd)	LH (ER% ± sd)	FH (ER% ± sd)	HH (ER% ± sd)
Bromazepam	85.3 ± 0.5	91.6 ± 0.9	93.9 ± 0.5	89.7 ± 0.8	92.9 ± 1.6	95.2 ± 0.8
Temazepam	90.6 ± 0.4	92.6 ± 0.9	91.7 ± 1.2	90.1 ± 0.6	94.4 ± 1.2	94.2 ± 0.2
Oxazepam	87.7 ± 0.6	89.7 ± 0.2	96.0 ± 0.7	88.4 ± 0.8	93.3 ± 0.3	104 ± 0.6
Alprazolam	90.7 ± 0.4	100 ± 0.5	109 ± 1.7	86.3 ± 0.2	101 ± 1.0	106 ± 1.3
7-aminoflunitrazepam (7-AF)	86.7 ± 0.5	93.7 ± 0.2	87.3 ± 0.7	88.1 ± 1.2	96.1 ± 0.1	96.4 ± 0.5
Methadone	84.6 ± 0.8	105 ± 1.9	99.6 ± 0.7	86.0 ± 0.5	96.2 ± 0.9	94.8 ± 0.3
Cathine	85.7 ± 0.4	89.3 ± 0.4	97.4 ± 0.6	86.2 ± 0.5	88.2 ± 0.6	97.2 ± 0.3
Mescaline	88.4 ± 0.8	94.6 ± 0.6	102 ± 1.0	90.2 ± 0.7	89.8 ± 0.9	98.5 ± 0.7
25B-NB2OMe	89.3 ± 0.7	92.7 ± 0.3	92.9 ± 0.7	88.8 ± 0.6	94.7 ± 0.5	103 ± 1.3
25C-NB2OMe	86.2 ± 0.8	88.7 ± 1.0	94.6 ± 0.8	86.6 ± 0.2	96.6 ± 0.9	88.8 ± 0.8
25I-NB2OMe	86.8 ± 0.1	95.1 ± 0.3	96.9 ± 0.3	89.0 ± 0.4	86.3 ± 0.2	97.4 ± 0.8
Mephedrone	88.9 ± 0.3	89.2 ± 0.6	101 ± 0.8	87.9 ± 0.6	96.4 ± 0.4	95.0 ± 0.1
AH-7921	91.5 ± 0.6	92.7 ± 1.2	88.2 ± 0.4	93.6 ± 0.7	95.5 ± 1.3	99.9 ± 1.1
1-Benzylpiperazine	88.3 ± 1.5	92.6 ± 0.5	85.6 ± 0.3	86.6 ± 0.3	94.8 ± 0.6	93.9 ± 0.4
Methylone	86.5 ± 0.6	88.2 ± 0.4	90.1 ± 0.4	89.0 ± 0.3	92.1 ± 0.4	98.0 ± 0.2
3,4-Methylenedioxypyrovalerone (MDPV)	89.2 ± 0.7	91.5 ± 0.5	96.2 ± 0.3	90.8 ± 0.5	95.0 ± 0.6	97.7 ± 0.8
JWH-018	85.1 ± 0.4	95.8 ± 0.6	88.6 ± 0.4	85.5 ± 0.3	87.3 ± 0.3	90.6 ± 0.4
	(B)					
	Fortified Concentration (ng/mL)					
Analyte	**2.5 ng/mL**			**25 ng/mL**		
	LH (ER% ± sd)	FH (ER% ± sd)	HH (ER% ± sd)	LH (ER% ± sd)	FH (ER% ± sd)	HH (ER% ± sd)
Diazepam	92.3 ± 1.2	86.3 ± 0.2	108 ± 1.0	89.1 ± 0.8	90.5 ± 0.8	105 ± 0.8
Cocaine	88.4 ± 0.4	93.4 ± 0.9	102 ± 1.5	85.9 ± 0.5	87.7 ± 0.9	103 ± 1.0
Benzoylecgonine	86.6 ± 0.2	98.7 ± 0.8	94.2 ± 0.2	85.2 ± 0.4	89.5 ± 0.4	91.6 ± 0.8
Methylecgonine	87.9 ± 0.8	105 ± 0.9	94.6 ± 0.8	85.6 ± 0.5	106 ± 0.5	89.7 ± 0.4
Cocaethylene	91.2 ± 0.5	95.6 ± 1.3	91.7 ± 1.2	90.5 ± 0.3	98.4 ± 1.0	88.8 ± 0.8
3,4-methylenedioxymethamphetamine (MDMA)	85.7 ± 0.5	98.7 ± 0.6	96.6 ± 0.9	88.0 ± 0.2	95.6 ± 0.6	94.8 ± 0.3
Tetrahydrocannabinol (THC)	85.7 ± 0.8	87.8 ± 0.6	88.2 ± 0.4	84.8 ± 0.2	96.4 ± 0.2	89.9 ± 0.7
Cannabinol (CBN)	85.2 ± 0.6	89.9 ± 1.1	85.6 ± 0.3	86.3 ± 1.1	88.9 ± 0.5	86.7 ± 1.2
Cannabidiol (CBD)	84.9 ± 0.3	90.4 ± 0.7	87.9 ± 0.4	87.4 ± 0.5	87.3 ± 1.2	89.3 ± 0.6

2.3.3. Stability

Analyte stability was evaluated by analyzing fortified qDBS samples under three different storage conditions (benchtop 20 °C, freezer −20 °C, and oven 30 °C) over 5 sequential days. Results of stability were expressed as % relative error (% E_r), demonstrated in Table 3. As observed in Figure 3, all analytes were estimated to be within +15% E_r to

−15% E_r in all cases, indicating that the analytes are stable under the given conditions for almost a week allowing for a quite adequate time frame for their analysis.

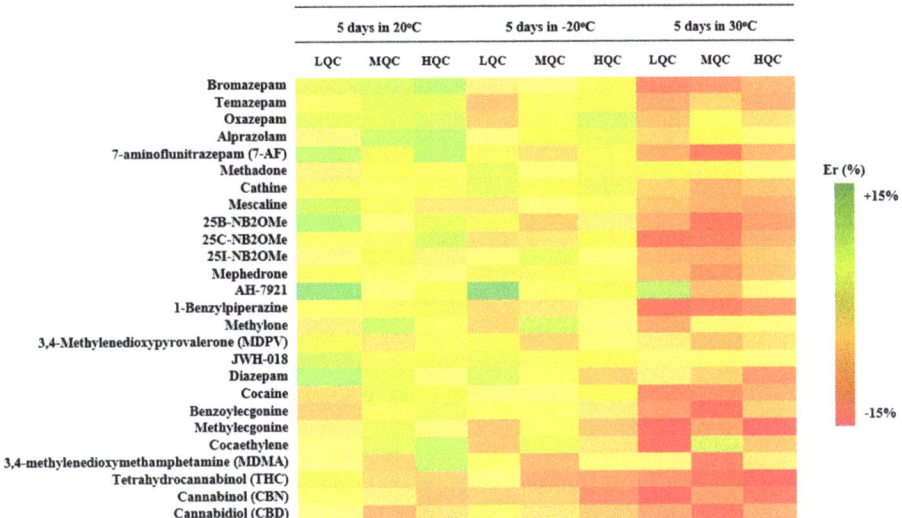

Figure 3. Heatmap illustrating the % E_r for all analytes, in three different levels, in three different storage conditions.

Table 3. Relative error (%E_r) found for all analytes in three different fortified levels, in three different storage conditions over 5 sequential days.

Analyte	5 Days in 20 °C			5 Days in −20 °C			5 Days in 30 °C		
	LQC (%Er)	MQC (%Er)	HQC (%Er)	LQC (%Er)	MQC (%Er)	HQC (%Er)	LQC (%Er)	MQC (%Er)	HQC (%Er)
Bromazepam	−2.10	−1.20	0.98	−6.77	−5.90	−3.83	−12.9	−12.1	−10.1
Temazepam	−4.40	−1.59	−1.94	−9.17	−4.18	−2.91	−10.8	−7.97	−10.4
Oxazepam	−1.86	−1.60	−1.50	−8.53	−3.80	−0.80	−9.28	−5.00	−7.60
Alprazolam	−6.51	0.20	0.99	−5.94	−3.40	−1.88	−7.85	−4.00	−5.35
7-aminoflunitrazepam (7-AF)	0.80	−2.97	1.20	−4.00	−7.59	−3.62	−10.3	−13.6	−9.92
Methadone	−6.75	−3.78	−4.61	−1.84	−5.78	−3.14	−2.86	−4.38	−5.88
Cathine	−4.33	−5.02	−3.96	−1.57	−3.82	−2.67	−8.01	−10.1	−9.04
Mescaline	−5.81	−4.18	−7.21	−7.25	−5.98	−4.62	−8.92	−10.2	−10.4
25B-NB2OMe	1.88	−5.99	−1.96	−2.97	−8.47	−6.63	−10.3	−13.8	−11.4
25C-NB2OMe	−4.82	−4.38	−0.99	−7.60	−7.17	−3.87	−13.6	−13.2	−10.2
25I-NB2OMe	−6.41	−2.83	−6.67	−5.48	−1.86	−5.73	−9.82	−10.1	−8.53
Mephedrone	−4.12	−5.01	−5.98	−3.16	−4.06	−5.04	−9.22	−11.6	−8.92
AH-7921	5.03	−5.99	−4.16	6.08	−5.05	−3.20	0.63	−9.18	5.94
1-Benzylpiperazine	−5.04	−4.37	−2.91	−7.80	−7.15	−5.74	−13.8	−13.2	−11.9
Methylone	−7.07	0.20	−4.90	−7.99	−0.79	−5.84	−10.5	−5.96	−6.00
3,4-Methylenedioxypyrovalerone (MDPV)	−2.90	−7.31	−4.55	−3.86	−8.23	−5.50	−6.46	−9.09	−7.43
JWH-018	−1.02	−3.54	−2.94	−2.00	−4.50	−3.90	−5.49	−5.31	−5.49
Diazepam	1.70	−1.98	−5.69	−1.26	−4.83	−8.43	−6.38	−7.91	−11.8
Cocaine	−7.81	−2.00	−3.48	−6.25	−6.00	−3.89	−12.4	−12.1	−10.2
Benzoylecgonine	−8.64	−4.38	−2.21	−4.94	−4.78	−6.84	−11.5	−14.3	−8.25

Table 3. Cont.

Analyte	5 Days in 20 °C			5 Days in −20 °C			5 Days in 30 °C		
	LQC (%Er)	MQC (%Er)	HQC (%Er)	LQC (%Er)	MQC (%Er)	HQC (%Er)	LQC (%Er)	MQC (%Er)	HQC (%Er)
Methylecgonine	−6.51	−2.02	−6.25	−9.24	−4.87	−8.98	−14.8	−11.1	−14.9
Cocaethylene	−5.49	−1.63	0.40	−9.02	−2.04	−7.24	−14.5	−1.22	−8.85
3,4-methylenedioxymethamphetamine (MDMA)	−5.83	−8.30	−0.20	−4.93	−9.88	−5.79	−5.83	−13.0	−6.59
Tetrahydrocannabinol (THC)	−2.90	−7.95	−8.33	−5.73	−10.63	−11.00	−11.9	−13.8	−14.8
Cannabinol (CBN)	−4.78	−6.64	−8.93	−8.37	−8.30	−12.30	−14.7	−11.2	−13.5
Cannabidiol (CBD)	−6.01	−9.88	−7.37	−4.72	−7.00	−8.37	−10.7	−14.0	−10.4

3. Materials and Methods

3.1. Reagents, Materials, and Chemicals

Methanol (MeOH) and acetonitrile (AcN), LC–MS grade, were purchased from HiPerSolv CHROMANORM®. LC–MS grade isopropanol (IPA) was obtained from Fisher Scientific International, Inc., Hampton, NH, USA. A Milli-Q purification system (18.2 MΩ cm^{-1}) was used to provide ultrapure water. Ammonium formate (AF) ≥99% and formic acid (FA) 98–100% mobile phase additives were purchased from Riedel-de Haën® (Sigma-Aldrich, Steinheim, Germany) and ChemLab, Zedelgem Belgium, respectively. Reference standards of diazepam, bromazepam, temazepam, oxazepam, alprazolam, 7-aminoflunitrazepam (7-AF), methadone, 3,4-methylenedioxymethamphetamine (MDMA), cathine, mescaline, cocaine, benzoylecgonine, methylecgonine, cocaethylene, 25B-NB2OMe, 25C-NB2OMe, 25I-NB2OMe, mephedrone, AH-7921, 1-Benzylpiperazine, methylone, 3,4-Methylenedioxy pyrovalerone (MDPV), JWH-018, tetrahydrocannabinol (THC), cannabinol (CBN), and cannabidiol (CBD) were of more than 98% purity and were purchased from Lipomed AG (Arlesheim, Switzerland).

Dried blood spots (qDBSs) devices were obtained from Capitainer AB® (Solna, Sweden).

3.2. Working Standard Solutions and Quality Control Samples (QCs)

All analytes' stock solutions (1 mg/mL) were prepared in methanol by dissolving an appropriate amount of each solid standard. Following dilutions with MeOH, working solutions of 0.1 mg/mL concentration were prepared for each drug. Subsequently, a mixture, containing a concentration of 10 μg/mL bromazepam, temazepam, oxazepam, alprazolam, 7-aminoflunitrazepam (7-AF), methadone, cathine, mescaline, 25B-NB2OMe, 25C-NB2OMe, 25I-NB2OMe, mephedrone, AH-7921, 1-benzylpiperazine, methylone, 3,4-methylenedioxypyrovalerone (MDPV), JWH-018, and at a concentration of 5 μg/mL diazepam, cocaine, benzoylecgonine, methylecgonine, cocaethylene, 3,4-methylenedioxymethamphetamine (MDMA), tetrahydrocannabinol (THC), cannabinol (CBN) and cannabidiol (CBD), was created in H$_2$O-MeOH, 50:50 (v/v). All working and stock solutions were stored at −20 °C.

For validation purposes, a sample prepared by pooling 10 whole blood samples was used (QC). By appropriate spiking of the latter at three different levels of the drugs LQC, MQC, and HQC, samples were prepared, which were then transferred by a syringe onto the qDBS disc. LQC, MQC, and HQC were spiked at 5 ng/mL, 50 ng/mL, and 100 ng/mL, respectively, with bromazepam, temazepam, oxazepam, alprazolam, 7-aminoflunitrazepam (7-AF), methadone, cathine, mescaline, 25B-NB2OMe, 25C-NB2OMe, 25I-NB2OMe, mephedrone, AH-7921, 1-benzylpiperazine, methylone, 3,4-methylenedioxypyrovalerone (MDPV), and JWH-018. For the rest of the drugs, namely, for diazepam, cocaine, benzoylecgonine, methylecgonine, cocaethylene, 3,4-methylenedioxymethamphetamine (MDMA), tetrahydrocannabinol (THC), cannabinol (CBN), cannabidiol (CBD), LQC, MQC, and HQC were prepared by spiking at 2.5 ng/mL, 25 ng/mL, and 50 ng/mL, respectively.

For the evaluation of extraction recovery, venous whole blood collected from three individuals with three different hematocrit levels (low 35%, medium 40%, and high 50%)

were used. Spiking at LQC and MQC, as described above for QC, was performed to study the impact of hematocrit in extraction efficiency. The collection of the blood samples was performed under the approval of the Ethical Committee of the Aristotle University of Thessaloniki (protocol number 62883/2023).

3.3. Instrumentation and Analytical Conditions

A reversed-phase liquid chromatography–tandem mass spectrometry (RPLC–MS/MS) method was developed for the determination of the 26 drugs in qDBS extracts using an Elute LC chromatographic system coupled to an EVOQ Elite triple quadrupole mass spectrometer (Bruker Daltonics, Bremen, Germany). Separation was carried out on an Intensity Solo 2 C18 (2.1 × 100 mm, 2 µm) column and the mobile phases consisted of A: H_2O-MeOH, 90:10 (v/v), 5 mM ammonium formate, 0.01% formic acid and B: MeOH, 5 mM ammonium formate, 0.01% formic acid. Elution was performed by a 15 min gradient as follows: 0–0.5 min: 15–30% B (flow rate 0.2 mL/min), 0.5–10 min: 30–80% B (flow rate 0.2 mL/min); 10–10.5 min: 80–100% B (flow rate 0.4 mL/min); 10.5–12 min 100% B. At 12.01 min, the composition was returned to the initial conditions and column re-equilibration was applied for 3 min. Column temperature was set at 50 °C, and autosampler's temperature at 4 °C. Injection volume was 5 µL.

3.4. qDBS Sample Extraction Optimization

Different extraction conditions were tested to examine the extraction efficiency of the analytes from the qDBS disc. Specifically, extraction recovery and repeatability of various solvents or mixtures, including AcN, MeOH, and AcN: MeOH, 1:1 (v/v), were assessed. The extraction procedure started by carefully removing one disc (1 × 10 µL) from the Capitainer device, and then by transferring it into an Eppendorf tube. One milliliter of the extraction solvent was added. Vortex-mixing for 10 min, sonication for 10 min, and/or homogenization by bead beater were tested. For the latter, the disc was placed in a tube that contained approximately 20 ceramic bead media balls, vortex-mixed for 10 min, and then was homogenized with solvent for 30 s at a speed of 6.0 m/s; this was repeated twice. In all cases, centrifugation for 10 min at 6700× g was thereafter held. Finally, 500 µL of the supernatant were transferred to a tube and evaporated until dryness. The dry residue was reconstituted with 50 µL of H_2O-MeOH, 85:15 (v/v). The procedure was performed three times for the different extraction conditions. The solvent system that provided better results was also tested at a smaller volume (200 µL); in this condition, one hundred microliters of supernatant were directly transferred to an LC–MS vial and subjected to analysis.

3.5. Final qDBS Sample Treatment Protocol

In a tube that had previously been filled with about 20 ceramic balls (1.4 mm ceramic bead media), one qDBS sample was placed. Then, 1 mL of can-MeOH, 1:1 (v/v) was added. After 10 min of vortex-mixing, two cycles of beat-beater homogenization lasting 30 s each were carried out at a speed of 6.0 m/s. Five hundred microliters of the supernatant was transferred to a 1.5 mL Eppendorf tube after centrifuged at 6700× g for 10 min and evaporated to dryness. The dry residue was reconstituted with 50 mL of H_2O-MeOH, 85:15 (v/v).

3.6. qDBS Drugs Screening

3.6.1. Sensitivity, LOD, and LOQ

The sensitivity of the method was estimated through the limits of detection (LODs) for the studied analytes. Limit of quantification (LOQ) values were estimated experimentally by analyzing the spiked qDBS HQC sample after serial dilutions. LODs were established as the concentration where the chromatographic peaks-to-noise ratio was 3:1, whereas for LOQ, a 10:1 ratio was considered.

3.6.2. Extraction Recovery (ER%) and Hematocrit Effect

Extraction recovery (ER%) and hematocrit effect were evaluated for the employed extraction protocol. Three blood samples with different hematocrit spanning from low to high levels (low hematocrit, LH 35%; fixed hematocrit, FH 40%; and high hematocrit, HH 50%) were obtained from volunteers to assess the impact of the hematocrit on the extraction efficiency. Two different levels of the analytes standard mixture were added in qDBS samples (LQC, MQC) before and after extraction. Extraction recovery, ER%, was determined based on Equation (1). Hematocrit effect was evaluated as part of extraction recovery efficiency at the three different samples of different hematocrit levels (LH, FH, HH).

$$\%ER = \frac{\text{Peak area spiked before extraction}}{\text{Peak area spiked after extraction}} \times 100 \quad (1)$$

3.6.3. Stability of qDBS Samples

Stability of the analytes in the qDBS samples was studied under three different storage conditions: at benchtop (20 °C), in the oven (30 °C), and in the freezer (−20 °C). Three concentrations (LQC, MQC, and HQC) were examined. Evaluation of short-term stability was carried out by analyzing the spiked qDBS samples (LQC, MQC, HQC) stored under three different conditions for 5 days. The same spiked qDBS samples were analyzed after being freshly prepared to estimate the % relative error (%Er). The three different freshly prepared spiked qDBS QC samples were used to plot a calibration curve, aiding in generating concentration data.

4. Conclusions

The applied UHPLC–MS/MS method was developed with the aim to detect 26 illicit drugs in qDBS samples by analyzing only 10 µL of capillary blood. Herein, a simple, rapid, and trustworthy extraction protocol was achieved, reaching high sensitivity levels for all analytes. This is the first approach reported for the detection of frequently screened drugs utilizing a qDBS device, using just a small drop of blood. Stability experiments showed negligible bias that suggests the validity of the method within a 5-day time period, even at RT storage conditions. Therefore, the method offers a great promise for future applications in drug screening for toxicological and forensic analysis purposes.

Author Contributions: Conceptualization, H.G. and G.T.; methodology, T.M.; software, T.M.; validation, T.M.; formal analysis, T.M.; investigation, T.M. and H.G.; resources, H.G., G.T. and O.B.; data curation, T.M.; writing—original draft preparation, T.M. and H.G.; writing—review and editing, T.M., H.G., G.T. and O.B.; visualization, T.M.; supervision, H.G.; project administration, H.G.; funding acquisition, H.G. and G.T. All authors have read and agreed to the published version of the manuscript.

Funding: This research received no external funding.

Institutional Review Board Statement: The collection of the blood samples was under the approval of the Ethical Committee of the Aristotle University of Thessaloniki (protocol number 62883/2023, 7 March 2023).

Informed Consent Statement: Informed consent was obtained from all subjects involved in the study. Written informed consent has been obtained from the patient(s) to publish this paper.

Data Availability Statement: Data are contained within the article.

Acknowledgments: All authors would like to acknowledge Bruker Daltonics (Germany) and Capitainer® AB (Sweden) for the collaboration and technical guidance. The authors also acknowledge Aristea Papaioannou for her technical assistance in the laboratory.

Conflicts of Interest: The authors declare no conflicts of interest.

References

1. Schmidt, V. Ivar Christian Bang (1869–1918), Founder of Modern Clinical Microchemistry. *Clin. Chem.* **1986**, *32*, 213–215. [CrossRef] [PubMed]
2. Guthrie, R.; Susi, A. A Simple Phenylalanine Method for Detecting Phenylketonuria in Large Populations of Newborn Infants. *Pediatrics* **1963**, *32*, 338–343. [CrossRef] [PubMed]
3. Li, W.; Tse, F.L.S. Dried Blood Spot Sampling in Combination with LC-MS/MS for Quantitative Analysis of Small Molecules. *Biomed. Chromatogr.* **2010**, *24*, 49–65. [CrossRef] [PubMed]
4. Hannon, W.H.; Therrell, B.L. *Overview of the History and Applications of Dried Blood Samples*; Li, W., Lee, M.S., Eds.; John Wiley & Sons, Inc.: Hoboken, NJ, USA, 2014; pp. 1–15.
5. Eshghi, A.; Pistawka, A.J.; Liu, J.; Chen, M.; Sinclair, N.J.T.; Hardie, D.B.; Elliott, M.; Chen, L.; Newman, R.; Mohammed, Y.; et al. Concentration Determination of >200 Proteins in Dried Blood Spots for Biomarker Discovery and Validation. *Mol. Cell. Proteom.* **2020**, *19*, 540–553. [CrossRef] [PubMed]
6. Edelbroek, P.M.; van der Heijden, J.; Stolk, L.M.L. Dried Blood Spot Methods in Therapeutic Drug Monitoring: Methods, Assays, and Pitfalls. *Ther. Drug Monit.* **2009**, *31*, 327–336. [CrossRef] [PubMed]
7. Stove, C.P.; Ingels, A.-S.M.E.; De Kesel, P.M.M.; Lambert, W.E. Dried Blood Spots in Toxicology: From the Cradle to the Grave? *Crit. Rev. Toxicol.* **2012**, *42*, 230–243. [CrossRef] [PubMed]
8. Kolmonen, M.; Leinonen, A.; Pelander, A.; Ojanperä, I. A General Screening Method for Doping Agents in Human Urine by Solid Phase Extraction and Liquid Chromatography/Time-of-Flight Mass Spectrometry. *Anal. Chim. Acta* **2007**, *585*, 94–102. [CrossRef] [PubMed]
9. Lee, H.K.; Ho, C.S.; Iu, Y.P.H.; Lai, P.S.J.; Shek, C.C.; Lo, Y.-C.; Klinke, H.B.; Wood, M. Development of a Broad Toxicological Screening Technique for Urine Using Ultra-Performance Liquid Chromatography and Time-of-Flight Mass Spectrometry. *Anal. Chim. Acta* **2009**, *649*, 80–90. [CrossRef]
10. Bjørk, M.K.; Simonsen, K.W.; Andersen, D.W.; Dalsgaard, P.W.; Sigurðardóttir, S.R.; Linnet, K.; Rasmussen, B.S. Quantification of 31 Illicit and Medicinal Drugs and Metabolites in Whole Blood by Fully Automated Solid-Phase Extraction and Ultra-Performance Liquid Chromatography–Tandem Mass Spectrometry. *Anal. Bioanal. Chem.* **2013**, *405*, 2607–2617. [CrossRef]
11. Gomes, D.; de Pinho, P.G.; Pontes, H.; Ferreira, L.; Branco, P.; Remião, F.; Carvalho, F.; Bastos, M.L.; Carmo, H. Gas Chromatography–Ion Trap Mass Spectrometry Method for the Simultaneous Measurement of MDMA (Ecstasy) and Its Metabolites, MDA, HMA, and HMMA in Plasma and Urine. *J. Chromatogr. B* **2010**, *878*, 815–822. [CrossRef]
12. Spooner, N.; Lad, R.; Barfield, M. Dried Blood Spots as a Sample Collection Technique for the Determination of Pharmacokinetics in Clinical Studies: Considerations for the Validation of a Quantitative Bioanalytical Method. *Anal. Chem.* **2009**, *81*, 1557–1563. [CrossRef] [PubMed]
13. Barfield, M.; Spooner, N.; Lad, R.; Parry, S.; Fowles, S. Application of Dried Blood Spots Combined with HPLC-MS/MS for the Quantification of Acetaminophen in Toxicokinetic Studies. *J. Chromatogr. B* **2008**, *870*, 32–37. [CrossRef] [PubMed]
14. Alfazil, A.A.; Anderson, R.A. Stability of Benzodiazepines and Cocaine in Blood Spots Stored on Filter Paper. *J. Anal. Toxicol.* **2008**, *32*, 511–515. [CrossRef] [PubMed]
15. Jantos, R.; Vermeeren, A.; Sabljic, D.; Ramaekers, J.G.; Skopp, G. Degradation of Zopiclone during Storage of Spiked and Authentic Whole Blood and Matching Dried Blood Spots. *Int. J. Leg. Med.* **2013**, *127*, 69–76. [CrossRef]
16. Ambach, L.; Menzies, E.; Parkin, M.C.; Kicman, A.; Archer, J.R.H.; Wood, D.M.; Dargan, P.I.; Stove, C. Quantification of Cocaine and Cocaine Metabolites in Dried Blood Spots from a Controlled Administration Study Using Liquid Chromatography-Tandem Mass Spectrometry. *Drug Test. Anal.* **2019**, *11*, 709–720. [CrossRef]
17. Chepyala, D.; Tsai, I.-L.; Liao, H.-W.; Chen, G.-Y.; Chao, H.-C.; Kuo, C.-H. Sensitive Screening of Abused Drugs in Dried Blood Samples Using Ultra-High-Performance Liquid Chromatography-Ion Booster-Quadrupole Time-of-Flight Mass Spectrometry. *J. Chromatogr. A* **2017**, *1491*, 57–66. [CrossRef] [PubMed]
18. Moretti, M.; Visonà, S.D.; Freni, F.; Tomaciello, I.; Vignali, C.; Groppi, A.; Tajana, L.; Osculati, A.M.M.; Morini, L. A Liquid Chromatography-Tandem Mass Spectrometry Method for the Determination of Cocaine and Metabolites in Blood and in Dried Blood Spots Collected from Postmortem Samples and Evaluation of the Stability over a 3-Month Period. *Drug Test. Anal.* **2018**, *10*, 1430–1437. [CrossRef]
19. Moretti, M.; Freni, F.; Tomaciello, I.; Vignali, C.; Groppi, A.; Visonà, S.D.; Tajana, L.; Osculati, A.M.M.; Morini, L. Determination of Benzodiazepines in Blood and in Dried Blood Spots Collected from Post-Mortem Samples and Evaluation of the Stability over a Three-Month Period. *Drug Test. Anal.* **2019**, *11*, 1403–1411. [CrossRef]
20. Kacargil, C.U.; Daglioglu, N.; Goren, I.E. Determination of Illicit Drugs in Dried Blood Spots by LC–MS/MS Method: Validation and Application to Real Samples. *Chromatographia* **2020**, *83*, 885–892. [CrossRef]
21. Sadler Simões, S.; Castañera Ajenjo, A.; Dias, M.J. Dried Blood Spots Combined to an UPLC-MS/MS Method for the Simultaneous Determination of Drugs of Abuse in Forensic Toxicology. *J. Pharm. Biomed. Anal.* **2018**, *147*, 634–644. [CrossRef]
22. de Lima Feltraco Lizot, L.; da Silva, A.C.C.; Bastiani, M.F.; Hahn, R.Z.; Bulcão, R.; Perassolo, M.S.; Antunes, M.V.; Linden, R. Simultaneous Determination of Cocaine, Egonine Methyl Ester, Benzoylecgonine, Cocaethylene and Norcocaine in Dried Blood Spots by Ultra-Performance Liquid Chromatography Coupled to Tandem Mass Spectrometry. *Forensic Sci. Int.* **2019**, *298*, 408–416. [CrossRef] [PubMed]

23. Saussereau, E.; Lacroix, C.; Gaulier, J.M.; Goulle, J.P. On-Line Liquid Chromatography/Tandem Mass Spectrometry Simultaneous Determination of Opiates, Cocainics and Amphetamines in Dried Blood Spots. *J. Chromatogr. B* **2012**, *885–886*, 1–7. [CrossRef] [PubMed]
24. Kyriakou, C.; Marchei, E.; Scaravelli, G.; García-Algar, O.; Supervía, A.; Graziano, S. Identification and Quantification of Psychoactive Drugs in Whole Blood Using Dried Blood Spot (DBS) by Ultra-Performance Liquid Chromatography Tandem Mass Spectrometry. *J. Pharm. Biomed. Anal.* **2016**, *128*, 53–60. [CrossRef] [PubMed]
25. Odoardi, S.; Anzillotti, L.; Strano-Rossi, S. Simplifying Sample Pretreatment: Application of Dried Blood Spot (DBS) Method to Blood Samples, Including Postmortem, for UHPLC-MS/MS Analysis of Drugs of Abuse. *Forensic Sci. Int.* **2014**, *243*, 61–67. [CrossRef] [PubMed]
26. Ellefsen, K.N.; da Costa, J.L.; Concheiro, M.; Anizan, S.; Barnes, A.J.; Pirard, S.; Gorelick, D.A.; Huestis, M.A. Cocaine and Metabolite Concentrations in DBS and Venous Blood after Controlled Intravenous Cocaine Administration. *Bioanalysis* **2015**, *7*, 2041–2056. [CrossRef] [PubMed]
27. Stöth, F.; Martin Fabritius, M.; Weinmann, W.; Luginbühl, M.; Gaugler, S.; König, S. Application of Dried Urine Spots for Non-Targeted Quadrupole Time-of-Flight Drug Screening. *J. Anal. Toxicol.* **2023**, *47*, 332–337. [CrossRef]
28. Scheidweiler, K.B.; Barnes, A.J.; Huestis, M.A. A Validated Gas Chromatographic–Electron Impact Ionization Mass Spectrometric Method for Methamphetamine, Methylenedioxymethamphetamine (MDMA), and Metabolites in Mouse Plasma and Brain. *J. Chromatogr. B* **2008**, *876*, 266–276. [CrossRef]
29. Abarca, R.; Gerona, R. Development and Validation of an LC-MS/MS Assay for the Quantitative Analysis of Alprazolam, α-Hydroxyalprazolam and Hydrocodone in Dried Blood Spots. *J. Chromatogr. B* **2023**, *1220*, 123639. [CrossRef]
30. Gaugler, S.; Al-Mazroua, M.K.; Issa, S.Y.; Rykl, J.; Grill, M.; Qanair, A.; Cebolla, V.L. Fully Automated Forensic Routine Dried Blood Spot Screening for Workplace Testing. *J. Anal. Toxicol.* **2019**, *43*, 212–220. [CrossRef]
31. Joye, T.; Sidibé, J.; Déglon, J.; Karmime, A.; Sporkert, F.; Widmer, C.; Favrat, B.; Lescuyer, P.; Augsburger, M.; Thomas, A. Liquid Chromatography-High Resolution Mass Spectrometry for Broad-Spectrum Drug Screening of Dried Blood Spot as Microsampling Procedure. *Anal. Chim. Acta* **2019**, *1063*, 110–116. [CrossRef]
32. Gaugler, S.; Rykl, J.; Grill, M.; Cebolla, V. Fully Automated Drug Screening of Dried Blood Spots Using Online LC-MS/MS Analysis. *J. Appl. Bioanal.* **2018**, *4*, 7–15. [CrossRef]
33. Stelmaszczyk, P.; Gacek, E.; Wietecha-Posłuszny, R. Optimized and Validated DBS/MAE/LC-MS Method for Rapid Determination of Date-Rape Drugs and Cocaine in Human Blood Samples-A New Tool in Forensic Analysis. *Separations* **2021**, *8*, 249. [CrossRef]
34. Liu, Q.; Liu, L.; Yuan, Y.; Xie, F. A Validated UHPLC–MS/MS Method to Quantify Eight Antibiotics in Quantitative Dried Blood Spots in Support of Pharmacokinetic Studies in Neonates. *Antibiotics* **2023**, *12*, 199. [CrossRef] [PubMed]
35. Li, W.; Chace, D.H.; Garrett, T.J. Quantitation of Phenylalanine and Tyrosine from Dried Blood/Plasma Spots with Impregnated Stable Isotope Internal Standards (SIIS) by FIA-SRM. *Clin. Chim. Acta* **2023**, *549*, 117551. [CrossRef]
36. Deprez, S.; Van Uytfanghe, K.; Stove, C.P. Liquid Chromatography-Tandem Mass Spectrometry for Therapeutic Drug Monitoring of Immunosuppressants and Creatinine from a Single Dried Blood Spot Using the Capitainer® qDBS Device. *Anal. Chim. Acta* **2023**, *1242*, 340797. [CrossRef]
37. Carling, R.S.; Barclay, Z.; Cantley, N.; Emmett, E.C.; Hogg, S.L.; Finezilber, Y.; Schulenburg-Brand, D.; Murphy, E.; Moat, S.J. Investigation of the Relationship between Phenylalanine in Venous Plasma and Capillary Blood Using Volumetric Blood Collection Devices. *JIMD Rep.* **2023**, *64*, 468–476. [CrossRef] [PubMed]
38. Meikopoulos, T.; Begou, O.; Theodoridis, G.; Gika, H. Ceramides Biomarkers Determination in Quantitative Dried Blood Spots by UHPLC-MS/MS. *Anal. Chim. Acta* **2023**, *1255*, 341131. [CrossRef]
39. Orfanidis, A.; Gika, H.G.; Theodoridis, G.; Mastrogianni, O.; Raikos, N. A UHPLC-MS-MS Method for the Determination of 84 Drugs of Abuse and Pharmaceuticals in Blood. *J. Anal. Toxicol.* **2021**, *45*, 28–43. [CrossRef]
40. Sadones, N.; Capiau, S.; De Kesel, P.M.; Lambert, W.E.; Stove, C.P. Spot Them in the Spot: Analysis of Abused Substances Using Dried Blood Spots. *Bioanalysis* **2014**, *6*, 2211–2227. [CrossRef]
41. Pablo, A.; Breaud, A.R.; Clarke, W. Automated Analysis of Dried Urine Spot (DUS) Samples for Toxicology Screening. *Clin. Biochem.* **2020**, *75*, 70–77. [CrossRef]
42. De Kesel, P.M.; Sadones, N.; Capiau, S.; Lambert, W.E.; Stove, C.P. Hemato-Critical Issues in Quantitative Analysis of Dried Blood Spots: Challenges and Solutions. *Bioanalysis* **2013**, *5*, 2023–2041. [CrossRef] [PubMed]
43. Lee, H.; Park, Y.; Jo, J.; In, S.; Park, Y.; Kim, E.; Pyo, J.; Choe, S. Analysis of Benzodiazepines and Their Metabolites Using DBS Cards and LC-MS/MS. *Forensic Sci. Int.* **2015**, *255*, 137–145. [CrossRef] [PubMed]
44. Timmerman, P.; White, S.; Globig, S.; Lüdtke, S.; Brunet, L.; Smeraglia, J. EBF Recommendation on the Validation of Bioanalytical Methods for Dried Blood Spots. *Bioanalysis* **2011**, *3*, 1567–1575. [CrossRef] [PubMed]
45. Timmerman, P.; White, S.; Cobb, Z.; de Vries, R.; Thomas, E.; van Baar, B. European Bioanalysis Forum Update of the EBF Recommendation for the Use of DBS in Regulated Bioanalysis Integrating the Conclusions from the EBF DBS-Microsampling Consortium. *Bioanalysis* **2013**, *5*, 2129–2136. [CrossRef]
46. Velghe, S.; Stove, C.P. Evaluation of the Capitainer-B Microfluidic Device as a New Hematocrit-Independent Alternative for Dried Blood Spot Collection. *Anal. Chem.* **2018**, *90*, 12893–12899. [CrossRef]

47. Carling, R.S.; Emmett, E.C.; Moat, S.J. Evaluation of Volumetric Blood Collection Devices for the Measurement of Phenylalanine and Tyrosine to Monitor Patients with Phenylketonuria. *Clin. Chim. Acta* **2022**, *535*, 157–166. [CrossRef]
48. Guideline on Bioanalytical Method Validation. EMEA/CHMP/EWP/192217/2009 Rev. 1 Corr. 2** Committee for Medicinal Products for Human Use (CHMP). 2011. Available online: https://www.ema.europa.eu/en/documents/scientific-guideline/guideline-bioanalytical-method-validation_en.pdf (accessed on 15 February 2024).

Disclaimer/Publisher's Note: The statements, opinions and data contained in all publications are solely those of the individual author(s) and contributor(s) and not of MDPI and/or the editor(s). MDPI and/or the editor(s) disclaim responsibility for any injury to people or property resulting from any ideas, methods, instructions or products referred to in the content.

Article

Sol-Gel Graphene Oxide-Coated Fabric Disks as Sorbents for the Automatic Sequential-Injection Column Preconcentration for Toxic Metal Determination in Distilled Spirit Drinks

Natalia Manousi [1], Abuzar Kabir [2], Kenneth G. Furton [2] and Aristidis Anthemidis [2,*]

1. Laboratory of Analytical Chemistry, Department of Chemistry, Aristotle University of Thessaloniki, 54124 Thessaloniki, Greece
2. International Forensic Research Institute, Department of Chemistry and Biochemistry, Florida International University, Miami, FL 33131, USA
* Correspondence: anthemid@chem.auth.gr

Abstract: Sol-gel graphene oxide-coated polyester fabric platforms were synthesized and used for the on-line sequential injection fabric disk sorptive extraction (SI-FDSE) of toxic (i.e., Cd(II), Cu(II) and Pb(II)) metals in different distilled spirit drinks prior to their determination by electrothermal atomic absorption spectrometry (ETAAS). The main parameters that could potentially influence the extraction efficiency of the automatic on-line column preconcentration system were optimized and the SI-FDSE-ETAAS method was validated. Under optimum conditions, enhancement factors of 38, 120 and 85 were achieved for Cd(II), Cu(II) and Pb(II), respectively. Method precision (in terms of relative standard deviation) was lower than 2.9% for all analytes. The limits of detection for Cd(II), Cu(II) and Pb(II) were 1.9, 7.1 and 17.3 ng L^{-1}, respectively. As a proof of concept, the proposed protocol was employed for the monitoring of Cd(II), Cu(II), and Pb(II) in distilled spirit drinks of different types.

Keywords: fabric disk sorptive extraction; sol-gel graphene oxide; sequential injection; atomic absorption spectrometry; metals; alcoholic beverages

Citation: Manousi, N.; Kabir, A.; Furton, K.G.; Anthemidis, A. Sol-Gel Graphene Oxide-Coated Fabric Disks as Sorbents for the Automatic Sequential-Injection Column Preconcentration for Toxic Metal Determination in Distilled Spirit Drinks. *Molecules* 2023, 28, 2103. https://doi.org/10.3390/molecules28052103

Academic Editor: Ahmad Mehdi

Received: 1 February 2023
Revised: 16 February 2023
Accepted: 21 February 2023
Published: 23 February 2023

Copyright: © 2023 by the authors. Licensee MDPI, Basel, Switzerland. This article is an open access article distributed under the terms and conditions of the Creative Commons Attribution (CC BY) license (https://creativecommons.org/licenses/by/4.0/).

1. Introduction

A wide variety of metals and metalloids have been detected in alcoholic beverages, (e.g., As, Cd, Cr, Co, Cu, Fe, Mn, Ni, Sn, Pb, and Zn), the presence of which can be attributed to the raw materials used for their production; to the substances added during brewing; and to the equipment used during the storage, bottling, distillation and aging process [1,2]. Although elements such as Cu are essential for living organisms, long-term exposure to high concentrations can lead to chronic toxicosis [2]. On the other hand, metals such as Cd and Pb are considered to be highly toxic even at low concentration levels and they are associated with neurological disorders, anemia, kidney damage, liver damage and gastrointestinal damage [3,4]. Besides the impact of metals in human health, their presence in alcoholic beverages can significantly influence the quality of the product [1]. Depending on the type of the alcoholic beverage, a very wide range of metal concentration can be observed. Cd concentration can range from not detected up to 5.31 mg L^{-1}, Pb concentration can range from not detected up to 1.125 mg L^{-1} and Cu concentration can range from not detected up to 14.6 mg L^{-1} [1].

Undoubtedly, electro-thermal atomic absorption spectrometry (ETAAS) is a powerful analytical spectrometric technique for metal determination due to its excellent applicability, high sensitivity, reduced operational cost and method simplicity [5–7]. However, due to the low concentration of the trace elements in the distilled spirit samples, a preconcentration step is necessary for the sensitive and accurate metal determination in complex samples. Currently, a lot of attention is being paid to the development of on-line automated sample preparation systems as an alternative to batch-mode extraction procedures. According to the principles of green sample preparation that were recently introduced [8], the development of automatic sample preparation procedures can contribute to a reduction in the required amount of chemicals and waste generated while providing a larger sample throughput. At the same time, automation can also assist in the minimization of the operator's exposure to hazardous chemicals, thus reducing the risk of accidents. Although there are various on-line approaches for the extraction and preconcentration of metals, most of them focused on environmental and biological samples, and on-line systems for the monitoring of the metal content of alcoholic beverages are limited [9,10].

Fabric phase sorptive extraction (FPSE) is an evolutionary sample preparation technique that utilizes sol-gel organic–inorganic sorbent-coated fabric substrates with high thermal, chemical and solvent stability as extraction platforms [11]. A plethora of sol-gel coatings can be easily prepared, resulting in the fabrication of neutral, anion and cation exchangers, zwitterionic, and mixed mode zwitterionic sorptive phases [12]. FPSE has proved to be a significant sample preparation technique for the determination of a wide variety of organic and inorganic chemical species at trace- and ultra-trace-level concentrations [11,13]. Recently, the on-line flow injection–fabric disk sorptive extraction (FI-FDSE) technique was proposed as an automatic alternative to FPSE [14]. This technique offers various benefits such as high extraction efficiency, good reproducibility and sensitivity, and sorbent reusability [14,15]. Sol-gel sorbents have proved to be an important tool in environmental analysis [16], food analysis [17] and bioanalysis [18].

The second generation of flow injection techniques, known as sequential injection analysis (SIA), is based on the usage of programmable, bi-directional discontinuous flow, which is precisely coordinated and controlled by a computer. Due to its inherent characteristics, SIA exhibits multiple benefits including low chemical consumption, versatility, manifold simplicity, low operational cost, and low waste generation [19–21]. Although the FI-FDSE technique is a powerful analytical technique for on-line metal determination [14,15], its operation in on-line sequential injection (SI) mode has not yet been reported.

In this work, novel sol-gel graphene oxide-coated polyester FDSE membranes were synthesized and used to prepare a reusable and renewable on-line fabric disk sorbent extraction microcolumn. Graphene oxide has been proved to be a powerful material for the extraction of metals [22,23]. The proposed platform was employed for the first time for the effective preconcentration and determination of Cd(II), Cu(II) and Pb(II) as model analytes in distilled spirit drinks as a front end to ETAAS. The optimization of the main factors of the FDSE procedure was conducted to ensure the high extraction efficiency of the target analytes. The herein-developed protocol was validated and used for analysis of real samples.

2. Results and Discussion

2.1. Characterization of the Sol-Gel Graphene Oxide-Coated Polyester Fabric Membranes

The sol-gel graphene oxide-coated FPSE membranes were characterized by Fourier Transform Infrared Spectroscopy (FT-IR) and Scanning Electron Microscopy (SEM). FT-IR spectra provides important information about the functional makeup of the building blocks as well as the information regarding successful integration of the building blocks in the final product and its functional composition (Supplementary Materials, Figures S1–S3). Moreover, the SEM images shed light on the surface morphology of the fabric substrate and coated sol-gel sorbent. The scanning electron micrographs (SEM) images of (a) uncoated polyester fabric at 100× magnifications; (b) built-in pores in polyester fabric at 500× magnifications; (c) sol-gel graphene oxide-coated polyester FPSE membrane at 100× magnifications; (d) uniformity of the sol-gel graphene oxide coating at 500× magnifications are shown in Figure 1. The sol-gel coatings are uniformly distributed on the polyester substrate. The through pores of the polyester fabric remained intact even after the sol-gel sorbent coating, allowing for the rapid permeation of aqueous sample through the FPSE membrane bed during the analyte extraction. The rapid permeation of the sample through the extraction bed facilitates fast extraction kinetic.

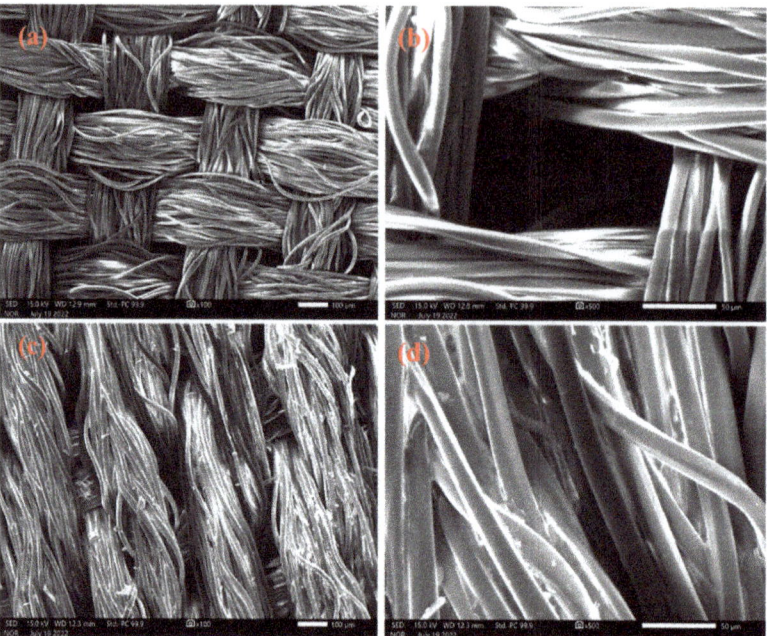

Figure 1. Scanning electron microscopy images of (**a**) uncoated polyester fabric at 100× magnifications; (**b**) demonstration of built-in pores in polyester fabric at 500× magnifications; (**c**) sol-gel graphene oxide-coated polyester FPSE membrane at 100× magnifications; (**d**) demonstration of the uniformity of the sol-gel graphene oxide coating at 500× magnifications.

2.2. Optimization of the On-Line Column Preconcentration Procedure

The main hydrodynamic and chemical factors of the SI-FDSE procedure were optimized by means of the one-variable-at-a-time approach (OVAT). In this case, each factor is individually examined, while the other factors remain constant. Method optimization was performed using aqueous standard solutions of 0.05 µg L^{-1} Cd(II), 0.15 µg L^{-1} Cu(II) and 0.5 µg L^{-1} Pb(II) at 0.01 mol L^{-1} HNO$_3$.

2.2.1. Effect of Chelating Agent and Sample Acidity

In this work, ammonium pyrrolidine dithiocarbamate (APDC) was used as chelating agent due to its well-known ability to form strong hydrophobic complexes in acidic solutions. Different concentrations (i.e., 0.01–0.5% m/v) of APDC solution were investigated to provide sufficient metal complexation and reduced chemical consumption. It was observed that an increase in the concentration up to 0.05% m/v had a positive effect, while any further increase in the concentration did not result in any additional benefits. However, since other potentially co-existing metals can consume the chelating reagent for metal complex formation, APDC was used in an excess concentration of 0.1% m/v.

The effect of sample acidity has been thoroughly investigated in earlier studies and it is evident that a stable metal complexation is observed using a nitric acid concentration of 1.0×10^{-3}–1.0×10^{-2} mol L^{-1} that corresponds to sample pH values of 2–3 [14]. This parameter plays a crucial role for the efficient on-line formation and retention of the analyte-APDC complexes onto the surface of the sol-gel sorbent. In this case, a concentration of HNO$_3$ of 0.01 mol L^{-1} was chosen to ensure the sufficient complexation.

2.2.2. Effect of Loading Flow Rate and Preconcentration Time

The sample loading flow rate and the preconcentration time in automatic time-based systems are considered to be two critical parameters since they determine the sample volume, the analyte retention and the preconcentration time. In addition, they affect the velocity of the liquid inside the column and, thus, the contact time between the target analyte and the active groups of the sorbent which is crucial for their interaction. These parameters are directly associated with the sorption equilibration and retention of the metals on the sorptive phase of the FDSE medium [24]. The sample loading flow rate was investigated between 60 µL s^{-1} and 180 µL s^{-1} for a preconcentration time of 60 s, maintaining a sample/APDC flow rate ratio of ca. 15. As shown in Figure S4, an almost linear increase in method sensitivity was reported up to 150 µL s^{-1}, demonstrating the efficient retention of the in situ-created metal complexes. At higher flow rates, a slight decrease in the response was observed. As a result, different loading flow rates can be employed in the on-line SI-FDSE procedure depending on the required sensitivity, the required sample throughput, and the sample availability. In this case, a sample loading flow rate of 150 µL s^{-1} was considered satisfactory for high method sensitivity, and it was finally adopted for further studies.

Subsequently, the preconcentration time was studied between 15 and 100 s. As it is presented in Figure S5, the increase in the absorbance for all analytes was practically proportional to the increase in preconcentration time. As a compromise between method high sensitivity and high sample throughput, a preconcentration time of 90 s was finally chosen.

2.2.3. Effect of Eluent Type and Elution Flow Rate

In this study, MIBK was chosen as an eluent. This solvent has been proven to be a powerful eluent compared to organic solvents with lower polarity due to the limited dispersion of the eluted analyte and the efficient desorption of the retained complexes from the sorbent [25]. The elution flow rate of MIBK within the SI-FDSE system was investigated between 5 and 20 µL s^{-1}. No significant response variations were observed within the examined range. Thus, an elution flow rate of 10 µL s^{-1} was chosen to ensure sufficient interaction between the adsorbed analytes and the MIBK.

2.2.4. Effect of Sample Volume Injected in Graphite Furnace

In order to achieve good method sensitivity, the effect of different sample volumes injected in the ETAAS system were studied. This parameter determines the amount of analyte that is finally analyzed. In this study, the first portion of the eluent zone was injected in the ETAAS system, since it is expected to contain the desorbed analytes at higher concentration in comparison with other parts along the zone of the eluent. For this purpose, different quantities between 25 and 45 µL were studied. Sufficient method sensitivity was obtained using 35 µL of eluent and, thus, this value was finally chosen.

2.2.5. Effect of Sample Ethanol Content

The effect of ethanol concentration on the absorbance was studied between 5% and 40% v/v ethanol in water. As shown in Figure 2, absorbance for all analytes was constant at ethanol contents varying between 5 and 20% v/v, while a slight decrease was observed at higher ethanol content. This observation is in accordance with other studies about alcoholic spirit drinks, in which sample dilution is recommended since high ethanol contents reduce the retention of the target analytes due to the disruption of the interaction between them and the active groups of the sorptive phase [26,27]. Thus, to ensure the efficient retention of the metal, the distilled spirit drinks should be at least 1:1 (v/v) diluted with water prior to SI-FDSE procedure.

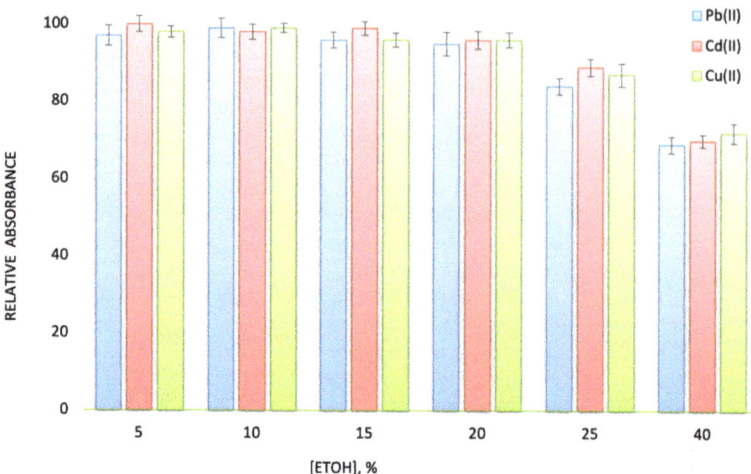

Figure 2. Effect of ethanol content (%) on the sensitivity of 0.05 µg L^{-1} Cd(II), 0.15 µg L^{-1} Cu(II), and 0.5 µg L^{-1} Pb(II). All parameters as presented in the operational sequence of the SI-FDSE-ETAAS method. The error bars were calculated based on standard deviation (±1 s).

2.3. Figures of Merit

Table 1 presents the analytical characteristics of the developed SI-FDSE-ETAAS protocol. As it can be observed, good linearity was obtained for all analytes within the studied linear range. The utilization of a preconcentration time of 60 s resulted in a high sample throughput of 24 samples h^{-1}. The enhancement factors for the target analytes were calculated as the ratio of the slope of the calibration curve for each metal after the on-line column preconcentration procedure versus the slope of the calibration curve for the same analyte without preconcentration (direct injection of 35 µL aqueous standard solution into the graphite furnace). The slopes of the batch ETAAS method were for 0.0605, 0.0052 and 0.0030 µg L^{-1} for Cd(II), Cu(II) and Pb(II), respectively.

Table 1. Figures of merit for the developed SI-FDSE-ETAAS method.

Parameter	Cd(II)	Cu(II)	Pb(II)
Slope, S (µg L^{-1})	2.3000 ± 0.1004	0.6272 ± 0.0256	0.2568 ± 0.0078
Intercept, i	0.0030 ± 0.0093	0.0022 ± 0.0100	0.0033 ± 0.0055
Linear range (µg L^{-1})	0.0064–0.160	0.0236–0.70	0.0578–1.5
Correlation coefficient, r^2	0.9985	0.9987	0.9993
LOD (ng L^{-1})	1.9	7.1	17.3
LOQ (ng L^{-1})	6.4	23.6	57.8
RSD (%)	2.4 (0.02 µg L^{-1})	2.2 (0.10 µg L^{-1})	2.9 (0.25 µg L^{-1})
Preconcentration time (s)	90	90	90
Sample throughput (h^{-1})	24	24	24
Enhancement factor	38	120	85
Sample consumption (mL)	13.5	13.5	13.5

Under optimum extraction conditions, the enhancement factors were found to be 38, 120 and 85 for Cd(II), Cu(II) and Pb(II), respectively. The limits of detection (LODs) and the limits of quantification (LOQs) were calculated according to the recommendation of the International Union of Pure and Applied Chemistry (IUPAC) following the 3 s and 10 s criteria, respectively.

The LOD values of the SI-FDSE-ETAAS method were 1.9 ng L^{-1} for Cd(II), 7.1 ng L^{-1} for Cu(II) and 17.3 ng L^{-1} for Pb(II). Moreover, the LOQ values were 6.4 ng L^{-1} for Cd(II),

23.6 ng L^{-1} for Cu(II) and 57.8 ng L^{-1} for Pb(II), demonstrating the applicability of the proposed method for trace metal analysis of distilled spirit drinks. The precision of the SI-FDSE-ETAAS method was calculated in terms of relative standard deviation (RSD) of ten repeated measurements of spiked distilled spirit drink samples containing 0.02 μg L^{-1} of Cd(II), 0.10 μg L^{-1} of Cu(II) and 0.25 μg L^{-1} of Pb(II). For all analytes, the RSD values were lower than 2.9%, indicating good precision.

The accuracy of the SI-FDSE-ETAAS method was evaluated by analyzing three different certified reference materials, i.e., NIST CRM 1643e (trace elements in water), IAEA-433 (marine sediment), and BCR 278-R (mussel tissue). Student's t-test (probability level of 95%) was conducted for the examination of the significance of the differences that occurred between the certified and the experimentally calculated values. As shown in Table 2, in all cases the t_{exp} values were lower than the $t_{crit,\,95\%}$ = 4.30, showing no statistical differences and good method accuracy.

Table 2. Determination of Cd(II), Cr(VI) and Pb(II) in certified reference materials Mean value ± standard deviation (n = 3), $t_{crit,\,95\%}$ = 4.30.

Certified Reference Material	Cd	Cu	Pb
CRM 1643e (Trace Element Water)			
Certified value (μg L^{-1})	6.568 ± 0.073	22.76 ± 0.31	19.63 ± 0.21
Found (μg L^{-1})	6.35 ± 0.29	22.26 ± 0.66	18.9 ± 1.12
Relative Error	−3.3	−2.2	−3.7
t_{exp}	1.302	1.312	1.129
IAEA-433 (Marine sediment)			
Certified value (μg L^{-1})	0.153 ± 0.033	30.8 ± 2.6	26.0 ± 2.7
Found (μg L^{-1})	0.149 ± 0.015	29.5 ± 1.2	26.4 ± 1.8
Relative Error	−2.6	−4.2	1.5
t_{exp}	0.462	1.876	−0.385
BCR 278-R (Mussel tissue)			
Certified value (μg L^{-1})	0.348 ± 0.007	9.45 ± 0.13	2.00 ± 0.04
Found (μg L^{-1})	0.33 ± 0.02	9.32 ± 0.35	1.90 ± 0.09
Relative Error	−5.2	−1.4	−5.0
t_{exp}	1.559	0.643	1.925

2.4. Interferences Studies

For a complete evaluate of the herein-developed analytical method, the effect of potentially interfering ions on the extraction performance was studied. For this purpose, a standard solution containing 0.02 μg L^{-1} of Cd(II), 0.10 μg L^{-1} of Cu(II), 0.20 μg L^{-1} of Pb(II) and different quantities of individual potential interferents were used, while the criterion of ≥5% of the analyte response was used to evaluate the existence of interferences. The results showed that the proposed SI-FDSE-ETAAS method can tolerate Al(III), Co(II), Fe(III), Mn(II), Ni(II) and Zn(II) at concentrations up to 10 mg L^{-1}, Hg(II) at concentrations up to 2.5 mg L^{-1} and Ca(II), Mg(II), Ba(II), Na(I), K(I), Br$^-$, Cl$^-$, I$^-$, SO$_4^{2-}$, NO$_3^-$ at concentrations up to 1000 mg L^{-1}. As a result, the tolerance ratio for toxic metals was higher than 12,500 for all cases.

2.5. Real Sample Analysis

The proposed SI-FDSE-ETAAS method was utilized for the elemental assessment of different distilled spirit drinks. Prior to the analysis, a 1:1 (v/v) dilution of the samples with water was performed. For the accuracy evaluation, spiked solutions at appropriate concentration levels were also analyzed. The results of real sample analysis are shown in Table 3. Cadmium, copper and lead concentration levels ranged up to 0.122 μg L^{-1}, up to 22.6 μg L^{-1}, and up to 101.5 μg L^{-1}, respectively. Moreover, the relative recoveries were 94.0–103.0%, demonstrating good method applicability.

Table 3. Analytical results for Cd, Cu, and Pb in distilled spirit drinks samples ($n = 3$).

Sample	Cd				Cu [1]				Pb			
	Added (µg L^{-1})	Found (µg L^{-1})	Recovery (%)	t_{exp}	Added (µg L^{-1})	Found (µg L^{-1})	Recovery (%)	t_{exp}	Added (µg L^{-1})	Found (µg L^{-1})	Recovery (%)	t_{exp}
Rum-1	-	0.090 ± 0.007	-	-	-	6.20 ± 0.90	-	-	-	2.12 ± 0.15	-	-
	0.200	0.284 ± 0.017	97.0	0.611	0.500	6.67 ± 0.50	94.0	0.104	0.500	2.60 ± 0.18	96.0	0.192
Rum-2	-	<LOD	-	-	-	7.90 ± 0.60	-	-	-	1.48 ± 0.10	-	-
	0.200	0.190 ± 0.014	95.0	1.237	0.500	8.41 ± 0.80	102.0	−0.022	0.500	2.000 ± 0.11	104.0	−0.315
Vodka-1	-	<LOD	-	-	-	10.30 ± 0.90	-	-	-	<LOD	-	-
	0.200	0.188 ± 0.012	94.0	1.732	0.500	10.78 ± 0.80	96.0	0.043	0.500	0.488 ± 0.661	97.6	0.031
Vodka-2	-	<LOD	-	-	-	4.60 ± 0.30	-	-	-	0.220 ± 0.011	-	-
	0.200	0.192 ± 0.015	96.0	0.924	0.500	5.09 ± 0.30	98.0	0.058	0.500	0.700 ± 0.032	96.0	1.155
Gin-1	-	<LOD	-	-	-	5.90 ± 0.20	-	-	-	<LOD	-	-
	0.200	0.210 ± 0.017	105.0	−1.019	0.500	6.38 ± 0.40	96.0	0.087	0.500	0.475 ± 0.033	95.0	1.443
Gin-2	-	0.120 ± 0.009	-	-	-	9.20 ± 0.70	-	-	-	1.50 ± 0.12	-	-
	0.200	0.325 ± 0.030	102.5	−0.289	0.500	9.73 ± 0.60	106.0	−0.087	0.500	1.97 ± 0.11	94.0	0.472
Tsipouro-1	-	0.100 ± 0.009	-	-	-	12.30 ± 1.00	-	-	-	30.2 ± 1.6 [1]	-	-
	0.200	0.293 ± 0.020	96.5	0.606	0.500	12.83 ± 1.10	106.0	−0.047	5.0	35.1 ± 1.8	98.0	0.096
Tsipouro-2	-	<LOD	-	-	-	12.10 ± 0.90	-	-	-	25.4 ± 5.0 [1]	-	-
	0.200	0.205 ± 0.015	102.5	−0.577	-	12.59 ± 0.80	98.0	0.022	5.0	30.5 ± 1.7	101.0	−0.551

[1] 1:25 v/v dilution.

The results for the real sample analysis were compared with the concentrations found in previous works. As shown in Table 4, similar concentrations levels were observed between the equivalent types of alcoholic beverages.

Table 4. Comparison of the elemental concentration levels ($\mu g\ L^{-1}$) in distilled spirit drinks with other studies.

Sample	Cd	Cu	Pb	Ref.
Gin	0.08–1.12	-	-	[28]
Rum	ND–0.70	-	-	[28]
Rum	3.0–30	ND–640	50–220	
Gin	10–30	ND–70	100–130	[29]
Vodka	10–30	ND–90	80–380	
Rum	-	3–45	23–65	[30]
Fruit spirits	ND–6.6	-	-	[31]
Gin	-	-	ND–35.70	[32]
Rum	-	-	ND–70.00	[32]
Pomace brandy (Tsikoudia)	ND–1	55–105,000	<1–1200	[33]
Rum	ND	ND–276,000	ND–16	[34]
Rum	-	ND–400	-	[35]
Rum	-	-	1.25	[36]
Rum, vodka, gin, tsipouro	ND–0.120	4.60–12.30	ND–30.2	This work

ND: Not detected.

2.6. Comparison with Other Studies

The developed SI-FDSE-ETAAS method was compared with other previously reported on-line ETAAS methods and the results of the comparative study are summarized in Table 5.

As it can be observed, the proposed method resulted in enhanced sensitivity for all the target analytes. At the same time, the SI-FDSE-ETAAS method is associated with adequate sample consumption, fast kinetics, high enhancement factors and good method precision. It is noteworthy that the herein-reported packed microcolumn is also characterized by fabrication simplicity, negligible back pressure, and high reusability (at least 500 cycles), making it an ideal option for on-line ETAAS methodologies for routine analysis.

Table 5. Comparison of the developed SI-FDSE-ETAAS method with previously reported on-line ETAAS methods.

Analyte	On-Line Procedure	P.T. (s)	S.C. (mL)	E.F.	RSD (%)	LOD (µg L^{-1})	Ref.
Cd(II)/ Pb(II)	SI-DLLME	90	8.1	34/ 80	4.1/ 3.8	0.002/ 0.01	[37]
Cd(II)	SI-SPE based on octadecylsilane functionalized maghemite particles	250	5.0	19	3.9	0.003	[38]
Cd(II)	SI–bead injection–lab-on-valve platform equipped with a microcolumn packed with PTFE beads	52	1.25	17.2	4.3	0.015	[39]
Cu(II)	SI-SPE based on silk fibroin sorbent	90	0.9	27.3	2.2	0.008	[40]
Cd(II)	SI-single-drop micro-extraction	600	15	10	3.9	0.01	[41]
Pb(II)	FI-SPE based on a chelating resin immobilized on aminopropyl-controlled pore glass	90	3.3	20.5	3.2	0.012	[42]
Cd(II)	SI-solvent extraction-back extraction	130	13.0	21.4	0.4	0.0027	[43]
Cu(II)	SI-SPE based on PTFE-beads-packed column	17	1.0	20	1.8	0.015	[44]
Cd(II)/ Cu(II)/ Pb(II)	SI-FDSE	90	13.5	38/ 120/ 85	2.4/ 2.2/ 2.9	0.0019/ 0.0071/ 0.0173	This work

2.7. Evaluation of the Green Character of the Proposed Method

In order to evaluate the green character of the on-line SI-FDSE-ETAAS method, complexGAPI index was used [45]. This tool enables the evaluation of the main parameters used for the analytical determination (i.e., sample collection, sample storage, sample transfer, sample preparation, chemicals and reagents, instrumentation, and method type), as well as the synthetic procedure used for the preparation of novel extraction phases, stationary phases, solvents, etc. Thus, this index was selected to provide a complete assessment of the proposed method from the production of FDSE membranes to analytes determination by ETAAS. The obtained pictogram is shown in Figure 3. It can be observed that regarding the production of the novel sorptive phases most of the requirements of complexGAPI index are met (green color in the hexagon). The synthetical route is characterized by a low E-factor, thus supporting the green economy. At the same time, a high process yield is obtained, relatively mild conditions are used, and a low amount of waste is generated. Regarding the analytical part of the SI-FDSE-ETAAS protocol, micro-scale extraction is used resulting in reduced consumption of hazardous chemicals. All things considered; the proposed method exhibits a green character in compliance with the principles of Green Analytical Chemistry [46].

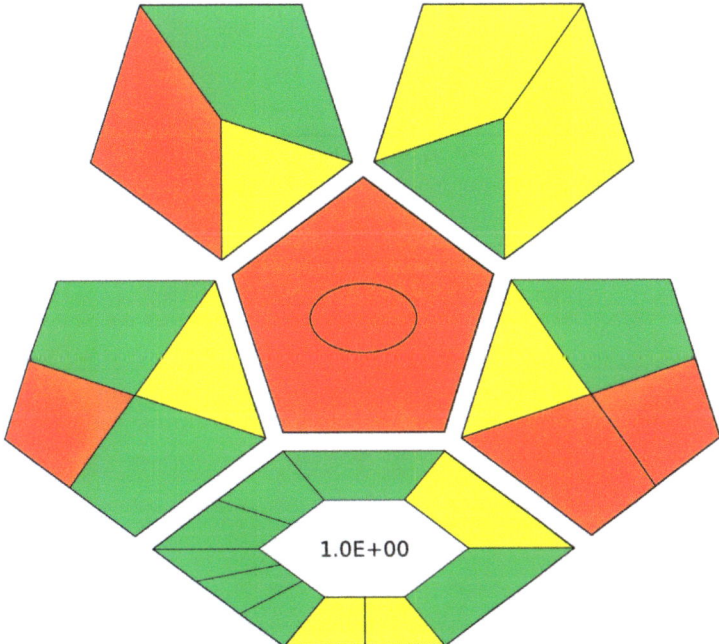

Figure 3. ComplexGAPI pictogram for the SI-FDSE-ETAAS method.

3. Materials and Methods

3.1. Reagents, Materials and Samples

Concentrated nitric acid (65%), methyl isobutyl ketone (MIBK) and stock standard solutions (1000 mg L^{-1}) of Cd(II), Cu(II) and Pb(II) were purchased from Merck (Merck, Darmstadt, Germany). Ultra-pure water produced by a Milli-Q system, also by Merck (Merck, Darmstadt, Germany), was used throughout the study. Working solutions for the target analytes were prepared on a daily basis through appropriate dilution of the stock standards. Ammonium pyrrolidine dithiocarbamate (APDC) and sol-gel precursors, methyl trimethoxysilane (MTMS) and tetramethyl orthosilicate (TMOS), were purchased from Sigma Aldrich (St. Louis, MO, USA). Aqueous solutions of APDC were prepared daily at appropriate concentration levels. The polyester fabric substrate was purchased from Joanne Fabrics (Miami, FL, USA). Isopropanol, HCl, NH$_4$OH, and methylene chloride were purchased from Fisher Scientific (Milwaukie, WI, USA). A single-layer graphene oxide dispersion in ethanol (5 mg mL^{-1}) was purchased from ACS Material (Pasadena, CA, USA). The preparation of the coated polyester fabric membranes is described in the Supplementary Materials. The proposed method was applied to the analysis of different types of alcoholic beverages (i.e., rum, vodka, gin, and tsipouro) with an ethanol content up to 40%. All samples were randomly purchased during April 2022 from the local market of Thessaloniki.

3.2. Instrumentation

A Perkin Elmer, Norwalk, Connecticut, USA (http://www.perkinelmer.com, accessed on 22 February 2023) model 5100 PC flame atomic absorption spectrometer with Zeeman-effect background correction equipped with AS-71 furnace auto-sampler was employed throughout study. Pyrolytically coated THGA graphite tubes (Perkin Elmer) with integrated L'vov platform were used. Argon 99.996% was applied as purge and protective gas. A hollow cathode lamp operated at 4 mA for Cd, a hollow cathode lamp operated at 30 mA for Cu, and an electrodeless discharge lamp operated at 10 W for Pb were used as light sources. The wavelength was set at 228.8 nm resonance line, at 324.7 nm and 283.3 nm for cadmium, copper and lead, respectively, while the monochromator spectral bandpass (slit) was set at 0.7 nm. Integrated absorbance (peak area) was used for signal evaluation throughout the study. The graphite furnace temperature/time program is presented in Table S1 (Supplementary Materials).

The online FDSE procedure took place utilizing a sequential injection (SI) system model FIALab®-3000 (FIAlab, Alitea, USA) consisting of a syringe pump (SP) with a glass barrel capacity of 1000 µL (Cavro, Sunnyvale, CA, USA), one six-port selection valve (SV), and one peristaltic pump (P) for sample or standard solution propulsion. The schematic diagram of the SI-FDSE-ETAAS procedure is depicted in Figure 4.

The SI system was controlled by a personal computer and the FIAlab application software for windows v.5.9.245 (http://www.flowinjection.com, accessed on 22 February 2023). The complete system was commanded by the computer that controlled the SIA system.

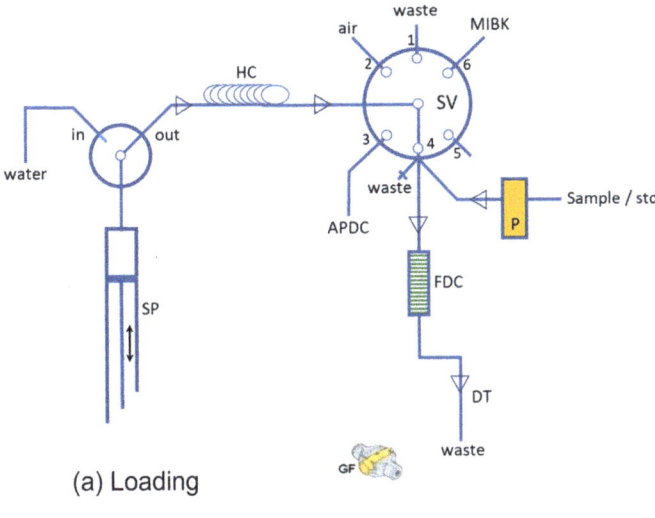

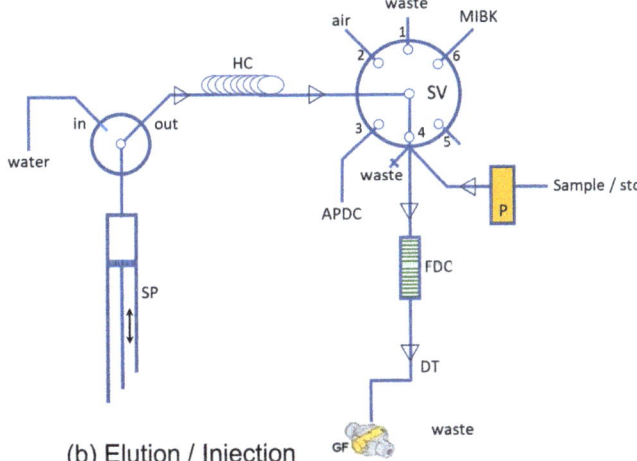

Figure 4. Schematic diagram of the proposed SI-FDSE-ETAAS manifold (**a**) loading, (**b**) elution/injection.

For the characterization of the FDSE disks, a Philips XL30 Scanning Electron Microscope (Cambridge, MA, USA) equipped with an EDAX detector and an Agilent Cary 670 FTIR Spectrometer (Santa Clara, CA, USA) were used.

3.3. Fabrication of the FDSE Microcolumn

The FDSE disks were obtained by cutting the fabric material into 40 disks of 4.0 mm i.d. using an appropriate hollow punch tool. For the fabrication of the FDSE microcolumn, a 1000 μL disposable polypropylene syringe of 10.0 cm length and 4.0 mm i.d. was shortened to 1.5 cm and packed with the FDSE disks. Prior to its use, the freshly prepared FDSE microcolumn was flashed with a solution of 1.0 mol L^{-1} HNO$_3$ solution, followed by rinsing with Milli-Q water to remove any undesired impurities from the surface of the FDSE disks. The microcolumn fabrication procedure is characterized by simplicity, and the obtained microcolumn can be easily repacked with new FDSE membranes when necessary.

Moreover, no frits or glass wool is necessary to block the FDSE media at the upper and lower part of the syringe. Due to the easy permeation of the incoming flow, through the pores of the FDSE disks, negligible back-pressure was observed, while high flow rates could be utilized resulting in rapid extraction and high extraction efficiency. It is noteworthy that the prepared FDSE microcolumns were proven to be reusable for at least 500 sample loading/elution cycles.

3.4. On-Line FDSE Analytical Procedure

The herein-developed automatic on-line analytical SI-FDSE-ETAAS procedure was operated in 12 individual steps categorized under four main sequences *a–d*. Table 6 (Supplementary Materials) summarizes the operational sequences used for the determination of Cd(II), Cu(II) and Pb(II). Using the proposed program, the sample analysis time was 150 s, and the sampling frequency (f) was 24 h^{-1}. For all instances, five replicates were performed. The detailed sequences are given in the Supplementary Materials.

Table 6. Operational sequence of the SI-FDSE-ETAAS method for metal determination.

Step [a]	V Position	SV Position	SP Flow Rate (µL s^{-1})	SP Operation	SP Volume (µL)	P Operation	Commentary
1	IN	2	80	Aspirate	50	OFF (*)	Water into SP
2	OUT	2	5	Aspirate	10	OFF	Air segment into HC
3	OUT	3	80	Aspirate	910	OFF	APDC into HC
4	OUT	4	10	Dispense	900	ON (*)	Sample loading, preconcentration for 90 s
5	OUT	1	80	Dispense	70	OFF	Emptying the SP
6	IN	1	80	Aspirate	300	OFF	Water into SP
7	OUT	6	10	Aspirate	500	OFF	MIBK into HC
8	OUT	4	5	Dispense	280	OFF	Elution– Transportation of eluent up to the exit of DT
9							DT into GF
10	OUT	4	5	Dispense	35	OFF	35 µL of MIBK into GF DT back to waste.
11							Starting the ETAAS program/measuring
12	OUT	4	50	Dispense	485	OFF	Cleaning of micro-column, Emptying the SP

[a] Sample propulsion with flow rate, (150 µL s^{-1}); SP, syringe pump; V, valve of SP; SV, multi-position selection valve; P, peristaltic pump; GF, graphite furnace of ETAAS. * ON means that the pump is operating and OFF means that the pump is not operating.

4. Conclusions

A novel SI-FDSE on-line automatic system was proposed as a front-end to ETAAS for the monitoring of trace amounts of toxic elements in distilled spirit drinks. The novel manifold showed ease in fabrication, low cost, fast extraction kinetics, high extraction performance, high reusability and low back-pressure. Moreover, the developed analytical method exhibited good accuracy, precision, linearity and low LOD and LOQ values. The proposed SI-FDSE-ETAAS method was successfully employed for the determination of Cd(II), Cu(II) and Pb(II) in a wide range of alcoholic beverages. The proposed system could be also used for the monitoring of other toxic elements with similar complexing reagents in other beverage samples (e.g., beer, wine). This system could be also expanded to extract other analytes (e.g., nickel, cobalt, etc.). For this purpose, different sol-gel-coated FDSE media can also be examined to ensure optimum extraction performance for the new analytes. Finally, the herein-developed SI-FDSE platform could also be used as a front-end to inductively coupled plasma atomic emission spectrometry after back-extraction, aiming to perform multi-elemental analysis.

Supplementary Materials: The following supporting information can be downloaded at: https://www.mdpi.com/article/10.3390/molecules28052103/s1, Fourier Transform Infra-Red Spectroscopy (FT-IR), Preparation of the of the coated polyester fabric membranes [47,48]. Main sequences of the on-line procedure, Figure S1: FT-IR spectra of (a) pristine graphene oxide; (b) sol-gel graphene oxide-coated polyester fabric phase sorptive extraction membrane; Figure S2: FT-IR spectra of uncoated polyester substrate; Figure S3: FT-IR spectra of methyl trimethoxysilane; Figure S4: Effect of sample flow rate on the extraction efficiency of 0.05 μg L^{-1}, Cd(II), 0.15 μg L^{-1} Cu(II), and 0.5 μg L^{-1} Pb(II). All parameters as presented in Table 6. The error bars were calculated based on standard deviation (± 1 s); Figure S5: Effect of loading time on the extraction efficiency of 0.05 μg L^{-1} Cd(II), 0.15 μg L^{-1} Cu(II), and 0.5 μg L^{-1} Pb(II). All parameters as presented in Table 6. The error bars were calculated based on standard deviation (± 1 s).; Table S1: Graphite furnace temperature/time program for Cd, Cu and Pb determination in 35 μL of MIBK.

Author Contributions: Conceptualization, A.A.; methodology, N.M., A.K. and A.A.; validation, N.M. and A.A.; formal analysis, N.M., A.K. and A.A.; investigation, N.M., A.K., K.G.F. and A.A. resources, A.K. and A.A.; writing—original draft preparation, N.M.; writing—review and editing, A.K., K.G.F. and A.A.; supervision, A.A.; project administration, A.A. All authors have read and agreed to the published version of the manuscript.

Funding: This research received no external funding.

Institutional Review Board Statement: Not applicable.

Informed Consent Statement: Not applicable.

Data Availability Statement: Not applicable.

Conflicts of Interest: The authors declare no conflict of interest.

Sample Availability: Samples of the compounds/analytes are available from the authors.

References

1. Ibanez, J.G.; Carreon-Alvarez, A.; Barcena-Soto, M.; Casillas, N. Metals in alcoholic beverages: A review of sources, effects, concentrations, removal, speciation, and analysis. *J. Food Compos. Anal.* **2008**, *21*, 672–683. [CrossRef]
2. Tobiasz, A.; Walas, S. Solid-phase-extraction procedures for atomic spectrometry determination of copper. *TrAC-Trends Anal. Chem.* **2014**, *62*, 106–122. [CrossRef]
3. Jaishankar, M.; Tseten, T.; Anbalagan, N.; Mathew, B.B.; Beeregowda, K.N. Toxicity, mechanism and health effects of some heavy metals. *Interdiscip. Toxicol.* **2014**, *7*, 60–72. [CrossRef]
4. Ebrahimi, M.; Khalili, N.; Razi, S.; Keshavarz-Fathi, M.; Khalili, N.; Rezaei, N. Effects of lead and cadmium on the immune system and cancer progression. *J. Environ. Health Sci. Eng.* **2020**, *18*, 335–343. [CrossRef] [PubMed]
5. Miró, M.; Estela, J.M.; Cerdà, V. Application of flowing stream techniques to water analysis: Part III. Metal ions: Alkaline and alkaline-earth metals, elemental and harmful transition metals, and multielemental analysis. *Talanta* **2004**, *63*, 201–223. [CrossRef]
6. Ferreira, S.L.C.; Bezerra, M.A.; Santos, A.S.; dos Santos, W.N.L.; Novaes, C.G.; de Oliveira, O.M.C.; Oliveira, M.L.; Garcia, R.L. Atomic absorption spectrometry—A multi element technique. *TrAC-Trends Anal. Chem.* **2018**, *100*, 1–6. [CrossRef]
7. Giakisikli, G.; Anthemidis, A.N. An automatic stirring-assisted liquid–liquid microextraction system based on lab-in-syringe platform for on-line atomic spectrometric determination of trace metals. *Talanta* **2017**, *166*, 364–368. [CrossRef]
8. López-Lorente, Á.I.; Pena-Pereira, F.; Pedersen-Bjergaard, S.; Zuin, V.G.; Ozkan, S.A.; Psillakis, E. The ten principles of green sample preparation. *TrAC-Trends Anal. Chem.* **2022**, *148*, 116530. [CrossRef]
9. Manousi, N.; Kabir, A.; Furton, K.G.; Stathogiannopoulou, M.; Drosaki, E.; Anthemidis, A. An automatic on-line sol-gel pyridylethylthiopropyl functionalized silica-based sorbent extraction system coupled to flame atomic absorption spectrometry for lead and copper determination in beer samples. *Food Chem.* **2022**, *394*, 133548. [CrossRef]
10. Ribeiro, G.C.; Coelho, L.M.; Coelho, N.M.M. Determination of nickel in alcoholic beverages by faas after online preconcentration using Mandarin Peel (Citrus reticulata) as biosorbent. *J. Braz. Chem. Soc.* **2013**, *24*, 1072–1078. [CrossRef]
11. Kabir, A.; Samanidou, V. Fabric Phase Sorptive Extraction: A Paradigm Shift Approach in Analytical and Bioanalytical Sample Preparation. *Molecules* **2021**, *26*, 865. [CrossRef]
12. Manousi, N.; Kabir, A.; Zachariadis, G.A. Green bioanalytical sample preparation: Fabric phase sorptive extraction. *Bioanalysis* **2021**, *13*, 693–710. [CrossRef]
13. Kazantzi, V.; Anthemidis, A. Fabric sol–gel phase sorptive extraction technique: A review. *Separations* **2017**, *4*, 20. [CrossRef]
14. Anthemidis, A.; Kazantzi, V.; Samanidou, V.; Kabir, A.; Furton, K.G. An automated flow injection system for metal determination by flame atomic absorption spectrometry involving on-line fabric disk sorptive extraction technique. *Talanta* **2016**, *156–157*, 64–70. [CrossRef]

15. Kazantzi, V.; Samanidou, V.; Kabir, A.; Furton, K.G.; Anthemidis, A. On-line fabric disk sorptive extraction via a flow preconcentration platform coupled with atomic absorption spectrometry for the determination of essential and toxic elements in biological samples. *Separations* **2018**, *5*, 34. [CrossRef]
16. Celeiro, M.; Acerbi, R.; Kabir, A.; Furton, K.G.; Llompart, M. Development of an analytical methodology based on fabric phase sorptive extraction followed by gas chromatography-tandem mass spectrometry to determine UV filters in environmental and recreational waters. *Anal. Chim. Acta X* **2020**, *4*, 100038. [CrossRef]
17. Kalogiouri, N.P.; Ampatzi, N.; Kabir, A.; Furton, K.G.; Samanidou, V.F. Development of a capsule phase microextraction methodology for the selective determination of coumarin in foodstuff analyzed by HPLC-DAD. *Adv. Sample Prep.* **2022**, *3*, 100026. [CrossRef]
18. Mazaraki, K.; Kabir, A.; Furton, K.G.; Fytianos, K.; Samanidou, V.F.; Zacharis, C.K. Fast fabric phase sorptive extraction of selected β-blockers from human serum and urine followed by UHPLC-ESI-MS/MS analysis. *J. Pharm. Biomed. Anal.* **2021**, *199*, 114053. [CrossRef]
19. Dvořák, M.; Miró, M.; Kubáň, P. Automated Sequential Injection-Capillary Electrophoresis for Dried Blood Spot Analysis: A Proof-of-Concept Study. *Anal. Chem.* **2022**, *94*, 5301–5309. [CrossRef]
20. Horstkotte, B.; Miró, M.; Solich, P. Where are modern flow techniques heading to? *Anal. Bioanal. Chem.* **2018**, *410*, 6361–6370. [CrossRef]
21. Zacharis, C.K.; Theodoridis, G.A.; Voulgaropoulos, A.N. Coupling of sequential injection with liquid chromatography for the automated derivatization and on-line determination of amino acids. *Talanta* **2006**, *69*, 841–847. [CrossRef]
22. Manousi, N.; Rosenberg, E.; Deliyanni, E.A.; Zachariadis, G.A. Sample preparation using graphene-oxide-derived nanomaterials for the extraction of metals. *Molecules* **2020**, *25*, 2411. [CrossRef] [PubMed]
23. Sun, J.; Liang, Q.; Han, Q.; Zhang, X.; Ding, M. One-step synthesis of magnetic graphene oxide nanocomposite and its application in magnetic solid phase extraction of heavy metal ions from biological samples. *Talanta* **2015**, *132*, 557–563. [CrossRef] [PubMed]
24. Zhao, S.L.; Chen, F.S.; Zhang, J.; Ren, S.B.; Liang, H.D.; Li, S.S. On-line flame AAS determination of traces Cd(II) and Pb(II) in water samples using thiol-functionalized SBA-15 as solid phase extractant. *J. Ind. Eng. Chem.* **2015**, *27*, 362–367. [CrossRef]
25. Anthemidis, A.; Tzili, A. An automatic on-line sorptive extraction platform for palladium determination in automobile exhaust catalysts based on a PTFE-turnings packed column, flow injection analysis and flame atomic absorption spectrometry. *Int. J. Environ. Anal. Chem.* **2021**, 1–13. [CrossRef]
26. Rascón, A.J.; Azzouz, A.; Ballesteros, E. Use of semi-automated continuous solid-phase extraction and gas chromatography–mass spectrometry for the determination of polycyclic aromatic hydrocarbons in alcoholic and non-alcoholic drinks from Andalucía (Spain). *J. Sci. Food Agric.* **2019**, *99*, 1117–1125. [CrossRef] [PubMed]
27. Regueiro, J.; Wenzl, T. Determination of bisphenols in beverages by mixed-mode solid-phase extraction and liquid chromatography coupled to tandem mass spectrometry. *J. Chromatogr. A* **2015**, *1422*, 230–238. [CrossRef]
28. Mena, C.; Cabrera, C.; Lorenzo, M.L.; López, M.C. Cadmium levels in wine, beer and other alcoholic beverages: Possible sources of contamination. *Sci. Total Environ.* **1996**, *181*, 201–208. [CrossRef]
29. Iwegbue, C.M.A.; Overah, L.C.; Bassey, F.I.; Martincigh, B.S. Trace metal concentrations in distilled alcoholic beverages and liquors in Nigeria. *J. Instig. Brew.* **2014**, *120*, 521–528. [CrossRef]
30. Barbeira, P.J.S.; Stradiotto, N.R. Simultaneous determination of trace amounts of zinc, lead and copper in rum by anodic stripping voltammetry. *Talanta* **1997**, *44*, 185–188. [CrossRef]
31. Tatarková, M.; Baška, T.; Ulbrichtová, R.; Kuka, S.; Sovičová, M.; Štefanová, E.; Malobická, E.; Hudečková, H. Determination of Cadmium and Chromium in Fruit Spirits Intended for Own Consumption Using Graphite Furnace Atomic Absorption Spectrometry. *Acta Med.* **2021**, *64*, 213–217. [CrossRef]
32. Mena, C.M.; Cabrera, C.; Lorenzo, M.L.; Lopez, M.C. Determination of Lead Contamination in Spanish Wines and Other Alcoholic Beverages by Flow Injection Atomic Absorption Spectrometry. *J. Agric. Food Chem.* **1997**, *45*, 1812–1815. [CrossRef]
33. Galani-Nikolakaki, S.; Kallithrakas-Kontos, N.; Katsanos, A.A. Trace element analysis of Cretan wines and wine products. *Sci. Total Environ.* **2002**, *285*, 155–163. [CrossRef]
34. Sampaio, O.M.; Reche, R.V.; Franco, D.W. Chemical profile of rums as a function of their origin. The use of chemometric techniques for their identification. *J. Agric. Food Chem.* **2008**, *56*, 1661–1668. [CrossRef]
35. Cardoso, D.R.; Andrade-Sobrinho, L.G.; Leite-Neto, A.F.; Reche, R.V.; Isique, W.D.; Ferreira, M.M.C.; Lima-Neto, B.S.; Franco, D.W. Comparison between cachaça and rum using pattern recognition methods. *J. Agric. Food Chem.* **2004**, *52*, 3429–3433. [CrossRef]
36. Elçi, L.; Arslan, Z.; Tyson, J.F. Determination of lead in wine and rum samples by flow injection-hydride generation-atomic absorption spectrometry. *J. Hazard. Mater.* **2009**, *162*, 880–885. [CrossRef]
37. Anthemidis, A.N.; Ioannou, K.I.G. Development of a sequential injection dispersive liquid-liquid microextraction system for electrothermal atomic absorption spectrometry by using a hydrophobic sorbent material: Determination of lead and cadmium in natural waters. *Anal. Chim. Acta* **2010**, *668*, 35–40. [CrossRef]
38. Giakisikli, G.; Anthemidis, A.N. Automated magnetic sorbent extraction based on octadecylsilane functionalized maghemite magnetic particles in a sequential injection system coupled with electrothermal atomic absorption spectrometry for metal determination. *Talanta* **2013**, *110*, 229–235. [CrossRef]
39. Miró, M.; Jończyk, S.; Wang, J.; Hansen, E.H. Exploiting the bead-injection approach in the integrated sequential injection lab-on-valve format using hydrophobic packing materials for on-line matrix removal and preconcentration of trace levels of

cadmium in environmental and biological samples via fomation of non-charged chelates prior to ETAAS detection. *J. Anal. At. Spectrom.* **2003**, *18*, 89–98. [CrossRef]
40. Chen, X.W.; Huang, L.L.; He, R.H. Silk fibroin as a sorbent for on-line extraction and preconcentration of copper with detection by electrothermal atomic absorption spectrometry. *Talanta* **2009**, *78*, 71–75. [CrossRef]
41. Anthemidis, A.N.; Adam, I.S.I. Development of on-line single-drop micro-extraction sequential injection system for electrothermal atomic absorption spectrometric determination of trace metals. *Anal. Chim. Acta* **2009**, *632*, 216–220. [CrossRef] [PubMed]
42. Vereda Alonso, E.; Siles Cordero, M.T.; García De Torres, A.; Cano Pavón, J.M. Lead ultra-trace on-line preconcentration and determination using selective solid phase extraction and electrothermal atomic absorption spectrometry: Applications in seawaters and biological samples. *Anal. Bioanal. Chem.* **2006**, *385*, 1178–1185. [CrossRef] [PubMed]
43. Wang, J.; Hansen, E.H. Development of an automated sequential injection on-line solvent extraction-back extraction procedure as demonstrated for the determination of cadmium with detection by electrothermal atomic absorption spectrometry. *Anal. Chim. Acta* **2002**, *456*, 283–292. [CrossRef]
44. Yu, Y.L.; Du, Z.; Wang, J.H. Determination of Copper in Seawater Using a Sequential Injection System Incorporating a Sample Pretreatment Module Coupled to Electrothermal Atomic Absorption Spectrometry. *Chin. J. Anal. Chem.* **2007**, *35*, 431–434. [CrossRef]
45. Płotka-Wasylka, J.; Wojnowski, W. Complementary green analytical procedure index (ComplexGAPI) and software. *Green Chem.* **2021**, *23*, 8657–8665. [CrossRef]
46. Gałuszka, A.; Migaszewski, Z.; Namieśnik, J. The 12 principles of green analytical chemistry and the SIGNIFICANCE mnemonic of green analytical practices. *TrAC-Trends Anal. Chem.* **2013**, *50*, 78–84. [CrossRef]
47. Samanidou, V.; Filippou, O.; Marinou, E.; Kabir, A.; Furton, K.G. Sol–gel-graphene-based fabric-phase sorptive extraction for cow and human breast milk sample cleanup for screening bisphenol A and residual dental restorative material before analysis by HPLC with diode array detection. *J. Sep. Sci.* **2017**, *40*, 2612–2619. [CrossRef] [PubMed]
48. Parvinzadeh, M.; Ebrahimi, I. Influence of atmospheric-air plasma on the coating of a nonionic lubricating agent on polyester fiber. *Radiat. Eff. Defects Solids* **2011**, *166*, 408–416. [CrossRef]

Disclaimer/Publisher's Note: The statements, opinions and data contained in all publications are solely those of the individual author(s) and contributor(s) and not of MDPI and/or the editor(s). MDPI and/or the editor(s) disclaim responsibility for any injury to people or property resulting from any ideas, methods, instructions or products referred to in the content.

Article

A Low-Cost Colorimetric Assay for the Analytical Determination of Copper Ions with Consumer Electronic Imaging Devices in Natural Water Samples

Argyro G. Gkouliamtzi, Vasiliki C. Tsaftari, Maria Tarara and George Z. Tsogas *

Laboratory of Analytical Chemistry, School of Chemistry, Faculty of Sciences, Aristotle University of Thessaloniki, GR-54124 Thessaloniki, Greece
* Correspondence: gtsogkas@chem.auth.gr

Abstract: This study reports a new approach for the determination of copper ions in water samples that exploits the complexation reaction with diethyldithiocarbamate (DDTC) and uses widely available imaging devices (i.e., flatbed scanners or smartphones) as detectors. Specifically, the proposed approach is based on the ability of DDTC to bind to copper ions and form a stable Cu-DDTC complex with a distinctive yellow color detected with the camera of a smartphone in a 96-well plate. The color intensity of the formed complex is linearly proportional to the concentration of copper ions, resulting in its accurate colorimetric determination. The proposed analytical procedure for the determination of Cu^{2+} was easy to perform, rapid, and applicable with inexpensive and commercially available materials and reagents. Many parameters related to such an analytical determination were optimized, and a study of interfering ions present in the water samples was also carried out. Additionally, even low copper levels could be noticed by the naked eye. The assay performed was successfully applied to the determination of Cu^{2+} in river, tap, and bottled water samples with detection limits as low as 1.4 μM, good recoveries (89.0–109.6%), adequate reproducibility (0.6–6.1%), and high selectivity over other ions present in the water samples.

Keywords: colorimetric determination; DDTC complexation; copper ions; simple imaging devices; natural water samples

Citation: Gkouliamtzi, A.G.; Tsaftari, V.C.; Tarara, M.; Tsogas, G.Z. A Low-Cost Colorimetric Assay for the Analytical Determination of Copper Ions with Consumer Electronic Imaging Devices in Natural Water Samples. *Molecules* 2023, 28, 4831. https://doi.org/10.3390/molecules28124831

Academic Editor: Gavino Sanna

Received: 23 May 2023
Revised: 15 June 2023
Accepted: 15 June 2023
Published: 17 June 2023

Copyright: © 2023 by the authors. Licensee MDPI, Basel, Switzerland. This article is an open access article distributed under the terms and conditions of the Creative Commons Attribution (CC BY) license (https://creativecommons.org/licenses/by/4.0/).

1. Introduction

Copper (Cu) is a metal that belongs in group 11 of the periodic table, and it is found naturally in the earth's solid crust. It is a soft, malleable, and ductile metal with great chemical and physical properties such as remarkably high thermal and electrical conductivity. It is such an important metal that it gave its name to an era of the prehistoric ages, "the copper age". For centuries, humans have used it to produce copper alloys including brass and bronze [1]. Copper is one of the most common metals used to manufacture many products, including electrical and electronics equipment; building construction, machinery, and consumer products; and spare parts for hydraulic systems and plumbing materials such as household water pipes [2,3]. It can be found in drinking water, when the water passes through domestic plumbing that contains copper piping or fittings, in its most stable oxidation state of Cu^{2+}. In many countries worldwide, the monitoring of copper in tap water systems [4] and urban wastewater [5] and the possible health effects of its concentration is of great importance.

The human body needs an amount of copper as an essential micronutrient to stay healthy, but in high concentrations, it is considered to be harmful. According to the Food and Drug Administration, a small daily dosage of copper of 1.4 mg for men and 1.1 mg for women is essential to maintain good human health [6]. Major food sources of copper are beef liver, shellfish, nuts, vegetables, mushrooms, and chocolate [7]. Exposure to high daily doses of copper or metal accumulation can cause serious health problems, including Wilson's disease [8]. Short-term exposure to high levels of copper can cause gastrointestinal distress, and long-term exposure and severe cases of copper poisoning may cause anemia, disruption of liver and kidney functions, and finally cell death—a recently studied disease called cuproptosis [9,10]. Therefore, the development of rapid, accurate, and easily applicable methods for the quantitative determination of copper ions in water samples is considered of great importance.

In terms of analytical chemistry, a vast number of recent analytical methods have been developed and validated for the quantification of Cu^{2+} in several samples, including biochemical, food, water, and wastewater samples. These methods are mainly spectroscopic and require expensive instrumentation and specialized staff, including batch UV-Vis spectrometry [11–14], flame atomic absorption spectroscopy (FAAS) [15–18], graphite furnace atomic absorption spectrometry (GFAAS) [19,20], inductively coupled plasma (ICP) [21,22], flow injection [23], and fluorescence [24–27].

Diethyldithiocarbamate (DDTC) is a dithiocarbamate pesticide and has been used for the complexation of many metal ions such as Cr^{3+}, Zn^{2+}, Ni^{2+}, Co^{2+}, Fe^{3+}, and Mn^{2+}, resulting in the formation of stable metal complexes [28]. One of the various metal ions that interacts strongly with DDTC is Cu^{2+}, forming a stable Cu(II)-DDTC complex [29–32].

Although the aforementioned analytical methods are accurate, provide low detection limits, and are reproducible, they, unfortunately, require trained personnel, expensive instrumentation and consumables, and high solvent volumes for the analytical detection of copper ions. In remote areas or in facilities with reduced budgets, these drawbacks are major obstacles to the development of precise analytical methods when multiple samples must be analyzed in short time periods. Motivated by these problems, we developed a new colorimetric analytical method for the determination of Cu^{2+} in natural water samples. Our research team has taken a step forward in the creation and development of low-cost and simple analytical colorimetric sensors, which can be applied with minimal resources and instrumentation to provide rapid and reliable results in the determination of various analytes carried out by a smartphone detector or a flatbed scanner [33–35].

In an effort to further simplify the overall analytical process, we report herein a 96-well plate assay for the analytical determination of Cu^{2+} relying on the colorimetric alteration that is analogous to copper concentration because of the complexation reaction between copper cations and DDTC molecules in a neutral environment. The analytical method designed is cheap, easy, and fast to implement, and the analytical protocol used for this determination is easily feasible with minimal technical expertise and is instrument-free. Based on this protocol, the determination of copper ions can be performed by the sequential addition of the pH-conditioning reagent (distilled water or the necessary acid or base), the complexation reagent (DDTC), and the analyte in the 96-well plate cells. We waited for a short time period at room temperature for the full development of the complex color, taking a photograph with a smartphone camera or a flatbed scanner and measurement with the ImageJ program. The color change during the development of this method was also evident with the naked eye (concentrations up to 10 µM had a pale yellow color, concentrations from 10 to 25 µM had a relatively bright yellow color, and concentrations above 25 µM had a very intense yellow color). These three areas can be distinguished by an appropriate experimental design and a semi-quantitative determination of copper ions can be accomplished. Finally, this method's applicability was tested for the determination of Cu^{2+} with promising results in terms of accuracy, sensitivity, and reproducibility.

2. Results and Discussion

2.1. Optimization Parameters

The possible effect of many parameters for the optimization of the proposed analytical procedure for the colorimetric determination of Cu^{2+} in natural water samples was studied in detail.

2.1.1. Effect of Reagent Series

After the preliminary experiments for the successful formation of the colored complex of Cu^{2+} and DDTC, the first parameter studied was the sequence of reagents added to the cells of the plate. We tested three different approaches for the addition of the reagents to complex the copper ions and thus to form the colored complex. On the first try, we added DDTC–deionized water–Cu^{2+} (1), on the second try DDTC–Cu^{2+}–deionized water (2), and on the third try deionized water–DDTC–Cu^{2+} (3) for 10 and 30 µmol L^{-1} Cu^{2+} concentrations. From Figure 1a, it is obvious that at the lower concentration, the sequence of the reagents had no effect on the net color intensity (the color of the blank minus color of the sample), but at the higher concentration, a significant increase in the net color intensity was observed for the third reagent sequence. This is expected from the point of view of analytical chemistry because it is optimal for the analyte to be added last, which occurs in the analysis of real samples. Thus, the third sequence of reagent addition was used for the experiments.

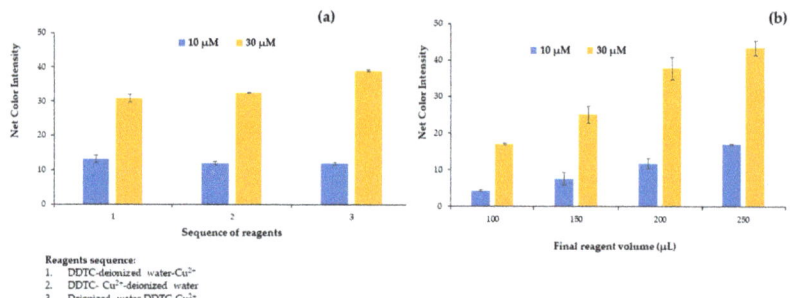

Figure 1. Effect of (**a**) reagent sequence and (**b**) final reagent volume on the net color signal for the proposed method.

2.1.2. Effect of Plate Volume

The next parameter studied was the influence of the final reagent volume contained in the plate cells for 10 and 30 µmol L^{-1} of Cu^{2+}. The plate cells had a maximum volume content of 250 µL, so 4 different volumes were tested from 100 to 250 µL. Volumes less than 100 µL were difficult to handle with minimum errors, and taking a photograph was impossible to accomplish. As depicted in Figure 1b, there is an upward trend of the signal up to the maximum possible volume that can be placed on the plate. The main disadvantage of using smaller volumes was the difficulty of taking the photo and correspondingly measuring the color variation due to the limited surface area visible to the smartphone lens because the solution was at a low level inside each plate cell.

2.1.3. Effect of DDTC Concentration

After determining the reagent sequence and the final volume added in each cell of the plate that we used in the analytical methodology, the influence of DDTC concentration in the evolution of the complexation reaction was studied in the range from 0.025 to 0.5 mmol L^{-1} by adding the appropriate diluted solution to the well. The DDTC concentrations of the solutions studied were 0.025, 0.04, 0.05, 0.075, 0.1, 0.25, and 0.5 mM. The experiments proved that maximum colorimetric values were achieved for a DDTC concentration of 0.25 mmol L^{-1}. Thus, this concentration was chosen for the subsequent experiments (Figure 2a). It is obvious that for lower concentrations there were not enough DDTC molecules to complex with Cu^{2+}, while for higher concentrations a flattening of the net colorimetric signal was observed. Even though the net signal was similar for DDTC concentrations higher than 0.075 mM, to ensure that Cu^{2+} quantitatively reacted with DDTC the 0.25 mmol L^{-1} DDTC concentration value was chosen for the experiments.

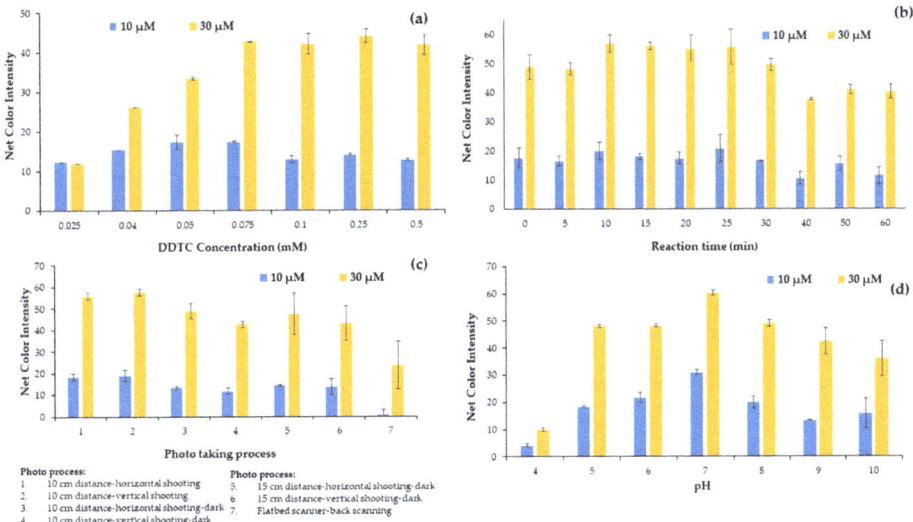

Figure 2. Effect of (**a**) DDTC concentration, (**b**) reaction time, (**c**) photo process and (**d**) pH on the net color signal for the proposed method.

2.1.4. Effect of Reaction Time

The complexation reaction between copper ions and DDTC molecules is rapid even at room temperature, and its intensity can be measured immediately after the addition of the DDTC molecules. Nevertheless, due to the quantitative formation of the complex during this study, it was necessary to study the reaction time for the colorimetric determination of the copper ions. The influence of the reaction time was studied in the range of 0 to 60 min with time intervals of 5 min from 0 to 30 min and intervals of 10 min from 30 to 60 min of the reaction time, as shown in Figure 2b. The slight decrease in the colorimetric signal for the higher internal times was attributed to the quenching of the color after a period of time. It is obvious from Figure 2b that a reaction time of 10 min is adequate for taking the photograph.

2.1.5. Effect of Photo Capture Apparatus

Next, the selection of the apparatus used to capture the photographs of the plates was studied. A smartphone and a flatbed scanner were used for this purpose. Because, unfortunately, in our lab there is a scanner that scans only from the bottom, we had the one and only option of taking the photo from the bottom of the plate, with disappointing results compared to the smartphone used. The smartphone photograph was tested with different layouts and alignments at a 10 cm distance from the plate, and we used a dark protective box to identify the effect of environmental conditions on the photo taking and consequently on the alarm signal (Figure 2c). The net signal intensity (sample minus blank signal) decreased when measured with the flatbed scanner, and although the scanner is considered to be less affected by external factors such as solar radiation, we had to rely on the mobile phone due to the large signal difference (Figure 2c).

2.1.6. pH Effect

DDTC molecules and Cu^{2+} ions reacted quickly for the generation of a yellow $Cu(DDTC)_2$ complex. It has been proved in previous reports that this complexation reaction takes place at neutral pH values [31,32]. In contrast, the chelation procedure of DDTC and Cu^{2+} was found in previous reports to be sensitive to pH [36], and the pH value of the solutions or the water samples might affect the efficiency of the proposed method. Motivated by these findings, we decided that the next parameter we should study in detail was the influence of pH on the formation of the metal complex. We tested the effect of different concentrations of hydrochloric acid and sodium hydroxide; in order to achieve higher colorimetric values, pH values from 4 to 10 were studied. Dilute HCl and NaOH solutions were prepared and used to achieve different pH values. As shown in Figure 2d, the color signal of the Cu(DDTC)$_2$ complex reached a peak as the pH raised from 4 to 7, but the color intensity decreased as the pH increased to 10. The maximum color intensity was determined at pH 7, which was conducive to the effective formation of the complex. For possible highly acidic or basic samples, buffer solutions such as phosphate buffer can replace deionized water and regulate the pH at 7, without interfering in the DDTC-Cu complexation reaction. When the pH exceeded 7, the hydrolysis of Cu^{2+} played an important role in the solution. In acidic pH, the DDTC is unstable [37], because Na-DDTC decomposed to release the free amine and carbon disulfide. Therefore, a pH of 7 was chosen for the experimental process.

2.1.7. Effect of Ionic Strength

Ionic strength occasionally affects the formation, the stability of metal complexes, and the robustness of analytical methods. Therefore, the effect of ionic strength was considered as an important parameter to study and carried out by the addition of different concentrations of NaNO$_3$ solutions to the well cells in the range between 0.01 and 0.75 mol L^{-1}. As can be seen from Figure 3, the alteration of ionic strength had practically no influence on the stability of the Cu-DDTC complex and consequently on the net colorimetric signal in the range between 0.01 and 0.75 mol L^{-1}, which is in agreement with previous reports [38]. Additionally, these findings show that the proposed method is suitable for the detection of copper ions in samples with high ionic strength, such as seawater samples, and also prove the robustness of the method to important changes such as the effect of salinity. In conclusion, it was chosen that no salt would be added in the next experiments.

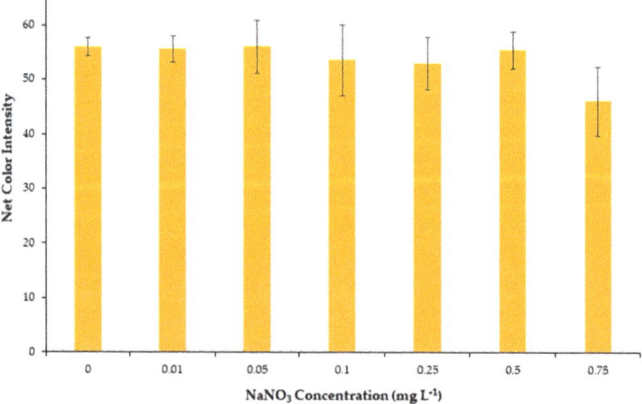

Figure 3. Effect of salinity on the net color signal for the proposed method.

2.2. Method Validation

Regarding the validation of the method in terms of analytical chemistry, the proposed colorimetric method was studied in terms of accuracy, linearity, limits of detection (LOD) and quantification (LOQ), precision, and selectivity.

2.2.1. Linearity, Precision, and Limits of Detection (LOD) and Quantification (LOQ)

The developed method showed adequate linearity of Cu^{2+} concentrations in the range between 2.5 and 40 µmol L^{-1} (Figure 4). At higher concentrations, color saturation occurred, and a plateau was reached. The values of the color signal were almost identical for concentrations above 40 µM due to the color saturation. The regression equation was obtained by integrating the results from 49 standard solutions measured on different working days (n = 7) for the validation of the method. In this way, the calibration curve is far more representational, including possible variations from day-to-day measurements, and the regression equation obtained was:

$$CI = 2.03\ (\pm 0.04)\ [Cu^{2+}] + 6.40\ (\pm 0.88),\ with\ R^2 = 0.986$$

where CI is the color intensity measured by the method.

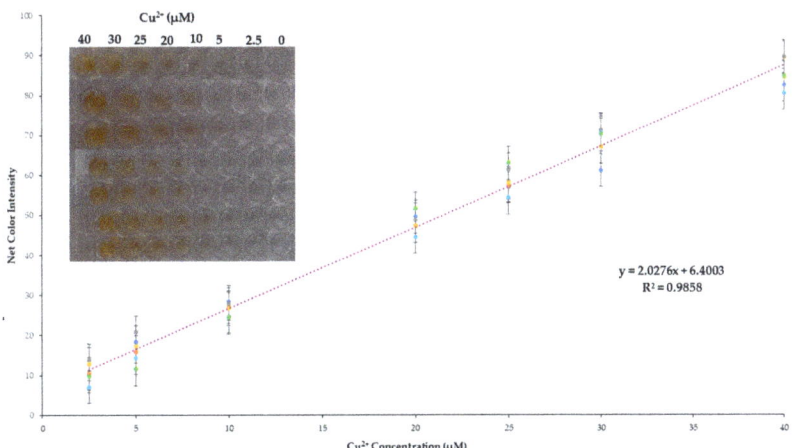

Figure 4. Cumulative calibration curve for the proposed method.

The within-day precision (intra-day) and between-days precision (inter-day) were validated at copper concentrations of 10 and 30 µmol L^{-1} by repetitive measurements of different samples (n = 7) on 7 different days. The intra-day relative standard deviation obtained (RSD) was 2.3% and 0.6%, while the inter-day was 6.1% and 3.7% for 10 and 30 µmol L^{-1} Cu^{2+}, respectively.

Finally, the limit of detection (LOD) and limit of quantification (LOQ) were calculated as LOD = 3.3 × SDi/s and LOQ = 10 × SDi/s, where SDi is the standard deviation of the intercept, and s is the slope of all regression lines. The LOD/LOQ calculated for the determination of Cu^{2+} was 1.4 and 4.3 µmol L^{-1}, respectively.

2.2.2. Interference Study Selectivity

The selectivity of the developed colorimetric method was validated against representative cations and anions that are the most frequently encountered and in higher concentrations in the natural water samples. All potential interferents were analyzed at different concentrations representative of their existence in natural surface waters and are shown in Table 1, whereas Cu^{2+} was at 30 µmol L^{-1}. The influence of the anions and cations is depicted in Figure 5. None of the selected ions formed the DDTC complex in the specific experimental conditions, and this could be observed even by the naked eye, as shown in the integrated photo of Figure 5. We also studied the selectivity of our method towards the transition metal ions. As expected, Cr^{3+}, Zn^{2+}, Mn^{2+}, and Ni^{2+} did not produce a colored complex with DDTC [29], and thus no interference was detected for concentrations up to 20 mg L^{-1} for each metal. Additionally, for Fe^{3+} and Co^{2+} a pale yellow color was formed at concentrations of 10 mg L^{-1}, while for concentrations up to 2.0 mg L^{-1} no colored complex was achieved for each metal (Figure 6 and integrated photo). This concentration is 10 times higher than that permitted by the European Union (0.2 mg L^{-1}), so no interference can be anticipated in surface water and drinking water samples [39]. It is clear that none of the most abundant ions in natural waters has a significant influence on the analytical signal of the proposed method, and as a result the method can be characterized as selective for the analytical determination of copper ions.

Table 1. Interfering ion concentrations used for the selectivity study.

Interfering Cations	Concentration (mg L^{-1})	Interfering Anions	Concentration (mg L^{-1})
Na$^+$	50	NO$_3^-$	50
Ca^{2+}	100	HCO$_3^-$	500
Mg^{2+}	100	SO$_4^{2-}$	25
K$^+$	25	Cl$^-$	100

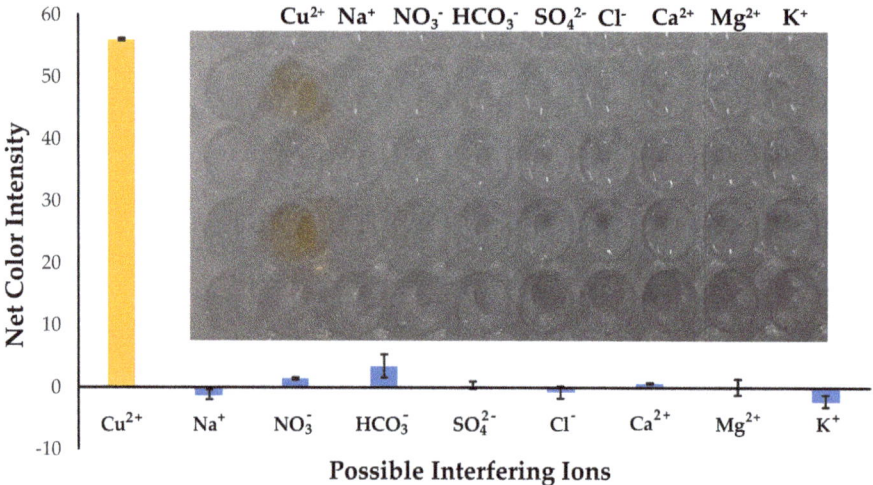

Figure 5. Selectivity of Cu^{2+} determination under the optimum experimental conditions for 30 µmol L^{-1}. Error bars are the standard deviation for n = 3.

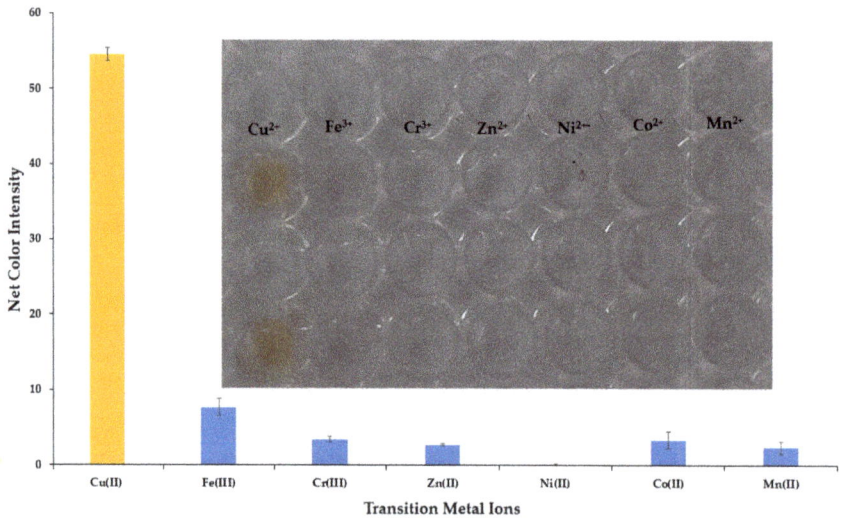

Figure 6. Selectivity of Cu^{2+} determination versus transition metal ions under the optimum experimental conditions for 30 µmol L^{-1}. Error bars are the standard deviation for n = 3.

2.3. Real Sample Applicability

Seven real samples—four bottled water samples from northern and northwestern Greece; two river water samples from the region of Epirus (northwestern Greece); and tap water from the city center of Thessaloniki—were measured with the developed colorimetric method. The samples were handled as described in the Experimental Section, and the results are presented in Table 2. The recoveries of the determined spiked levels of Cu^{2+} were satisfactory and ranged between 89.0 and 109.6%, with a good standard deviation.

Table 2. Applicability of the colorimetric method for Cu^{2+} analysis in water samples.

Samples	Spiked (μmol L^{-1})	Found (μmol L^{-1})	% Recovery ($\pm$RSD, n = 5)
Bottled water 1	10	10.1	101.5 ± 0.3
	30	28.6	95.2 ± 1.5
Bottled water 2	10	10.8	107.6 ± 6.3
	30	29.1	96.9 ± 3.5
Bottled water 3	10	10.3	102.8 ± 6.5
	30	29.1	96.9 ± 3.2
Bottled water 4	10	8.9	89.0 ± 5.0
	30	27.0	89.9 ± 5.8
River water 1	10	11.0	109.6 ± 7.1
	30	28.7	95.6 ± 4.6
River water 2	10	9.9	98.8 ± 8.4
	30	29.4	97.8 ± 5.0
Tap water	10	9.2	92.3 ± 4.7
	30	28.1	93.6 ± 0.6

3. Materials and Methods

3.1. Reagents and Solutions

Copper sulfate pentahydrate ($CuSO_4 \times 5H_2O$) was purchased from Penta chemicals unlimited (Prague, Czech Republic). Sodium chloride, sodium sulfate anhydrous, hydrochloric acid, and sodium hydrogen carbonate were provided by Merck (Darmstadt, Germany). Magnesium nitrate hexahydrate and sodium nitrate were provided by Panreac (Madrid, Spain), while calcium nitrate tetrahydrate and potassium nitrate were bought from Chem Lab NV (Zedelgem, Belgium). Diethyldithiocarbamate sodium salt (Na-DDTC, $C_5H_{10}NNaS_2 \times 3H_2O$) was provided by Carl Roth GmbH (Karlsruhe, Germany). For the selectivity of the proposed method towards transition metal ions (Fe^{3+}, Ni^{2+}, Zn^{2+}, Cr^{3+}, Co^{2+}, and Mn^{2+}), 6 salts were used, and stock solutions of 100 mg L^{-1} were prepared. Nickel (II) sulfate hexahydrate $NiSO_4 \times 6H_2O$ was purchased from Penta chemicals unlimited (Prague, Czech Republic). Chromium (III) nitrate nonahydrate $Cr(NO_3)_3 \times 9H_2O$, Zinc sulfate heptahydrate $ZnSO_4 \times 7H_2O$, and Cobalt (II) nitrate hexahydrate $Co(NO_3)_2 \times 6H_2O$ were provided by Merck (Darmstadt, Germany). Iron (III) nitrate nonahydrate $Fe(NO_3)_3 \times 9H_2O$ was bought from Chem Lab NV (Zedelgem, Belgium), and finally Manganous chloride tetrahydrate $MnCl_2 \times 4H_2O$ was purchased from BDH Chemicals (Poole, England).

All chemical substances were of analytical grade, and all the solutions were prepared with deionized water. The standard stock Cu^{2+} solution (1.0 mmol L^{-1}) was prepared weekly in deionized water, while the standard stock DDTC solution (1.0 mmol L^{-1}) was prepared daily in deionized water. Working solutions of Cu^{2+} and DDTC were prepared every day by diluting the stock solutions in deionized water. Standard stock solutions of 2.0 mol L^{-1} HCl and NaOH were used for the pH study. Additionally, for the selectivity study, different cation and anion stock solutions of 1000 mg L^{-1} for each ion were prepared in deionized water. Finally, sodium nitrate stock solution (2.0 mol L^{-1}) used for the salinity study was prepared by dissolving the appropriate amount in deionized water.

3.2. Apparatus

The instrumentation used for this study is clearly minimal. The pH values of the tested acidic and alkaline solutions were measured with a pH meter (Orion). The images of the devices were captured at specific time periods and at the same distance and angle of the photograph taken, using a mobile smartphone (Samsung Galaxy A21s, Samsung, Ridgefield Park, NJ, USA) or a common flatbed scanner (HP Scanjet 4850, Hewlett Packard Inc., Palo Alto, CA, USA).

3.3. Experimental Process

The experimental process was easy to perform, with absolutely no demand for laboratory instrumentation. In brief, a few μL of deionized water (pH = 7), DDTC (62.5 μL, 0.25 mmol L^{-1}), and the appropriate volumes (e.g., 75 μL for 30 μmol L^{-1} Cu^{2+}) of sample/standard solutions were sequentially added in the wells of the plate for a final volume of 250 μL. For the blank samples, deionized water was used, replacing copper ions. After a short time of 10 min, at room temperature, a photograph of the plate was taken. The photographs taken after this analytical procedure were saved as JPEG files (minimum analysis of 300 dpi), and the ImageJ program (Version 1.53.k) in RGB mode in the blue area was used for the measurement of the mean color intensity (Figure 7). From the ImageJ software, the blue type of images were chosen, and an elliptical shape was selected with the same area in every measurement carried out. The mean color intensity was determined, and the results were transferred to Excel and formed into the figures presented herein.

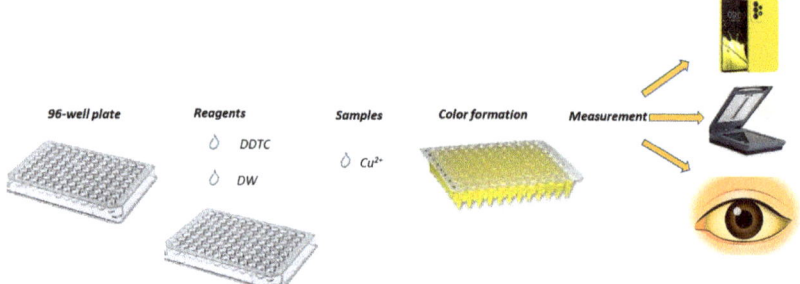

Figure 7. Experimental procedure of the proposed colorimetric method.

3.4. Real Samples

The method was tested on seven different natural water samples. The origin of the samples was divided into three distinct categories: river samples, bottled samples, and tap water samples. Four bottled water samples were purchased from local stores in the Thessaloniki city center. Two river water samples were collected in the Epirus region (NW Greece), from the Louros and Acheron rivers, stored in glass containers of 500 mL each, and filtered through a Whatman 0.45 μm filter to remove any suspended solids. Finally, the tap water sample was collected from the water supply system of the Aristotle University of Thessaloniki campus. All samples were stored at 4 °C in the refrigerator.

4. Conclusions

A fast, equipment-free, and reliable colorimetric method for the analytical determination of Cu^{2+} in natural water samples was developed and validated. The developed sensor utilized only two readily available reagents and is desirable for real-time applications. The developed analytical method relied on the colorimetric differences generated at neutral conditions by the different copper ion concentrations, and the color intensity was recorded by a domestic smartphone camera. The calibration curve obtained was linearly proportional to the concentration of the analyte. The maximum allowable concentration in surface waters according to Directive (EU) 2020/2184 of the European Parliament and of the

Council on the quality of water intended for human consumption is 2.0 mg/L [39]. Thus, in µM concentration units, this corresponds to 31.5 µM, which is much higher than the limit of detection accomplished by this method. Furthermore, the analytical methodology developed is robust, can analyze many samples in a short time period (up to 96 different samples), and permits the analysis of Cu^{2+} at low µM levels (LOD: 1.4 µmol L^{-1}). The levels of Cu^{2+} in the spiked analyzed real samples were within acceptable recovery limits from 89.0 to 109.6%.

Author Contributions: Conceptualization, G.Z.T.; methodology, M.T. and G.Z.T.; validation, A.G.G., V.C.T. and M.T.; investigation, A.G.G.; data curation, A.G.G., V.C.T. and G.Z.T.; writing—original draft preparation, A.G.G.; writing—review and editing, G.Z.T.; supervision, G.Z.T. All authors have read and agreed to the published version of the manuscript.

Funding: This research received no external funding.

Institutional Review Board Statement: Not applicable.

Informed Consent Statement: Not applicable.

Data Availability Statement: Not applicable.

Conflicts of Interest: The authors declare no conflict of interest.

Sample Availability: DDTC, Cu2+ and all other compounds used in this work are available from the authors.

References

1. Park, J.-S.; Voyakin, D.; Kurbanov, B. Bronze-to-brass transition in the medieval Bukhara oasis. *Archaeol. Anthropol. Sci.* **2021**, *13*, 32. [CrossRef]
2. D'Antonio, L.; Fabbricino, M.; Panico, A. Monitoring copper release in drinking water distribution systems. *Water Sci. Technol.* **2008**, *57*, 1111–1115. [CrossRef] [PubMed]
3. Lee, Y. An evaluation of microbial and chemical contamination sources related to the deterioration of tap water quality in the household water supply system. *Int. J. Environ. Res. Public Health* **2013**, *10*, 4143–4160. [CrossRef] [PubMed]
4. Zietz, B.P.; Dassel de Vergara, J.; Dunkelberg, H. Copper concentrations in tap water and possible effects on infant's health—Results of a study in Lower Saxony, Germany. *Environ. Res.* **2003**, *92*, 129–138. [CrossRef]
5. Sörme, L.; Lagerkvist, R. Sources of heavy metals in urban wastewater in Stockholm. *Sci. Total Environ.* **2002**, *298*, 131–145. [CrossRef]
6. Klevay, L.M. Copper. In *Encyclopedia of Dietary Supplements*, 2nd ed.; Coates, P.M., Betz, J.M., Blackman, M.R., Eds.; Informa Healthcare: London, UK; New York, NY, USA, 2010; pp. 604–611.
7. Zhao, D.; Huang, Y.; Wang, B.; Chen, H.; Pan, W.; Yang, M.; Xia, Z.; Zhang, R.; Yuan, C. Dietary intake levels of iron, copper, zinc, and manganese in relation to cognitive function: A cross-sectional study. *Nutrients* **2023**, *15*, 704. [CrossRef]
8. Ala, A.; Walker, A.P.; Ashkan, K.; Dooley, J.S.; Schilsky, M.L. Wilson's disease. *Lancet* **2007**, *369*, 397–408. [CrossRef]
9. Tsvetkov, P.; Coy, S.; Petrova, B.; Dreishpoon, M.; Verma, A.; Abdusamad, M.; Rossen, J.; Joesch-Cohen, L.; Humeidi, R.; Spangler, R.D.; et al. Copper induces cell death by targeting lipoylated TCA cycle proteins. *Science* **2022**, *375*, 1254–1261. [CrossRef]
10. Chen, L.; Min, J.; Wang, F. Copper homeostasis and cuproptosis in health and disease. *Signal Transduct. Target. Ther.* **2022**, *7*, 378. [CrossRef]
11. Thongkam, T.; Apilux, A.; Tusai, T.; Parnklang, T.; Kladsomboon, S. Thy-AuNP-AgNP hybrid systems for colorimetric determination of Copper (II) ions using UV-Vis spectroscopy and smartphone-based detection. *Nanomaterials* **2022**, *12*, 1449. [CrossRef]
12. Böck, F.C.; Helfer, G.A.; da Costa, A.B.; Dessuy, M.B.; Ferrão, M.F. Low cost method for copper determination in sugarcane spirits using Photometrix UVC® embedded in smartphone. *Food Chem.* **2022**, *367*, 130669. [CrossRef] [PubMed]
13. Giokas, D.L.; Paleologos, E.K.; Veltsistas, P.G.; Karayannis, M.I. Micellar enhanced analytical application of Bismuthiol-II for the spectrophotometric determination of trace copper in nutritional matrices. *Microchim. Acta* **2002**, *140*, 81–86. [CrossRef]
14. Sahu, S.; Sikdar, Y.; Bag, R.; Drew, M.G.B.; Cerón-Carrasco, J.P.; Goswami, S. A Quinoxaline–Naphthaldehyde conjugate for colorimetric determination of copper ion. *Molecules* **2022**, *27*, 2908. [CrossRef]
15. Anthemidis, A.N.; Zachariadis, G.A.; Stratis, J.A. On-line preconcentration and determination of copper, lead and chromium(VI) using unloaded polyurethane foam packed column by flame atomic absorption spectrometry in natural waters and biological samples. *Talanta* **2002**, *58*, 831–840. [CrossRef]
16. Silveira, J.R.K.; Brudi, L.C.; Waechter, S.R.; Mello, P.A.; Costa, A.B.; Duarte, F.A. Copper determination in beer by flame atomic absorption spectrometry after extraction and preconcentration by dispersive liquid–liquid microextraction. *Microchem. J.* **2023**, *184*, 108181. [CrossRef]

17. Manousi, N.; Kabir, A.; Furton, K.G.; Zachariadis, G.A.; Anthemidis, A. Automated solid phase extraction of Cd(II), Co(II), Cu(II) and Pb(II) coupled with flame atomic absorption spectrometry utilizing a new sol-gel functionalized silica sorbent. *Separations* **2021**, *8*, 100. [CrossRef]
18. Yilmaz, E.; Soylak, M. Development a novel supramolecular solvent microextraction procedure for copper in environmental samples and its determination by microsampling flame atomic absorption spectrometry. *Talanta* **2014**, *126*, 191–195. [CrossRef]
19. Han, Q.; Yang, X.; Huo, Y.; Lu, J.; Liu, Y. Determination of ultra-trace amounts of copper in environmental water samples by dispersive liquid-liquid microextraction combined with graphite furnace atomic absorption spectrometry. *Separations* **2023**, *10*, 93. [CrossRef]
20. Burylin, M.Y.; Kopeyko, E.S.; Bauer, V.A. Determination of Cu and Mn in seawater by high resolution continuum source graphite furnace atomic absorption spectrometry. *Anal. Lett.* **2022**, *55*, 1663–1671. [CrossRef]
21. Manousi, N.; Kabir, A.; Furton, K.G.; Zachariadis, G.A.; Anthemidis, A. multi-element analysis based on an automated on-line microcolumn separation/preconcentration system using a novel sol-gel thiocyanatopropyl-functionalized silica sorbent prior to ICP-AES for environmental water samples. *Molecules* **2021**, *26*, 4461. [CrossRef]
22. Kalogiouri, N.P.; Manousi, N.; Mourtzinos, I.; Zachariadis, G.A. Multielemental inductively coupled plasma–optical emission spectrometric (ICP-OES) method for the determination of nutrient and toxic elements in wild mushrooms coupled to unsupervised and supervised chemometric tools for their classification by species. *Anal. Lett.* **2022**, *55*, 2108–2123. [CrossRef]
23. Granado-Castro, M.D.; Díaz-de-Alba, M.; Chinchilla-Real, I.; Galindo-Riaño, M.D.; García-Vargas, M.; Casanueva-Marenco, M.J. Coupling liquid membrane and flow-injection technique as an analytical strategy for copper analysis in saline water. *Talanta* **2019**, *192*, 374–379. [CrossRef] [PubMed]
24. Jiang, S.; Lu, Z.; Su, T.; Feng, Y.; Zhou, C.; Hong, P.; Sun, S.; Li, C. High sensitivity detection of copper ions in oysters based on the fluorescence property of cadmium selenide quantum dots. *Chemosensors* **2019**, *7*, 47. [CrossRef]
25. Sayin, S. Synthesis of new quinoline-conjugated calixarene as a fluorescent sensor for selective determination of Cu^{2+} ion. *J. Fluoresc.* **2021**, *31*, 1143–1151. [CrossRef] [PubMed]
26. Mahnashi, M.H.; Mahmoud, A.M.; Alkahtani, S.A.; Ali, R.; El-Wekil, M.M. A novel imidazole derived colorimetric and fluorometric chemosensor for bifunctional detection of copper (II) and sulphide ions in environmental water samples. *Spectrochim. Acta A Mol. Biomol. Spectrosc.* **2020**, *228*, 117846. [CrossRef]
27. Zong, W.; Cao, S.; Xu, Q.; Liu, R. The use of outer filter effects for Cu^{2+} quantitation: A unique example for monitoring nonfluorescent molecule with fluorescence. *Luminescence* **2012**, *27*, 292–296. [CrossRef]
28. Cvek, B.; Milacic, V.; Taraba, J.; Ping, D.Q. Ni(II), Cu(II), and Zn(II) diethyldithiocarbamate complexes show various activities against the proteasome in breast cancer cells. *J. Med. Chem.* **2008**, *51*, 6256–6258. [CrossRef]
29. Ly, N.H.; Nguyen, T.D.; Zoh, K.-D.; Joo, S.-W. Interaction between diethyldithiocarbamate and Cu(II) on gold in non-cyanide wastewater. *Sensors* **2017**, *17*, 2628. [CrossRef]
30. Wang, T.; Fu, Y.; Huang, T.; Liu, Y.; Wu, M.; Yuan, Y.; Li, S.; Li, C. Copper ion attenuated the antiproliferative activity of di-2-pyridylhydrazone dithiocarbamate derivative; However, there was a lack of correlation between ROS generation and antiproliferative activity. *Molecules* **2016**, *21*, 1088. [CrossRef]
31. Sant'Ana, O.D.; Jesuino, L.S.; Cassella, R.J.; Carvalho, M.S.; Santelli, R.E. Solid phase extraction of Cu(II) as diethyldithiocarbamate (DDTC) complex by polyurethane foam. *J. Braz. Chem. Soc.* **2003**, *14*, 728–733. [CrossRef]
32. Li, L.; Xu, K.; Huang, Z.; Xu, X.; Iqbal, J.; Zhao, L.; Du, Y. Rapid determination of trace Cu^{2+} by an in-syringe membrane SPE and membrane solid-phase spectral technique. *Anal. Methods* **2021**, *13*, 4691. [CrossRef] [PubMed]
33. Kappi, F.A.; Papadopoulos, G.A.; Tsogas, G.Z.; Giokas, D.L. Low-cost colorimetric assay of biothiols based on the photochemical reduction of silver halides and consumer electronic imaging devices. *Talanta* **2017**, *172*, 15–22. [CrossRef] [PubMed]
34. Bizirtsakis, P.A.; Tarara, M.; Tsiasioti, A.; Tzanavaras, P.D.; Tsogas, G.Z. Development of a paper-based analytical method for the selective colorimetric determination of bismuth in water samples. *Chemosensors* **2022**, *10*, 265. [CrossRef]
35. Tarara, M.; Tzanavaras, P.D.; Tsogas, G.Z. Development of a paper-based analytical method for the colorimetric determination of calcium in saliva samples. *Sensors* **2023**, *23*, 198. [CrossRef]
36. Li, W.; Wang, L.; Tong, P.; Iqbal, J.; Zhang, X.; Wang, X.; Du, Y. Determination of trace analytes based on diffuse reflectance spectroscopic techniques: Development of a multichannel membrane filtration-enrichment device to improve repeatability. *RSC Adv.* **2014**, *4*, 52123–52129. [CrossRef]
37. Hogarth, G. Transition Metal Dithiocarbamates: 1978–2003. *Prog. Inorg. Chem.* **2005**, *53*, 383–411.
38. Zarei, A.R.; Mardi, K. A green approach for photometric determination of copper β-resorcylate in double base solid propellants. *Chem. Methodol.* **2021**, *5*, 513–521.
39. European Union. Directive (Eu) 2020/2184 of the European Parliament and of the Council, on the quality of water intended for human consumption. *Off. J. Eur. Union* **2020**, *435*, 36.

Disclaimer/Publisher's Note: The statements, opinions and data contained in all publications are solely those of the individual author(s) and contributor(s) and not of MDPI and/or the editor(s). MDPI and/or the editor(s) disclaim responsibility for any injury to people or property resulting from any ideas, methods, instructions or products referred to in the content.

Review

Two-Dimensional High-Performance Liquid Chromatography as a Powerful Tool for Bioanalysis: The Paradigm of Antibiotics

Christina Papatheocharidou and Victoria Samanidou *

Laboratory of Analytical Chemistry, School of Chemistry, Aristotle University of Thessaloniki, GR-54124 Thessaloniki, Greece; papatheocharidouchristina@gmail.com
* Correspondence: samanidu@chem.auth.gr

Abstract: The technique of two-dimensional high-performance liquid chromatography has managed to gain the recognition it deserves thanks to the advantages of satisfactory separations it can offer compared to simple one-dimensional. This review presents in detail key features of the technique, modes of operation, and concepts that ensure its optimal application and consequently the best possible separation of even the most complex samples. Publications focusing on the separation of antibiotics and their respective impurities are also presented, providing information concerning the analytical characteristics of the technique related to the arrangement of the instrument and the chromatographic conditions.

Keywords: two-dimensional liquid chromatography; modes; instrumentation; antibiotics

Citation: Papatheocharidou, C.; Samanidou, V. Two-Dimensional High-Performance Liquid Chromatography as a Powerful Tool for Bioanalysis: The Paradigm of Antibiotics. *Molecules* **2023**, *28*, 5056. https://doi.org/10.3390/molecules28135056

Academic Editor: Gavino Sanna

Received: 12 June 2023
Revised: 25 June 2023
Accepted: 26 June 2023
Published: 28 June 2023

Copyright: © 2023 by the authors. Licensee MDPI, Basel, Switzerland. This article is an open access article distributed under the terms and conditions of the Creative Commons Attribution (CC BY) license (https://creativecommons.org/licenses/by/4.0/).

1. Introduction

The innovative idea of applying multidimensional techniques to challenging complex matrices such as blood, urine, biological cells, and environmental or forensic samples emerged due to the inability of simple methods to achieve high resolution in a short time [1]. The first application of two-dimensional chromatography was proposed by Consden, Gordon, and Martin when they used paper chromatography to separate 22 protein amino acids, within 120 h [2]. Then, Kirchner implemented thin-layer chromatography in 2D mode, turning it directly into a promising analytical separation tool. These applications were followed by the evolution of two-dimensional gas and liquid chromatography, the very primary separation techniques of recent years.

Two-dimensional liquid chromatography has deservedly won a dominant position in the family of LC techniques [3]. Even though liquid chromatography has great flexibility, regarding the different forms in which it has prevailed and the variety of samples it can analyze, it is unable to provide high resolving power in a short period of time. It is not able to deal with two types of mixtures: those consisting of thousands of analytes such as biological samples and those consisting of closely related compounds which cannot be separated by their physicochemical properties, such as enantiomers [3].

In fact, peaks of one-dimensional chromatograms overlap either with other compounds of related samples or with matrix components, although selectivity between different analyzers can be improved. There is not enough room to separate similar compounds, and max capacity is not sufficient for complex samples [1]. As a rule, a simple liquid chromatography system analyzes samples of 20 components in 2 h, and the time increases in proportion to the compounds. In addition, the incompatibility between the LC techniques and mass spectroscopy, which leads to the ion suppression phenomenon due to co-eluting compounds, is another issue that discourages LC application. This is particularly the case in the domain of pharmaceutical analysis, where the detection of impurities is mandatory [4]. Therefore, multidimensional separations can deal with complex and difficult mixtures if they are carried out by qualified operators [5].

In 2D-LC, two different separation modes are exploited to increase the maximum peak capacity and achieve the optimal separation efficiency. The first dimension separates the sample based on one set of physicochemical properties, related to molecular size and polarity, while the co-eluted compounds are subsequently transferred to the second dimension for a further separation [6]. To achieve this goal, the total analysis time is clearly increased compared to conventional liquid chromatography. The procedure of method development, the assembly of two different stationary phases, and the data processing stage are extremely time-consuming [7]. It should be mentioned that the 1D separation time influences the total analysis time, as 2D separations must be shorter [8].

2. General Aspects of 2D-LC

2.1. Historical Data

Two-dimensional separations are not a recent invention. The first mention of this type of chromatography was made in 1941 by Martin and Synge. They argued that two-dimensional separation of liquid samples, known as partition chromatography, depends on two factors: the rate at which the liquid mobile phase flows and the square of the diameter of the particles fixed in the separation column. After this paper, Dent and his colleagues used 2D paper for the isolation and separation of 19 amino acids from potatoes. The results of various subsequent attempts raised several questions in the scientific community, which was unable to provide a conclusive explanation of the divisions until 1967, when scientists discovered high-pressure liquid chromatography (HPLC), providing documented answers to key questions. In 1978, Erni and Frei conducted a very important experiment in which they tried to combine a gel permeation stationary phase with a reverse phase one, separating sienna glycoside [9]. In 1990, Bushey and Jorgenson applied for the first time the functionally complete technique of two-dimensional liquid chromatography. They separated 14 components contained in a protein mixture, combining size-exclusion and ion-exchange chromatography. The applications that followed were based on homemade heart-cutting mode for most samples, except for proteomics. In proteomics, scientists attempted to apply the comprehensive technique. The first decade of 2000 was followed by papers that highlighted some weaknesses of the technique, such as undersampling. Others highlighted features that upgrade the analysis in terms of both accuracy and time. Among the properties investigated were that of maximum capacity, the application of gradient elution to the second column, and the statistical processing of data.

2.2. Modes of 2D-LC

The heart-cutting mode is a simple and easy application of 2D-LC, which provides targeted analysis in a complex matrix (e.g., proteins in blood serum). The analytical process involves transferring one or even a small number of 1D fractions to the second column for further separation [3]. By focusing on the compounds of interest, selectivity and sensitivity are enhanced, thus improving detection limits. The operating cost is relatively low as it does not require specialized instruments compared to comprehensive mode. This setup increases 2D sampling time from 1D runtime. Consequently, the separation power of the total chromatograph increases, and higher separation efficiency is achieved (Figure 1). One disadvantage is the possible loss of important information due to selective transfer of compounds. The heart-cutting mode is very useful for identifying peaks, a very important tool for forensics [1]. Conventionally it is denoted as LC–LC.

The multiple heart-cutting mode is an intermediate mode between comprehensive and heart-cutting mode which provides broader flexibility to the simple heart-cutting mode. More specifically, this mode focuses on multiple peaks of interest which are transferred sequentially to the 2D column for further separation (the total separation process lasts a few seconds or a minute). The looping systems are connected to specific valves to store and purify the 1D eluents before their reinjection. When only the 2D column is ready for a separation, the sampling loop will permit the transfer among columns [3]. The multiple heart-cutting mode is the preferable method for the nonvolatile mobile phases with the ion source in mass spectroscopy [10]. However multiple heart-cutting increases complexity [11]. Conventionally, it is denoted as mLC–LC.

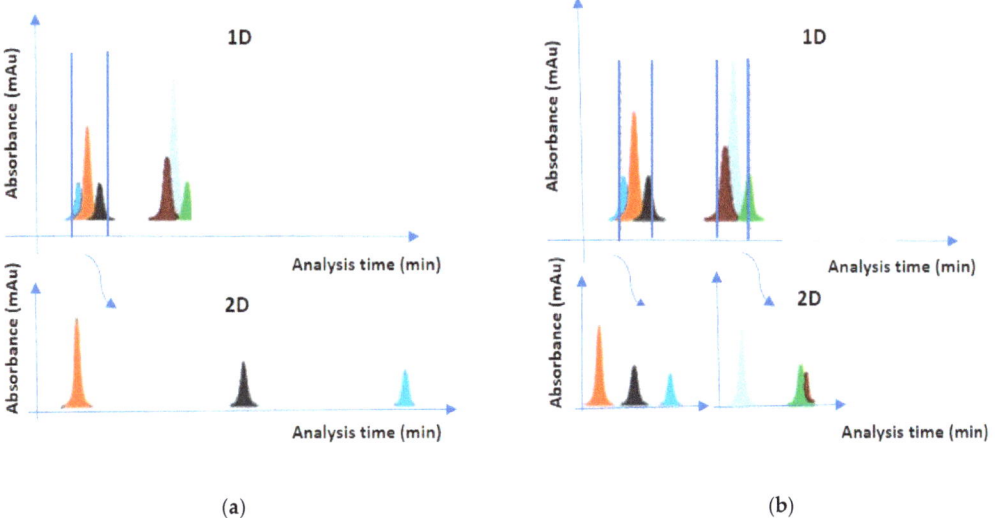

Figure 1. Heart-cutting mode (**a**) and multiple heart-cutting mode (**b**).

The comprehensive mode was proposed after the heart-cutting mode's innovation [12]. It expanded the field of applications and offered a total analysis of the sample, especially for the compounds consisting of thousands of metabolites or co-eluted 1D fractions. Every compound which is spotted in the sample is compulsorily passed through the whole analysis system [4]. The 1D eluents are temporally stored in loops (the most common form) [13] until the 2D column performs further separations (Figure 2). The 2D separations must be faster than those of 1D. The primary goal of a complex fraction is to obtain as much information as possible about the sample by extracting a two-dimensional chromatogram [3]. In simple case studies, the goal of the technique is the successful separation of all components. Sampling time is frequent to avoid remixing fractions that have been successfully separated in the first dimension. Thus, this setup requires short columns with a significant rate difference between columns, where the 2D rate must be higher. This technique is particularly widespread in the fields of omics technology [14]. It is very useful for composite samples consisting of nonvolatile samples. It involves a more extensive separation process, resulting in a higher overall analysis time [1]. It requires specialized operation and application of chemometric methods, increasing the operating cost [15]. A significant advantage of the comprehensive mode is the ability to be combined with mass spectroscopy, eliminating the problems arising from conventional LC–MS. LC × LC/MS is a three-dimensional separation system which prevents matrix effects, offering quantitative analysis and identification of even unknown compounds [16]. Conventionally, it is denoted as LC × LC.

A basic prerequisite for the application of the comprehensive technique is that the separation of 2D fractions and the sampling time of the next 1D eluent should be carried out simultaneously. Such limitations complicate the technique's implementation compared to one-off heart cutting. Time constraints become sampling limitations, since the volume of sample that can be hosted in loop is specific and limited [3]. Under perfect conditions, each 1D fraction should be transferred to the second dimension in three or four consecutive fractions. The maximum allowable pressure in both dimensions determines the time of a full LC × LC analysis. The selective comprehensive mode is a hybrid one that excellently combines the principles of the heart and comprehensive techniques. It can perform comprehensive separations, while still focusing on specific analytes. This specific feature has a double-positive impact: it restricts the problem of undersampling peaks in the first dimension, and it shortens the analysis time without taking under consideration the empty content of 1D separation. The selective comprehensive mode is used for quantitative purposes due to the ability to provide high resolution [17,18]. It cannot be compared with the multiple heart-cutting setup, although similar instrument hardware can be used [3]. Conventionally, it is denoted as sLC × LC.

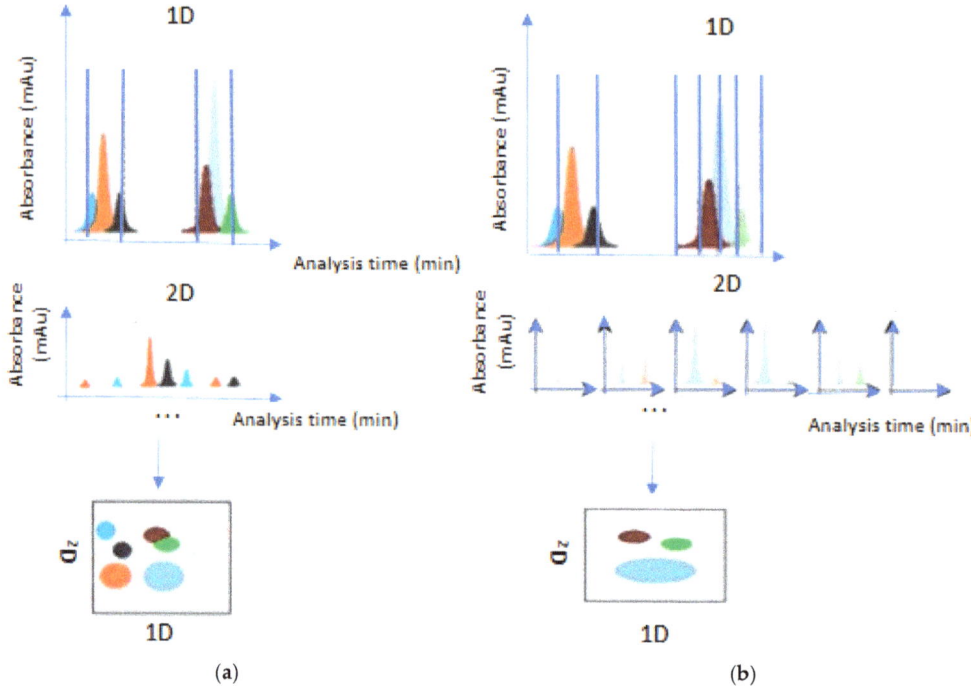

Figure 2. Comprehensive mode (**a**) and selective comprehensive mode (**b**).

Scientists developed the hybrid multiple heart-cutting and selective comprehensive mode to overcome the difficulties that arise when applying traditional methods. This approach can perform the basic concepts of a conventional 2D-LC method, while offering some extra advantages. However, its application is not always possible. Table 1 summarizes the benefits and drawbacks of two-dimensional liquid chromatography.

Table 1. Benefits and drawbacks of hybrid modes of two-dimensional liquid chromatography.

Advantages	Disadvantages
The ability to combine different separation mechanisms increases selectivity by focusing on different analyte properties.	The coordination of all parameters required for the multiple combinations increases system complexity.
The different columns are combined, leading to orthogonal systems that can achieve improved separation efficiency.	The system's complexity increases the overall analysis time, a primary limitation for time-sensitive applications.
The introduction of new orthogonal systems increases the peak capacity and the resolution power of the analytes of interest.	The use of multiple loops, columns, and valves necessitates upgrading the chromatographic system, whereby new modifications must be implemented.
Scientists broaden the fields of application to even the most complicated matrices.	Scientists consume time in searching for sources and, therefore, developing methods.

2.3. Classification of Two-Dimensional Techniques Based on Temporal Transfer of Fractions

In offline mode, fractions from the first column are processed before being injected into the second column. The storage of fractions in vials for reinjection is a factor contributing to contamination or sample loss. It can be performed with an LC system using the same instrument if there are complementary mechanisms among the two dimensions. The first dimension operates continuously while the analysis time of the second column does not affect the total analysis time [3,12]. This is a significant benefit over the other modes as it can be very convenient for a variety of applications. However, the offline mode is time-consuming and cannot be performed in an automated manner [4].

In online mode, the technique can be fully automated until the data processing stage and is much faster than offline mode. The two columns are combined via a specific interface (a two-position switching valve of eight or six ports equipped with two storage loops of the same volume), responsible for the direct/continuous transfer of eluents between columns. An important advantage of online modes is full control of the sample, as well as the fact that there is no possibility of loss. Nevertheless, the implementation of online mode requires complex settings for the utilization of data, and it offers lower peak capacity than offline mode [3,12].

The stop-and-go approach is a compromise in terms of time and power of analysis compared to the previous approaches. The analytical procedures among the two dimensions are carried out alternately. When a specific volume of mobile phase has passed through the 1D column, the flow of the 1D mobile phase is temporarily stopped, allowing the fractions to be retained on the 1D column. Then, the 2D column is ready to start the separation without using sampling loops [3].

3. Orthogonality

The word orthogonal is a combination of the two Greek acronyms "ortho" and "gon", meaning right and angled, respectively. In 2D-LC, orthogonality refers to the degree of independence achieved between the two separation dimensions [19]. The dimensions are uncorrelated and unable to influence each other. The orthogonal characteristic of 2D-LC improves the analytical procedure, enhancing peak capacity and achieving the maximum separation efficiency. Any attempt to apply these systems can be very challenging as scientists must choose carefully among the variety of different columns, which is also very important for accurate results [20]. The appropriate combinations of stationary phases are selected by considering the physicochemical properties of the relevant sample. The possible combinations that can be achieved are RPLC × RPLC, HILIC × RPLC, IEC × RPLC, SEC × RPLC, and NPLC × RPLC [3].

The final choice of separation mechanisms depends on the analytical purpose, the types of analytes, and the matrix. The frequency of column combinations is illustrated in Figure 3.

Some of the above combinations are theoretically possible but practically impossible [3]. When the solvents of mobile phases are immiscible, then the relative mobile phases are incompatible. Generally, mobile phases with different viscosities result in unstable flows, due to the inability of the stationary phases to hold the compounds. A typical example of solvent incompatibility is the normal/reversed phase combination, which has a limited number of practical applications, mostly in offline modes. On the other hand, there are other case studies in which the limited peak capacity is compensated for by the orthogonality achieved. For example, the implementation of ion-exchange or size-exclusion chromatography with reversed-phase chromatography. These combinations are widely used in trace analysis and chiral analysis. Essentially, IEX and SEC columns are used for impurity removal, while RP columns are used for component separation. However, column equilibration requires substantial time for analysis, and the mobile phase is not compatible with MS detectors, a very important tool in analytical separations [4]. The application of reversed-phase chromatography in both dimensions is most suitable in the two-dimensional field. The high separation efficiency, the generation of maximum peak capacity, the commercial availability of RP stationary phases, and the compatibility of the relative solvents with the MS detector are some of the attributes that enhance RP–RP applications [21]. However, the double RPLC combination lacks orthogonality compared to the above combinations [22]. A very promising combination is HILIC–RP, as the mobile phases used in HILIC are similar to those used in RP, resulting in orthogonality [4].

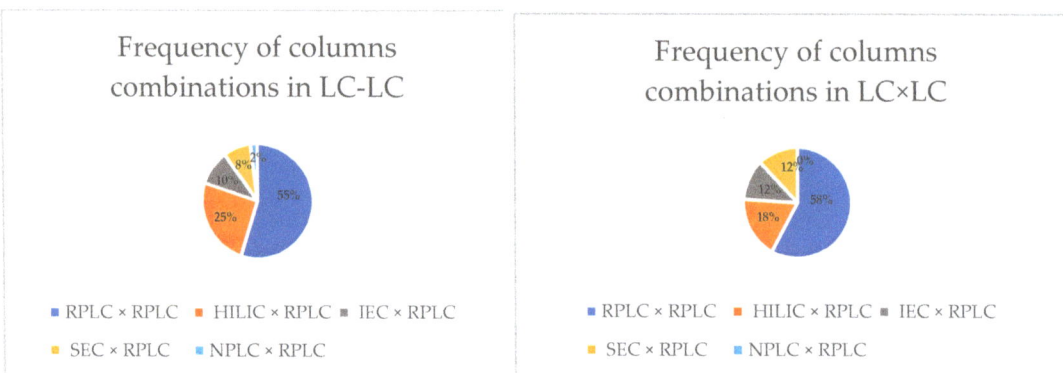

Figure 3. Frequency of column combinations in LC–LC and LC × LC analyses [1].

The simplest way to combine different separation mechanisms is passive modulation, which is based on an interface valve with two identical storage loops. When the 1D eluent is sampled by the first loop, the second one transfers the previous eluent into the second column. However, there are cases in which solvent incompatibility broadens or distorts peaks and leads to difficult quantitation [1]. Scientists have tried addressing this difficulty by making modifications to the modulation process. Some cases are distinguished below.

In active-solvent modulation, the 1D eluents are prediluted in the second mobile phase by a weak solvent before they proceed to the 2D column. There is an eight-port valve with four positions, two of which are functionally identical to those of passive modulation. The content of the second pump is separated by the other two positions and travels through the bypass capillary. This part of the flow joins the stream of fluid coming from the sampling loop before the sample returns in the second column. In this way, the separated flow portion of the second column acts as a diluent [23].

In stationary-phase-assisted modulation (SPAM), the sampling loops are replaced by small-volume trapping columns with which increase the sample's concentration while reducing problems that occur during injection. These columns can be combined mostly with a ten-port valve. Even though the sensitivity of the method can be highly increased, this modulation is not so simple to implement. The stationary phase is chosen according to the physicochemical properties of the sample and the types of mobile phases in both dimensions, choices that can be very challenging. In addition, it is considered the most used but unstable modulation due to the possible premature elution of the analytes. This technique is suitable for multiple heart cutting if the traps are identical [24].

Vacuum-evaporation modulation includes the evaporation of 1D eluents to remove the solvent of the first dimension [1]. During switching, the mobile phase of the second column redissolves the specified compounds and introduces them into the second column for further separation. This setup is often used in noncomprehensive modes, and, although its advantages have been demonstrated, there are still some questions. The first refers to the volatility of the evaporated solvent, the second concerns compounds that cannot be easily dissolved, and the last focuses on the potential loss of specific analytes [25].

4. Peak Capacity

According to Giddings, peak capacity is the maximum number of equal peaks that can be placed between an early and a late peak in a chromatogram [26]. In other words, it is a term that describes the ability of the analytical technique to separate multiple peaks in a single run. The chief advantage of 2D-LC is the second dimension's ability to increase peak capacity without increasing the overall analysis time. The orthogonality achieved between two dimensions increases separation's efficiency, resulting in maximum peak capacity. The relative calculation is as follows: peak capacity = number of peaks obtained from the 1D × number of compounds obtained from the 2D [27]. It is well known that the increase in resolution of each dimension contributes to the overall calculation of peak capacity. However, several parameters determine the peak capacity of a 2D-LC application, such as the number of theoretical plates (N) and the separation selectivity α.

To determine these parameters, despite the easy application and the high speed of isocratic elution, despite the instrument and the columns not needing re-equilibration after each run, gradient elution is the most applicable method [28,29]. The peaks of gradient chromatography are sharper and narrower with equal peak widths compared to the wider peaks of isocratic elution. This attribute provides better quantifications, especially for the late-eluting fractions. In gradient elution, the constant change in the characteristic retention factor broadens the range of analytes that can be separated. Its greatest advantage is that it avoids dilution during the outflow from the first column and the introduction of the sample to the top of the second column, which occurs during isocratic elution [30].

5. Undersampling

The problem of undersampling refers to a small set of 1D compounds transferred to the second column, degrading the quality of the results. The compounds that are not transferred are mixed in the sampling loop of 1D, with their immediate subsequent loss. The key parameter involved in the exemplary condition is the sampling frequency of the first column (when carried out late), as well as the inability to implement all necessary processes to ensure the full coverage of peak capacity [3]. The loss of specific fractions during transfer from one column to another is inevitable, but not undetectable. Therefore, the product rule of peak capacity is adjusted to add the correction factor Cf: peak capacity = number of peaks obtained from the 1D/Cf × number of compounds obtained from the 2D [19]. Consequently, the Giddings definition of maximum capacity cannot be fully applied, taking into consideration the losses that occur.

6. 2D-LC Instrumentation

The individual parts of a 2D-LC instrument are significantly different from those of a one-dimensional liquid chromatography system. Both components and data processing software have made significant progress, transforming homemade instrumentations into advanced systems that perform analyses depending on their assembly. The most important parts of the instrumentation are outlined below.

Pumping system: The pumping system is determined by two main attributes. The first one is the gradient delay volume, i.e., the solvent volume which is delivered by the pump before the 2D column accepts the solvent of the mobile phase. The second one is the flush-out volume, which refers to the use of an organic solvent that will wash the column after every analysis. Modern 2D-LC comprehensive separations have pumps with a volume of 100 mL, and they are capable of gradient elution in a short time. If the system does not retain such requirements, undersampling can occur. To improve the capabilities of pumps, the use of parallel identical columns is recommended in terms of efficiency and retention mechanisms. The systems have totally been improved compared to past pumps with a capacity of 1 mL. In heart cutting, the volume of the pumping system is not investigated, as the execution time of each 2D separation is sufficient [6].

Columns: The two different columns have different dimensions and stationary phases. The final choice is based on the analyte to be separated [12].

Interface valve: The interface valve is located between the separation columns, and it is responsible for the transformation of 1D eluent to 2D. It affects the performance of the separation column, the detection of the identified components, and therefore, the quantification of the results. Due to its usefulness, one- or two-position valves with six (only applicable for heart-cutting mode), eight, 10, or 12 loops (applicable in comprehensive mode), or even two-port valves with six loops have been invented. Every valve consists of a valve motor, stator, and rotor [31]. The most common type among all practical combinations is a two-position valve connected with two columns, while the most common model is asymmetrical instead of symmetrical, as shown in Figure 4 [31]. The results of Van der Horst's research, however, overturned the current data when he reported that the direction of samples within the sampling loops has a significant effect on the retention time of the determined components. Therefore, the use of asymmetric valves should be avoided in cases where the first column is characterized by a very slow speed.

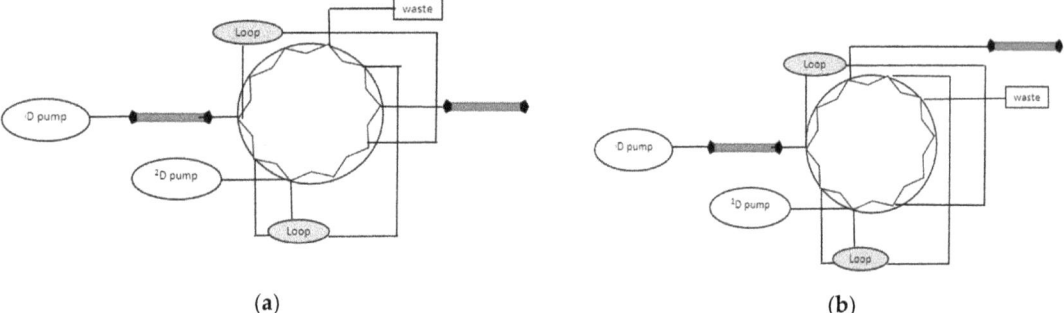

Figure 4. (a) Symmetrical arrangement. (b) Asymmetrical arrangement.

Detectors for data acquisition and analysis: The detector system is an integral part of the second dimension, while it is often optional for the first one. The most common detectors are UV/Vis detectors, mass spectrometers, diode array detectors (DAD), photo array detectors (PDA), and fluorescence detectors [32]. However, no one detector can be effective without the appropriate system responsible for converting the generated signal into chemical information. The processes that take place refer to the quantification and identification of chromatography peaks [3].

The visual representation of 2D data includes counter plots where the x- and y-axes represent the retention times of the two dimensions. The color of the plots is proportional to the intensity of the vertices. Therefore, the extracted chromatographic information develops an absorption area of specific values based on retention times, creating a three-dimensional data representation using special software (Fortner Transform) [33]. The quantification of a 2D chromatograph is a very challenging procedure, especially for the analytes of biological matrices. The utilized chemometric tools (e.g., ChromoSquare from Shimadzu), which include statistical models, have a fairly high cost and cannot be applied to all fields. To address these limitations, new approaches, which do not require specialized software, have been proposed. The strategies are based on the concept "region of interest "(ROI) and the design of multivariate calibration curves [34]. A typical 2D-LC system is illustrated in Figure 5.

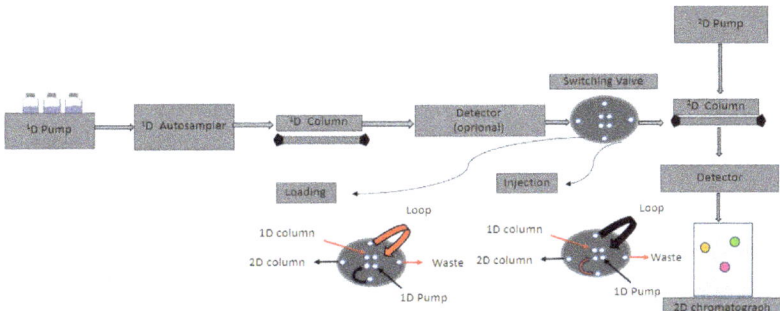

Figure 5. A typical 2D-LC system.

7. Automation of 2D-LC

The application of online 2D-LC was the first recorded type of automation that upgraded the overall analysis in terms of time and efficiency [8]. However, the scientific community is still trying to shorten the analytical procedure by focusing on the instrument's operation and the method's optimization [1]. Some relative examples are discussed. Jintgal et al. proposed the detection and qualification of monoclonal antibody therapeutics at pg/mL levels with clearly increased sensitivity compared to the conventional LC–MS. The automation referred to the design of 2D-LC instrumentation, including the dilution of a specific analyte using an organic solvent, while two RPLC columns with different pH values were used. The separation time (1D and 2D) was only about 20 min [35]. Another study described the automation of the online selective comprehensive and multiple heart-cutting mode for the quantification of different forms of amino acids, by derivatizing the analytes before their injection into the separation columns, in just 45 min [36].

8. The Application of 2D-LC in Antibiotics Analysis

Antibiotics are chemotherapeutic drugs that destroy bacterial infections through preventive or therapeutic actions. They are low-molecular weight compounds with different physicochemical characteristics. Based on their mechanism of action, antibiotics are used to treat a wide range of diseases such as pneumonia, syphilis, and septicemia. The bacterial cell wall consists of peptidoglycan, which is responsible for the basic functions of the bacterium. Hence, broad classes of antibiotics including beta-lactams, glycopeptides, cyclic peptides, and lipoglycopeptides act against the cell wall through peptidoglycan inhibition [37]. The consumption of antibiotics for a long time, in combination with the appearance of changes in the clinical picture of the patient, can lead to unpredictable concentrations of drugs in the human body. Their quantification has so far been highly focused on toxic antibiotics such as glycopeptides. However, therapeutic drug monitoring (TDM) involves quantitative testing of all possible classes, especially beta-lactams [38]. Interest is focused on the free fraction of pharmaceutical compounds that remains active, i.e., its free concentration in plasma, tissues, etc.

Study Cases

Here, we discuss papers that describe the identification or quantification of different antibiotics and their impurities using 2D-LC. Processes such as polymerization, isomerization, degradation reactions, or even the storage of antibiotics in packages result in impurities which are implicated in causing allergies.

The first group of papers focused on resolving the incompatibility between the nonvolatile mobile phase and mass spectroscopy. The use of a normal, reversed, hydrophilic, or ion-exchange stationary phase in the first dimension is combined with a nonvolatile mobile phase for better separation. However, it is well known that most LC methods use mass spectroscopy detectors with electrospray ionization, showing a preference for volatile mobile phases. Their combination is challenging in 2D-LC application. Multiple heart-cutting mode in coordination with the appearance of trap-free columns provides a fast transportation of 1D eluents to the second column. Cefpiramide and eight impurities were determined using trap-free 2D-LC, allowing the detection of even low concentrations through the injection of a large sample volume. The results showed seven new impurities [39]. The identification of impurities in cefonicid sodium followed the same method, resulting in the identification of seven unknown degradation products of the main compound [40]. The application of multiple heart-cutting mode in combination with demineralization offered a fast and convenient technique. The simultaneous successful 2D-LC separation of polymerized impurities in cephalosporins (cefodizime, cefmenoxime, and cefonicid) followed the same method [41].

The second group of studies applied the 2D-LC technique to provide accurate results. A heart-cutting method was used for the qualitative and quantitative determination of vancomycin. Through two-dimensional separation combined with a third intermediate column, it was possible to inject a large sample volume (200 µL) and reduce the total analysis time to 4 min in contrast with a 7.5 min UPLC–MS/MS method. The 1D column was used as a cleaner agent while the 2D column was used as a separator. Therefore, any pretreatment process such as SPE or LLE was not necessary. The lower limit of detection was 0.20 µg/mL [42]. Using 2D-LC, scientists tried to study the active ingredient of meropenem and quantify the ring-opened meropenem and meropenem dimer, known as A and B, with LODs of 0.07 $\mu g \cdot mL^{-1}$ and 0.22 $\mu g \cdot mL^{-1}$, respectively, and LOQs of 0.25 $\mu g \cdot mL^{-1}$ and 0.74 $\mu g \cdot mL^{-1}$, respectively. 2D-LC was also used to identify three new degradation impurities, and the results were evaluated using ESI-MS/MS data [43]. Another paper described the quantification of amoxicillin, benzylpenicillin, benzylpenicillin, flucloxacillin, and piperacillin, as well as the beta-lactamase inhibitors clavulanic acid and clindamycin, macrolide antibiotics, and tazobactam clindamycin [44]. Scientists studied plasma and tissue samples from surgical rehabilitation operations. All matrices were submitted to precipitation of their protein content, applying online extraction using the 1D column and separation using the 2D column. In plasma and plasma ultrafiltrate, the inaccuracy and imprecision for any analyte remained below 15%. In tissue, the accuracy and precision varied up to 16% and 20%, respectively, when various tissues were analyzed after water calibration. In addition, studies quantified oxacillin, cloxacillin, α-amoxicillin, and linezolid in human plasma using the same method. The innovation of this method was the application of 5-sulfosalicylic acid dihydrate (SSA) solution as a precipitation agent of substrate proteins [45]. Amoxicillin's impurities were also studied separately. To separate high-molecular-weight impurities in amoxicillin, a heart-cutting mode was used, providing an eco-friendly separation technique. Gel filtration chromatography was combined with reversed-phase LC [46]. In 2022, two different methods were developed to determine polymerized impurities in cefotaxime sodium and cefepime [47]: an HPSEC TSK gel column, which provided co-induced impurities with lower molecular weights than the identified compounds, and an RP–RP technique, which offered higher sensitivity and selectivity. For the separation of sulfonamides, beta-agonists, and (steroid) hormones in urine samples, a home-made device was initially developed, providing low limits of detection (LOD) ranging from 1 µg/L to 10 µg/L. A comprehensive 2D-LC was further used for more sensitive results, applying the pretreatment technique of 96-well plates for urine precipitation [48]. The experimental conditions, under which all the above applications were carried out, are described in Table 2.

Table 2. Experimental conditions for several antibiotics.

Antibiotic	Column Type	Mobile Phase	Program's Elution	Detector	References
Cefpiramide and relative impurities	1D: Kromasil C8 analytical column (250 mm × 4.6 mm, 5 µm). 2D: Shimadzu Shim-pack GISS C18 analytical column (124 mm × 2.1 mm, 1.9 µm).	1D: (A) 30 mM phosphate/methanol (75:25, v/v) and (B) phosphate buffer (30 mM, pH 7.5)/methanol (50:50, v/v) / 2D: (A) ammonium formate solution (10 mM) and (B) methanol.	1D: gradient conditions: Time (min) B% 0 0 12 0 32 100 33 0 The mobile phase flow rate was 0.80 mL/min. 2D: gradient conditions: Time (min) B% 0 5 5 95 5.5–9 5 The mobile phase flow rate was 0.30 mL/min.	Ion trap/time-of-flight mass spectrometer (Shimadzu Corp., Kyoto, Japan), equipped with an electrospray ionization (ESI) source in positive and negative mode.	[39]
Polymerized impurities of cephalosporins: cefodizime, cefmenoxime, and cefonicid	1D: Xtimate SEC-120 analytical column (7.8 mm × 30 cm, 5 m). 2D: Method A: Shimadzu Shim-pack GISS C18 analytical column (50 mm × 2.1 mm, 1.9 m). Method B/C: ZORBAX SB-C18 analytical column (4.6 × 150 mm, 3.5 m).	1D: (A) 0.005 mol/L dibasic sodium phosphate solution and (B) 0.005 mol/L sodium dihydrogen phosphate solution (61:39, v/v) [acetonitrile v/v)]. 2D: Method A: (A) ammonium formate solution (10 mM) and (B) acetonitrile. Method B: (A) acetic acid solution (0.1‰, v/v) and (B) acetonitrile. Method C: (A) ammonium formate solution (10 mM) and (B) ammonium formate (8 mM) in [acetonitrile/water (4:1, v/v)] solution.	1D: gradient conditions: Time (min) B% 0 5 18 20 20 5 The mobile phase's flow rate was 0.80 mL/min and the injection volume was 30 µL. 2D: gradient elution Method A: Time (min) B% 0 5 5 95 5.5–9 5 Method B: Time (min) B% 0 5 12 15 45 60 55 5 Method C: Time (min) B% 0 12 9 16 15 20 18 40 19 12 The mobile phase flow rate was 0.40 mL/min.	Ion trap/time-of-flight mass spectrometer (Shimadzu Corp., Kyoto, Japan), equipped with an electrospray ionization (ESI) source in positive and negative mode.	[41]

Table 2. Cont.

Antibiotic	Column Type	Mobile Phase	Program's Elution	Detector	References
Impurities in cefonicid sodium	1D: GRACE Alltima C18 analytical column (250 mm × 4.6 mm × 5 μm)/2D: Shimadzu Shim-pack GISS C18 analytical column (50 mm × 2.1 mm, 1.9 μm).	1D: (A) 0.02 mol·L^{-1} ammonium dihydrogen phosphate solution in H$_2$O with 40% aqueous ammonia solution) and (B) methanol. 2D: (A) 10 mmol·L^{-1} ammonium formate solution and (B) methanol.	1D gradient elution: Time (min) B% 0–10 16 10–30 60 41 16 The mobile phase flow rate was 0.80 mL/min. 2D gradient elution: Time (min) B% 0 5 5 95 5.5–9 5 The mobile phase flow rate was 0.30 mL·min^{-1}.	Ion trap/time-of-flight mass spectrometer (Shimadzu Corp, Kyoto, Japan), equipped with an electrospray ionization (ESI) source in positive and negative mode.	[40]
Meropenem	1D: Shim-Pack CLC ODS (6 mm × 150 mm, 5 μm, Shimadzu, Kyoto, Japan)/2D: Shim-pack GISS C18 column (50 mm × 2.1 mm, 1.9 μm, Shimadzu, Kyoto, Japan) used at 40 °C.	1D: (A) 0.1% triethylamine/acetonitrile (96:4, v/v) and (B) 0.1% triethylamine/acetonitrile (70:30, v/v) / 2D: (A) 10 mM ammonium formate and (B) methanol.	1D: Gradient elution Time (min) B% 0 0 18 10 55 60 56 0 The flow rate was 1.5 mL·min^{-1}. 2D: Gradient elution mode: Time (min) B% 0 5 5 95 5.5–9 5 The flow rate was 0.30 mL·min^{-1}.	Ion trap/time-of-flight mass spectrometer (Shimadzu Corp, Kyoto, Japan), equipped with an electrospray ionization (ESI) source in positive and negative mode.	[43]
Amoxicillin	1D: Shim-pack GIS C18 (4.6 mm × 250 mm, 5 μm)/2D: Shim-pack GIS C18(4.6 mm × 150 mm, 5 μm).	1D: ammonium dihydrogen phosphate/tetrahydrofuran/methanol (730:12.5:300) / 2D: 1% formic acid aqueous solution (A)/acetonitrile (B).	Isocratic conditions: the flow rate was 1 mL/min.	UV detector at 254 nm.	[46]
Benzylpenicillin, Flucloxacillin, Amoxicillin, Piperacillin, the beta-lactamase inhibitors Clavulanic acid and Clindamycin, Macrolide antibiotics, and Tazobactam clindamycin	1D: XBridge® C8 Direct Connect HP column (10 μm particle size, 2.1 × 30 mm)/2D: Acquity UPLC® BEH C18 (1.7 μm particle size, 2.1 × 100 mm i.d.; Waters) analytical column equipped with an Acquity UPLC® BEH C18 VanGuard precolumn (1.7 μm particle size, 2.1 × 5 mm i.d.; Waters).	1D: 100% H$_2$O with 0.1% formic acid/2D: (A) water containing 0.1% formic acid and (B) acetonitrile containing 0.1% formic acid.	1D: Isocratic elusion at a flow rate of 1.0 mL/min. 2D: Gradient elution Time (min) B% 0.0 10 1.8 10 4.5 45 5.0 100 6.0 100 6.1 10 The flow rate was 700 μL/min.	A triple-quadrupole mass spectrometer (TSQ ENDURA) (Thermo Scientific, Reinach, Switzerland), equipped with an electrospray ionization source in positive mode.	[44]

244

Table 2. Cont.

Antibiotic	Column Type	Mobile Phase	Program's Elution	Detector	References
Vancomycin	1D: Reversed-phase Diamonsil C18(2) column (100 mm × 4.6 mm, 5 μm, China; C1). Middle column: a strong cation-exchange column (20 mm × 4.6 mm, 5 μm, ANAX, China; MC). 2D: Inertsil ODS-3 column (150 mm × 4.6 mm, 5 μm, GL Science Inc., Japan; C2).	1D: (A) 20 mmol/L ammonium acetate buffer and (B) acetonitrile (88:12, v/v)/ 2D: (A) 50.0 mmol/L Ammonium acetate buffer (pH 5.0) and (B) acetonitrile (85:15, v/v), mobile phase C.	1D: isocratic elusion with a flow rate of (A) 1.2 mL/min, (B) 1.6 mL/min, and (C) 1.5 mL/min.	UV detector at 282 nm.	[42]
Amoxicillin, cloxacillin, oxacillin, and linezolid	1D: Perfusion column (POROS R1/20, 20 m, 2.1 mm × 30 mm, Applied Biosystems, Darmstadt, Germany)/2D: Pentafluorophenyl (PFP) analytical column (Phenomenex Kinetex, 2.6 m, 2 mm × 50 mm, Aschaffenburg, Germany).	1D: (A) water + 0.1% formic acid, (B) MeOH + 0.1% formic acid. (C) water/10 mM ammonium formate, with formic acid, and (D) ACN + C1% formic acid.	Isocratic conditions: a flow rate of 4.0 mL/min over 0.70 min. Gradient conditions: Time (min) D% 0 10 0.75 10 3.20 98 3.80 10 3.81 10	A TQD triple-quadrupole mass spectrometer) equipped with an electrospray ionization source (Waters, St Quentin, France) in positive mode.	[45]
Sulfonamides, Beta-agonists, and (Steroid) hormones	Self-made: 1D: Waters HSS Cyano (1.8 μm, 1 × 150 mm), Waters BEH C18 (1.8 μm, 1 × 150 mm), Waters Phenyl (1.8 μm, 1 × 150 mm) and a Phenomenex Kinetix (2.6 μm, 1.0 × 150 mm) column. 2D: Waters Phenyl column (1.7 μm, 2.1 × 50 mm). Commercial LC × LC	1D: (A) water/acetonitrile (90:10) containing 0.1% formic acid and (B) water/acetonitrile (10:90) containing 0.1% formic acid. 2D: (A) water/acetonitrile (90:10) containing 0.1% formic acid and (B) water/acetonitrile (10:90) containing 0.1% formic acid.	1D gradient conditions: Time (min) B% 0 0 0 80 27 80 28 0 2D gradient conditions: Time (min) B% 0 0 45 40 50 100	Ion trap/time-of-flight mass spectrometer (Waldbronn, Germany), was equipped with an electrospray ionization (ESI) source in positive mode.	[48]

Table 2. Cont.

Antibiotic	Column Type	Mobile Phase	Program's Elution	Detector	References
The polymerized impurities in cefotaxime sodium and cefepime 2D HPSEC/RP-HPLC system 2D RP-HPLC/RP-HPLC system	1D: a TSK-gel G2000SWxl column (7.8 mm × 30 cm, 5 μm) from TOSOH Corporation (Tokyo, Japan). 2D: Agilent ZORBAX SB-C18 analytical column (4.6 mm × 150 mm, 3.5 μm) (Santa Clara, CA, USA). 1D: Kromasil (Nouryon, Bohus, Sweden) 100-5-C18 analytical column (4.6 mm × 250 mm, 5 μm). 2D: Shimadzu Shim-pack GISS C18 analytical column (50 mm × 2.1 mm, 1.9 μm)	1D: phosphate buffer of dibasic sodium phosphate solution/0.005 mol/L sodium dihydrogen phosphate solution, 61:39 (v/v), and acetonitrile at 95:5 (v/v). 2D: (A) 10 mM ammonium formate solution and (B) acetonitrile. 1D: For cefotaxime sodium injection, the mobile phases were 0.05 M disodium hydrogen phosphate solution (A) and methanol (B). For cefepime, the mobile phases were 0.05 M ammonium dihydrogen phosphate solution (A) and acetonitrile (B). 2D: (A) 10 mM ammonium formate solution and (B) acetonitrile.	1D: Isocratic conditions: a flow rate of 0.50 mL/min 2D: Gradient elution Time (min) B% 0–20 5 20–21 40 21–29 5 The flow rate of the mobile phase was 0.40 mL/min. 1D for cefotaxime, gradient elution: Time (min) B% 0–10 15 10–30 15 30–40 70 40–41 70 41–50 15 The flow rate was 0.8 mL/min and the injection volume was 20 μL. 1D for cefepime: the gradient program was set as follows: Time (min) B% 0 10 10 10 30 70 40 70 41–50 10 The flow rate was 1.00 mL/min. 2D, gradient conditions: Time (min) B% 0–5 5–95 5.5–5.5 5	1D: PDA detector in the range of 200–400 nm. 2D: UV detection wavelength of 254 nm. 1D: PDA detector in the range of 200–400 nm. 2D: Ion trap/time-of-flight mass spectrometer (Shimadzu Corp., Kyoto, Japan), equipped with an electrospray ionization (ESI) source in positive and negative mode.	[41]

9. Conclusions

Undoubtedly, 2D-LC has set new standards in the domain of analytical chemistry. The ability to combine two different separation mechanisms has improved the resolution and detection limits of complex and closely related compounds. In some cases, pretreatment techniques have been replaced by using appropriate columns to clear and separate the matrix. This is easily combined with mass spectroscopy as a valuable comparison tool that can solve various problems during method development. On the other hand, there are some drawbacks that discourage the method's application, e.g., the acquisition of appropriate instrumentation and instrumental complexity. Moreover, the overall analysis time is increased in the case of multiple cuts. Nevertheless, 2D-LC will become one of the most applicable separations techniques as more and more scientists adjust their theoretical backgrounds with the practical implementation of 2D-LC modes.

Author Contributions: Conceptualization, V.S.; investigation, V.S. and C.P.; writing—original draft preparation, V.S. and C.P.; writing—review and editing, V.S. and C.P.; supervision, V.S. All authors have read and agreed to the published version of the manuscript.

Funding: This research received no external funding.

Institutional Review Board Statement: Not applicable.

Informed Consent Statement: Not applicable.

Data availability statement: Not applicable.

Conflicts of Interest: The authors declare no conflict of interest.

Sample Availability: Not available.

References

1. Pirok, B.W.J.; Stoll, D.R.; Schoenmakers, P.J. Recent Developments in Two-Dimensional Liquid Chromatography: Fundamental Improvements for Practical Applications. *Anal. Chem.* **2019**, *91*, 240–263. [CrossRef] [PubMed]
2. Guiochon, G.; Gonnord, M.F.; Zakaria, M.; Siouffi, A.M. Chromatography with a Two-Dimensional Column. *Chromatographia* **1983**, *17*, 121–124. [CrossRef]
3. Stoll, D.R.; Carr, P.W. Two-Dimensional Liquid Chromatography: A state of the art tutorial. *Anal. Chem.* **2017**, *89*, 519–531. [CrossRef]
4. Iguiniz, M.; Heinisch, S. Two-Dimensional Liquid Chromatography in Pharmaceutical Analysis. Instrumental Aspects, Trends and Applications. *J. Pharm. Biomed. Anal.* **2017**, *145*, 482–503. [CrossRef]
5. Davis, J.M.; Glddings, J.C. Statistical Theory of Component Overlap in Multicomponent Chromatograms. *Anal. Chem.* **1983**, *55*, 418–424. [CrossRef]
6. Stoll, D.R. Introduction to Two-Dimensional Liquid Chromatography—Theory and Practice. In *Handbook of Advanced Chromatography/Mass Spectrometry Techniques*; Academic Press: Cambridge, MA, USA, 2017; pp. 227–286. [CrossRef]
7. Hinzke, T.; Kouris, A.; Hughes, R.A.; Strous, M.; Kleiner, M. More Is Not Always Better: Evaluation of 1D and 2D-LC-MS/MS Methods for Metaproteomics. *Front. Microbiol.* **2019**, *10*, 238. [CrossRef]
8. Pandohee, J.; Stevenson, P.G.; Zhou, X.-R.; Spencer, M.J.S.; Jones, O.A.H. Current Metabolomics BENTHAM SCIENCE Send Orders for Reprints to Reprints@benthamscience.Ae Multi-Dimensional Liquid Chromatography and Metabolomics: Are Two Dimensions Better Than One? *Curr. Metab.* **2015**, *3*, 10–20. [CrossRef]
9. Erni, F.; Frei, R.W. Two-Dimensional Column Liquid Chromatographic Tech-Nique for Resolution of Complex Mixtures. *J. Chromatogr. A* **1978**, *149*, 561–569. [CrossRef]
10. Petersson, P.; Haselmann, K.; Buckenmaier, S. Multiple Heart-Cutting Two-Dimensional Liquid Chromatography Mass Spectrometry: Towards Real Time Determination of Related Impurities of Bio-Pharmaceuticals in Salt Based Separation Methods. *J. Chromatogr. A* **2016**, *1468*, 95–101. [CrossRef]
11. Fernández, A.S.; Rodríguez-González, P.; Álvarez, L.; García, M.; Iglesias, H.G.; García Alonso, J.I. Multiple Heart-Cutting Two-Dimensional Liquid Chromatography, and Isotope Dilution Tandem Mass Spectrometry for the Absolute Quantification of Proteins in Human Serum. *Anal. Chim. Acta* **2021**, *1184*, 339022. [CrossRef]
12. Stoll, D.R. Recent Advances in 2D-LC for Bioanalysis. *Bioanalysis* **2015**, *7*, 3125–3142. [CrossRef] [PubMed]
13. Martín-Pozo, L.; Arena, K.; Cacciola, F.; Dugo, P.; Mondello, L. Comprehensive Two-Dimensional Liquid Chromatography in Food Analysis. Is Any Sample Preparation Necessary? *Green Anal. Chem.* **2022**, *3*, 100025. [CrossRef]

14. Roca, L.S.; Gargano, A.F.G.; Schoenmakers, P.J. Development of Comprehensive Two-Dimensional Low-Flow Liquid-Chromatography Setup Coupled to High-Resolution Mass Spectrometry for Shotgun Proteomics. *Anal. Chim. Acta* **2021**, *1156*, 338349. [CrossRef]
15. Pierce, K.M.; Hoggard, J.C.; Mohler, R.E.; Synovec, R.E. Recent Advancements in Comprehensive Two-Dimensional Separations with Chemometrics. *J. Chromatogr. A* **2008**, *1184*, 341–352. [CrossRef] [PubMed]
16. Pól, J.; Hyötyläinen, T. Comprehensive Two-Dimensional Liquid Chromatography Coupled with Mass Spectrometry. *Anal. Bioanal. Chem.* **2008**, *391*, 21–31. [CrossRef] [PubMed]
17. Groskreutz, S.R.; Swenson, M.M.; Secor, L.B.; Stoll, D.R. Selective Comprehensive Multidimensional Separation for Resolution Enhancement in High Performance Liquid Chromatography. Part II: Applications. *J. Chromatogr. A* **2012**, *1228*, 41–50. [CrossRef] [PubMed]
18. Groskreutz, S.R.; Swenson, M.M.; Secor, L.B.; Stoll, D.R. Selective Comprehensive Multi-Dimensional Separation for Resolution Enhancement in High Performance Liquid Chromatography. Part I: Principles and Instrumentation. *J. Chromatogr. A* **2012**, *1228*, 31–40. [CrossRef]
19. Jones, O. *Two-Dimensional Liquid Chromatography Principles and Practical Applications*; Springer Briefs in Molecular Science; Springer Nature: Berlin, Germany, 2020.
20. Gilar, M.; Olivova, P.; Daly, A.E.; Gebler, J.C. Orthogonality of Separation in Two-Dimensional Liquid Chromatography. *Anal. Chem.* **2005**, *77*, 6426–6434. [CrossRef]
21. Li, D.; Jakob, C.; Schmitz, O. Practical Considerations in Comprehensive Two-Dimensional Liquid Chromatography Systems (LCxLC) with Reversed-Phases in Both Dimensions. *Anal. Bioanal. Chem.* **2015**, *407*, 153–167. [CrossRef]
22. Allen, R.C.; Barnes, B.B.; Haidar Ahmad, I.A.; Filgueira, M.R.; Carr, P.W. Impact of Reversed Phase Column Pairs in Comprehensive Two-Dimensional Liquid Chromatography. *J. Chromatogr. A* **2014**, *1361*, 169–177. [CrossRef]
23. Stoll, D.R.; Shoykhet, K.; Petersson, P.; Buckenmaier, S. Active Solvent Modulation: A Valve-Based Approach to Improve Separation Compatibility in Two-Dimensional Liquid Chromatography. *Anal. Chem.* **2017**, *89*, 9260–9267. [CrossRef]
24. den Uijl, M.J.; Roeland, T.; Bos, T.S.; Schoenmakers, P.J.; van Bommel, M.R.; Pirok, B.W.J. Assessing the Feasibility of Stationary-Phase-Assisted Modulation for Two-Dimensional Liquid-Chromatography Separations. *J. Chromatogr. A* **2022**, *1679*, 463388. [CrossRef] [PubMed]
25. Ding, K.; Xu, Y.; Wang, H.; Duan, C.; Guan, Y. A Vacuum Assisted Dynamic Evaporation Interface for Two-Dimensional Normal Phase/Reverse Phase Liquid Chromatography. *J. Chromatogr. A* **2010**, *1217*, 5477–5483. [CrossRef] [PubMed]
26. Giddings, J.C. Maximum Nlumber of Components Resolvable by Gel Filtration and Other Ellution Chromatographic Methods. *Anal. Chem.* **1967**, *39*, 1027–1028. [CrossRef]
27. Li, X.; Stoll, D.R.; Carr, P.W. Equation for Peak Capacity Estimation in Two-Dimensional Liquid Chromatography. *Anal. Chem.* **2009**, *81*, 845–850. [CrossRef]
28. Stoll, D.R.; Cohen, J.D.; Carr, P.W. Fast, Comprehensive Online Two-Dimensional High Performance Liquid Chromatography through the Use of High Temperature Ultra-Fast Gradient Elution Reversed-Phase Liquid Chromatography. *J. Chromatogr. A* **2006**, *1122*, 123–137. [CrossRef] [PubMed]
29. Pirok, B.W.J.; Gargano, A.F.G.; Schoenmakers, P.J. Optimizing Separations in Online Comprehensive Two-Dimensional Liquid Chromatography. *J. Sep. Sci.* **2018**, *41*, 68–98. [CrossRef] [PubMed]
30. Jandera, P. Can the Theory of Gradient Liquid Chromatography Be Useful in Solving Practical Problems? *J. Chromatogr. A* **2006**, *1126*, 195–218. [CrossRef]
31. Soliven, A.; Soliven, A.; Edge, T. *Considerations for the Use of LC x LC Active Flow Technology (AFT) Application in LC-MS Pesticide Residue Analysis View Project Selective Detection in Complex Samples View Project Considerations for the Use of LC x LC*; University of Sydney: Sydney, Australia, 2014.
32. Dugo, P.; Cacciola, F.; Kumm, T.; Dugo, G.; Mondello, L. Comprehensive Multidimensional Liquid Chromatography: Theory and Applications. *J. Chromatogr. A* **2008**, *1184*, 353–368. [CrossRef]
33. François, I.; Sandra, K.; Sandra, P. Comprehensive Liquid Chromatography: Fundamental Aspects and Practical Considerations—A Review. *Anal. Chim. Acta* **2009**, *641*, 14–31. [CrossRef]
34. Pérez-Cova, M.; Platikanov, S.; Tauler, R.; Jaumot, J. Quantification Strategies for Two-Dimensional Liquid Chromatography Datasets Using Regions of Interest and Multivariate Curve Resolution Approaches. *Talanta* **2022**, *247*, 123586. [CrossRef] [PubMed]
35. He, J.; Meng, L.; Ruppel, J.; Yang, J.; Kaur, S.; Xu, K. Automated, Generic Reagent and Ultratargeted 2D-LC-MS/MS Enabling Quantification of Biotherapeutics and Soluble Targets down to Pg/ML Range in Serum. *Anal. Chem.* **2020**, *92*, 9412–9420. [CrossRef] [PubMed]
36. Karongo, R.; Ge, M.; Geibel, C.; Horak, J.; Lämmerhofer, M. Enantioselective Multiple Heart Cutting Online Two-Dimensional Liquid Chromatography-Mass Spectrometry of All Proteinogenic Amino Acids with Second Dimension Chiral Separations in One-Minute Time Scales on a Chiral Tandem Column. *Anal. Chim. Acta* **2021**, *1180*, 338858. [CrossRef]
37. Pauter, K.; Szultka-Młyńska, M.; Buszewski, B. Determination and Identification of Antibiotic Drugs and Bacterial Strains in Biological Samples. *Molecules* **2020**, *25*, 2556. [CrossRef] [PubMed]
38. Muller, A.E.; Huttner, B.; Huttner, A. Therapeutic Drug Monitoring of Beta-Lactams and Other Antibiotics in the Intensive Care Unit: Which Agents, Which Patients and Which Infections? *Drugs* **2018**, *78*, 439–451. [CrossRef]

39. Wang, J.; Xu, Y.; Wen, C.; Wang, Z. Application of a Trap-Free Two-Dimensional Liquid Chromatography Combined with Ion Trap/Time-of-Flight Mass Spectrometry for Separation and Characterization of Impurities and Isomers in Cefpiramide. *Anal. Chim. Acta* **2017**, *992*, 42–54. [CrossRef]
40. Wang, J.; Xu, Y.; Zhang, Y.; Wang, H.; Zhong, W. Separation and Characterization of Unknown Impurities in Cefonicid Sodium by Trap-Free Two-Dimensional Liquid Chromatography Combined with Ion Trap Time-of-Flight Mass Spectrometry. *Rapid Commun. Mass Spectrom.* **2017**, *31*, 1541–1550. [CrossRef]
41. Xu, Y.; Wang, D.D.; Tang, L.; Wang, J. Separation and Characterization of Allergic Polymerized Impurities in Cephalosporins by 2D-HPSEC × LC-IT-TOF MS. *J. Pharm. Biomed. Anal.* **2017**, *145*, 742–750. [CrossRef]
42. Sheng, Y.; Zhou, B. High-Throughput Determination of Vancomycin in Human Plasma by a Cost-Effective System of Two-Dimensional Liquid Chromatography. *J. Chromatogr. A* **2017**, *1499*, 48–56. [CrossRef]
43. Ren, X.; Ye, J.; Chen, X.; Wang, F.; Liu, G.; Wang, J. Development of a Novel HPLC Method for the Analysis of Impurities in Meropenem and Identification of Unknown Impurities by 2D LC-IT-TOF MS. *Chromatographia* **2021**, *84*, 937–947. [CrossRef]
44. Rehm, S.; Rentsch, K.M. A 2D HPLC-MS/MS Method for Several Antibiotics in Blood Plasma, Plasma Water, and Diverse Tissue Samples. *Anal. Bioanal. Chem.* **2020**, *412*, 715–725. [CrossRef] [PubMed]
45. Palayer, M.; Chaussenery-Lorentz, O.; Boubekeur, L.; Urbina, T.; Maury, E.; Maubert, M.A.; Pilon, A.; Bourgogne, E. Quantitation of 10 Antibiotics in Plasma: Sulfosalicylic Acid Combined with 2D-LC-MS/MS Is a Robust Assay for Beta-Lactam Therapeutic Drug Monitoring. *J. Chromatogr. B Analyt. Technol. Biomed. Life Sci.* **2023**, *1221*, 123685. [CrossRef] [PubMed]
46. Shan-Ying, C.; Chang-Qin, H.; Ming-Zhe, X. Chromatographic Determination of High-Molecular Weight Impurities in Amoxicillin. *J. Pharm. Biomed. Anal.* **2003**, *31*, 589–596. [CrossRef] [PubMed]
47. Ren, X.; Zhu, B.; Gao, J.; Tang, K.; Zhou, P.; Wang, J. Study of the Polymerized Impurities in Cefotaxime Sodium and Cefepime by Applying Various Chromatographic Modes Coupled with Ion Trap/Time-of-Flight Mass Spectrometry. *Talanta* **2022**, *238*, 123079. [CrossRef]
48. Blokland, M.H.; Zoontjes, P.W.; Van Ginkel, L.A.; Van De Schans, M.G.M.; Sterk, S.S.; Bovee, T.F.H. Multiclass Screening in Urine by Comprehensive Two-Dimensional Liquid Chromatography Time of Flight Mass Spectrometry for Residues of Sulphonamides, Beta-Agonists and Steroids. *Food Addit. Contam. Part A Chem. Anal. Control. Expo Risk Assess* **2018**, *35*, 1703–1715. [CrossRef]

Disclaimer/Publisher's Note: The statements, opinions and data contained in all publications are solely those of the individual author(s) and contributor(s) and not of MDPI and/or the editor(s). MDPI and/or the editor(s) disclaim responsibility for any injury to people or property resulting from any ideas, methods, instructions or products referred to in the content.

MDPI
St. Alban-Anlage 66
4052 Basel
Switzerland
www.mdpi.com

Molecules Editorial Office
E-mail: molecules@mdpi.com
www.mdpi.com/journal/molecules

Disclaimer/Publisher's Note: The statements, opinions and data contained in all publications are solely those of the individual author(s) and contributor(s) and not of MDPI and/or the editor(s). MDPI and/or the editor(s) disclaim responsibility for any injury to people or property resulting from any ideas, methods, instructions or products referred to in the content.

www.ingramcontent.com/pod-product-compliance
Lightning Source LLC
LaVergne TN
LVHW070455100526
838202LV00014B/1730